SURFACE COMPLEXATION MODELING

SURFACE COMPLEXATION MODELING

Gibbsite

Athanasios K. Karamalidis
David A. Dzombak

Carnegie Mellon University
Pittsburgh, Pennsylvania

WILEY

A JOHN WILEY & SONS, INC., PUBLICATION

Published by John Wiley & Sons, Inc., Hoboken, New Jersey.
Published simultaneously in Canada.

Library of Congress Cataloging-in-Publication Data:

Karamalidis, Athanasios K.
Surface complexation modeling : gibbsite / Athanasios K. Karamalidis, David A. Dzombak.
p. cm.
Includes bibliographical references and index.
ISBN 978-0-470-58768-3 (cloth)
1. Aluminum oxide–Surfaces–Simulation methods. 2. Aluminum oxide–Solubility.
3. Surface chemistry–Simulation methods. 4. Coordination compounds. 5. Chemical models.
I. Dzombak, David A. II. Title.
QD181.A4.K37 2010
546′.6732–dc22

2010004435

Printed in the United States of America.

10 9 8 7 6 5 4 3 2 1

To Werner Stumm

Water chemist, surface chemist, and scientific leader
his insights and vision continue to inspire

CONTENTS

FOREWORD

Writing this foreword brings me several years back, when David Dzombak was a graduate student at MIT, a young man in whom one could already perceive the future eminent professor. This was a time when a great wave of experimental work on the adsorption of solutes on oxides had just peaked, the data being interpreted and reported with the help of a variety of models with multiple surface layers. Though differing from each other in the way they represented and parameterized the solid–water interface, all these models descended from Werner Stumm and Paul Schindler's original insight that surface complexation—that is, the formation of chemical bonds between solutes and atoms at the surface of solids—dominate adsorption phenomena in nature and that the corresponding free energy can be added to a coulombic term calculated from some version of the Gouy–Chapman–Stern–Grahame theory.

But the different models used by different research groups resulted in parameters that had different meanings, and different values for fitting the same experimental data. This made comparison among data sets difficult and also limited their use. David Dzombak, who always had a keen interest in making use of scientific advances for practical applications, embarked on the daunting task of interpreting all published adsorption data on oxides with a unique model. Refusing to be intimidated by the formidable triple layers emanating from the West Coast, he chose the unassuming two-layer model, which could simply account for the experimental observations. It turned out, of course, that assembling critically and re-interpreting all published adsorption data on oxides was an enormous undertaking, one that tested David's Augustinian patience and attention to details. (The truth be told, he found weekly solace with friends drinking in a louche bar on Massachusetts Avenue.) When it was time to graduate, he presented the first installment of his work, that on hydrous ferric oxide, as his doctoral thesis. It was accepted and published as a book.

Nearly 20 years later, in a demonstration of uncommon tenacity, David Dzombak and his postdoctoral associate Athanasios Karamalidis have brought to fruition another chapter of that original thesis idea; yet another book; this one on Gibbsite. This important addition to the available database on sorption reactions has great practical value, of course. Reactions at the solid–water interface play a key role in controlling the concentration and fate of solutes in natural and engineered aquatic systems. This book, like its predecessor, provides the means to quantify these reactions conveniently. But I believe that these books do more than reinterpreting data and compiling coherent thermodynamic parameters. By enabling comparison among sorption parameters, they also give insight into the variety of physical and chemical mechanisms responsible for the adsorption of solutes on solid surfaces. For example, by exposing the effects of ionic strength, they bring to light the contrasting behaviors and roles of activity coefficients in the bulk solution and at surfaces (i.e., the "coulombic term"). More importantly perhaps, the remarkable Linear Free Energy Relationships obeyed by consistent adsorption constants demonstrate an important underlying chemical regularity; they also provide the means to effectively make predictions where no data are available. We can now begin to reflect on the question of what coherence we should or should not expect between stoichiometries that underpin average thermodynamic quantification of sorption reactions and spectroscopic information on the coordination environment of atoms at surfaces.

Whether interested in making practical thermodynamic calculations, or in reflecting on the fundamental nature of the interactions at interfaces, we should be thankful to Athanasios Karamalidis and David Dzombak for this new opus.

François M. M. Morel

Allbert G. Blanke Professor of Geosciences
Princeton University
Princeton, New Jersey
December 2009

PREFACE

The development of this thermodynamic database for surface complexation of inorganic ions on gibbsite builds on the effort initiated in 1990 by David Dzombak and François Morel for amorphous iron oxyhydroxide, or hydrous ferric oxide (HFO). A primary objective in the publication of *Surface Complexation Modeling: Hydrous Ferric Oxide* (Wiley Interscience, 1990) was to advance the practical application of surface-complexation modeling. From the start of the development of surface-complexation models in the 1960s, a variety of different models have been developed, each with a particular physical description of the solid–water interface and hence with a different formulation for the electrostatic interaction submodel. The variations in descriptions of the solid–water interface result in different model parameters and types of surface-complexation reactions used to fit the same ion-sorption data. As a result, surface-complexation reactions and equilibrium constants extracted from sorption data sets with different models cannot be used collectively as a database for modeling sorption reactions.

Dzombak and Morel used a surface-complexation model with a simple solid–water interface model to interpret available data for inorganic ion sorption on HFO and thereby develop an internally consistent thermodynamic database for modeling sorption on this important sorbent in natural systems. While the simple two-layer model (one surface layer and a diffuse layer of counterions in solution) has its limitations as a description of reality for a complex solid like HFO in complex aqueous solutions, it has the benefit of being able to fit data for equilibrium sorption of ions on oxide surfaces across a range of solution conditions usually as well as more complex models and with fewer fitting parameters. The comparable performance and relative simplicity of the two-layer model is what guided its selection for the initial work with HFO.

The extensive use that the HFO database and the two-layer model have received has been gratifying. The model and HFO database have been used to gain quantitative insight into the role of sorption in natural aquatic systems of all types as well as in water, soil, and waste-treatment systems. The use of the model and HFO database has been facilitated by the incorporation of both into widely used general chemical equilibrium models derived from the original MINEQL model (John Westall, Joseph Zachary, and François Morel, Massachusetts Institute of Technology, 1976): MINEQL+, developed by Environmental Research Software, Inc.

(http://www.mineql.com); MINTEQ developed by the U.S. Environmental Protection Agency (http://www.epa.gov/ceampubl/mmedia/minteq/); MINTEQA2 for Windows developed by Allison Geoscience Consultants (http://www.allison-geoscience.com/); and VisualMINTEQ developed by the KTH Royal Institute of Technology in Sweden (http://www.lwr.kth.se/English/OurSoftware/vminteq/). While all models are subject to misuse and the HFO database and model no doubt have been applied inappropriately in some cases, the majority of applications have been appropriate and helpful to investigators. Results have been interpreted with consideration of the limitations associated with all applications of chemical-equilibrium models to complex aqueous systems encountered in natural and contaminated environments and treatment systems. The HFO model and database have certainly advanced the practical application of surface-complexation modeling.

Another goal in developing the HFO database was to provide the start of a larger database encompassing other important oxide sorbents for aqueous systems of interest in environmental science and engineering. Subsequent to the publication of the HFO database in 1990, the two-layer model and methodology used to develop the database were employed in follow-up database-development efforts for goethite (Samir Mathur and David Dzombak, 1995, 2006) and for hydrous manganese oxide (Jennifer Tonkin, Laurie Balistrieri, and James Murray, 2004). In this book, we present the fourth internally consistent database, for the very important natural sorbent, the aluminum hydroxide mineral gibbsite.

This book is organized based on the outline of the original HFO study. The first two chapters serve as a general introduction to aluminum (hydr)oxide chemistry, the following two chapters serve as the "materials and methods" section, and the final four are dedicated to surface complexation and modeling. In Chapter 1, we describe the importance of aluminum oxides and hydroxides in natural or engineered environments and their abundant occurrence. In Chapter 2, the formation and properties of gibbsite and its closely related minerals are presented. In this chapter, we try to clear the ambiguity in the various designations given for the aluminum (hydr)oxides and suggest a nomenclature. Chapters 3 and 4 describe the experimental data that constrain surface-complexation models and the systematic procedure followed to extract model constants from the data in the current and previous efforts. The historical development of surface-complexation modeling and the history and development of the generalized two-layer model are presented in detail by Dzombak and Morel (1990) and are not repeated in full here, but summarized information is provided. In addition, there is a discussion of surface spectroscopy constraints on surface-complexation modeling which has advanced since the initial work with HFO. In Chapter 5, we focus on potentiometric titration data fitting and surface acidity constant extraction for gibbsite. In this chapter, we also explore the influence of gibbsite dissolution on acid–base titration of gibbsite suspensions and the issue of accounting for the dissolution in formulating surface-complexation models for gibbsite. In Chapters 6 and 7, we present the available cation and anion sorption data for gibbsite and the results of fitting those data with the generalized two-layer model. In an attempt to constrain our model based on spectroscopic evidence, detailed information about available spectroscopic data for sorption of different ions on gibbsite is presented. The manner in which spectroscopic data were considered in fitting the experimental sorption data for each ion is discussed. Chapter 8 presents an investigation of linear free energy relationships for identifying trends in sorption data and predicting surface-complexation constants for conditions not yet studied experimentally. This chapter also examines the relative reactivity of the surface hydroxyl groups of gibbsite, goethite, HFO, and hydrous manganese oxide, through the comparison of the two-layer model surface-complexation constants extracted using the methodology specified in Dzombak and Morel (1990).

We gratefully acknowledge support for this work from Carnegie Mellon University through the Walter J. Blenko, Sr. Professorship, and assistance from the Department of Civil and Environmental Engineering. We thank Sabine Goldberg of the U.S. Salinity Laboratory for her time and comments on our work. Her extensive knowledge of the surface chemistry of aluminum oxide and surface-complexation modeling, and her generosity in answering our questions and reviewing various portions of our work were most helpful. We also thank Antonio Violante of Università degli Studi di Napoli Federico II for discussing various issues related to the crystallography of gibbsite, for reviewing Chapter 2, and for sharing with us information and parts of his voluminous work on gibbsite. We are grateful to several anonymous reviewers arranged by Wiley, whose comments improved our manuscript. We thank François Morel for reading parts of the manuscript and writing the Foreword, and for his vision which initiated this effort almost 30 years ago. We express special thanks to our families, and especially to Thanasis' family for putting up with the extended absences from home required to bring this effort to completion, particularly during the summer of 2009 when video conferences substituted for family travel!

Finally, we humbly thank all the experimentalists whose careful and diligent work was critical to the development of this gibbsite surface-complexation database.

Athanasios K. Karamalidis
David A. Dzombak

Carnegie Mellon University
Pittsburgh, Pennsylvania
December 2009

1

ALUMINUM OXIDES AND HYDROXIDES UNDER ENVIRONMENTAL CONDITIONS

1.1 INTRODUCTION

In natural aqueous systems, reactions at the solid–water interface are among the key processes controlling the transport and fate of metals and other inorganic chemicals. Hydrous metal oxides are abundant minerals in soils and sediments and are also important sorbents for inorganic species in these systems (Dzombak and Morel, 1990; Alloway, 1995; Goldberg et al., 1996a; Martinez and McBride, 1999; Trivedi and Axe, 2000, 2001).

The sorption of metal ions and other inorganic species on hydrous metal oxides across a range of solution chemistry conditions can be described with surface complexation models. To use such models for predictive simulation in chemical equilibrium modeling, it is necessary to have a consistent surface complexation model and a database of reactions and equilibrium constants extracted from experimental data with the particular surface complexation model (Dzombak and Morel, 1990; Goldberg, 1992).

With use of the generalized two-layer surface complexation model, internally consistent databases have been developed for sorption of inorganic ions on hydrous ferric oxide (Dzombak and Morel, 1990), on the common crystalline iron oxide goethite (Mathur and Dzombak, 2006), and on hydrous manganese oxide (Tonkin et al., 2004). Another class of metal oxides equivalently common to hydrous iron oxides is gibbsite, which is the subject of this book. Iron, aluminum, and manganese oxides and hydroxides are the most ubiquitous of the hydrous metal oxides.

Surface Complexation Modeling: Gibbsite By Athanasios K. Karamalidis and David A. Dzombak

Aluminum (Al) occurs ubiquitously in the terrestrial environment. It is the most abundant metal in the lithosphere, comprising about 8.2 percent of the Earth's crust (Bowen, 1979) and about 7.2 percent of soils (Schacklette and Boerngen, 1984; as reported by Sposito, 1989). Due to its reactivity, Al does not occur in elemental form in nature but is present predominantly in sparingly soluble oxides and aluminosilicates (Scancar and Milacic, 2006). Aluminum is a key component of clays, and also occurs in various oxide and hydroxide minerals as described below. Aluminum oxide and hydroxide solids are of great importance in the chemistry of soil, sediment, surface water, and groundwater systems because of their adsorptive role and dissolution properties.

When aluminum dissolves into water from clays and hydrous metal oxides, the liberated free aluminum ion Al^{3+} reacts with water to form various hydroxy species including $AlOH^{2+}$, $Al(OH)_2{}^{+}$, $Al(OH)_3{}^{0}$, and $Al(OH)_4{}^{-}$. The aqueous speciation of dissolved Al strongly depends on pH. The released mononuclear ionic Al species may undergo polymerization (Bi et al., 2004) or may be complexed by available organic or inorganic ligands (Martell et al., 1996; Smith, 1996; Exley et al., 2002; Scancar and Milacic, 2006). Polymerization of the ionic Al species leads to precipitation of aluminum hydroxide or oxyhydroxide solids. Thus, the abundance and form of aluminum in soil and water systems is dependent on the dissolution–precipitation cycle of clays and aluminum oxides and hydroxides.

Aluminum oxides and hydroxides are also used in a variety of industrial and technological applications, including as adsorbents in water and wastewater treatment. Aluminum sulfate (alum) is widely employed in drinking water treatment systems across the world to precipitate aluminum hydroxide as a coagulation agent for particle removal (Licsko, 1997; Letterman et al., 1999) and as an adsorbent for contaminants such as arsenate (McNeill and Edwards, 1999). In addition, aluminum oxide is used as a fixed-bed adsorbent for removal of arsenate, fluoride, and other ionic contaminants in drinking water treatment (Chowdhury et al., 1991; Huang and Shiu, 1996; Martell et al., 1996; Clifford, 1999; Viraraghavan et al., 1999; Dayton and Basta, 2005; Ayoob et al., 2008) and industrial wastewater treatment (Karthikeyan et al., 1997; De-Bashan and Bashan, 2004).

1.2 OCCURRENCE OF ALUMINUM OXIDES AND HYDROXIDES IN THE SUBSURFACE

The nonsilicate aluminum minerals that occur in soils and the shallow subsurface environment include the aluminum hydroxides, gibbsite ($Al(OH)_3(s)$), bayerite ($Al(OH)_3(s)$), nordstrandite ($Al(OH)_3(s)$), and the oxides or (oxy)hydroxides boehmite ($AlO(OH)(s)$), diaspore ($AlO(OH)(s)$), alunite ($KAl_3(SO_4)_2(OH)_6(s)$), and corundum ($Al_2O_3(s)$) (Eswaran et al., 1977). The most common aluminum hydroxides, oxyhydroxides, and oxides found in nature are shown in Table 1.1.

Weathering of minerals to form gibbsite occurs most intensely in humid tropic environments, but also in other environments. Gibbsite has been found in the clays of alpine soils (Reynolds, 1971), in volcanic ash soils in Japan (Wada and Aomine, 1966),

TABLE 1.1 Aluminum oxides and hydroxides occurring under environmental (surface and near-surface, low-temperature) conditions

Name	Chemical Formula	Occurrence in Nature	Frequency of Occurrence	Year Approved by IMA[a]	Year of Discovery
Gibbsite	$Al(OH)_3$	Occurs abundantly in humid tropical climates (usually in soils at high weathering stage, such as oxisols, ultisols, or ferrolsols).[b] Occurs ubiquitously in soils and structures of clays	Very common[c,d]	1822	1820 (Dewey)
Bayerite	$Al(OH)_3$	Precipitated from high aluminum concentration gels at pH >5.8; as weathered crusts on amphiboles and pyroxenes; in bauxites	Very rare[d,e]	1928	1925 (Böhm)
Nordstrandite	$Al(OH)_3$	Weathering product of bauxitic soils derived from limestone. Alteration product of aluminum carbonate minerals. Late-stage mineral in nepheline syenite pegmatites	Very rare[c,d]	1958	1956 (Van Nordstrand)
Doyleite	$Al(OH)_3$	Occurs in albitite veins in nepheline syenite and in silicocarbonatite sills	Very rare[c]	1985	1985 (E.J. Doyle)
Boehmite	AlO(OH)	Occurs abundantly in subtropical areas of high rainfall and commonly in soils elsewhere	Very common[c]	1927	1925 (Böhm and Niclassen)
Diaspore	AlO(OH)	Common in many environments	Very common[c]	1801	1801 (Hauy)
Corundum	Al_2O_3	Corundum occurs as a mineral in mica schist, gneiss, and some marbles in metamorphic terranes. It also occurs in low silica igneous syenite and nepheline syenite intrusions. Because of its hardness and resistance to weathering, it commonly occurs as a detrital mineral in stream and beach sands	Common[c]		1798 (Greville)

[a] The International Mineralogy Association, which maintains a public listing of all the approved mineral names for all minerals since 1959, where official determinations are on record.
[b] USDA (1996); FAO (1998); Brady and Weil (2002).
[c] Wefers and Misra (1987).
[d] Hsu (1977).
[e] Huneke et al. (1980).

in temperate areas in North Carolina (Cate and McCracken, 1972), France (Dejou et al., 1970), and Scotland (Wilson, 1970), and in hotter areas such as Zaire (Eswaran et al., 1977).

An accumulation of gibbsite with or without the other forms of aluminum hydroxides or oxides (e.g., diaspore) characterizes bauxite, the primary ore used for production of aluminum metal. Bauxite rich in gibbsite is usually found in areas characterized by a tropical climate with alternating rainy and dry periods. Bauxite with primarily boehmite appears to be more constrained to the subtropical areas (Mediterranean-type bauxite) (Kloprogge et al., 2006). In lateritic bauxites, gibbsite and boehmite are the most common minerals, diaspore occurs but not frequently, and corundum appears very rarely. In karst and sedimentary bauxites, diaspore is frequently found while corundum is occasionally observed (Trolard and Tardy, 1987). It has been shown (Bardossy, 1982; Trolard and Tardy, 1987) that the distribution of these minerals is a function of climate, petrografic organization, age of formation, degree of compaction, and, in some cases, temperature and degree of metamorphism.

Generally, bauxite is considered to have at least 45.5 percent by weight Al_2O_3 and less than 20 percent Fe_2O_3 and 3.5 percent combined silica (Valeton, 1972; Eswaran et al., 1977). Global bauxite resources are estimated to be 55–75 billion tons, located in Africa (33 percent), Oceania (24 percent), South America and the Caribbean (22 percent), Asia (15 percent), and elsewhere (6 percent) (U.S. Geological Survey, 2008).

In addition to its occurrence as a pure phase, gibbsite occurs ubiquitously as precipitates in the interlayer or as part of the structure of common clay minerals, especially the illite, kaolinite, chlorite, and montmorillonite/smectite groups (Barnishel and Rich, 1963; Weaver and Pollard, 1973; Violante and Jackson, 1981; Dixon et al., 1989; Bird et al., 1994; Sposito, 1996). The individual aluminum hydroxide layers in these clays are gibbsite. Gibbsite can be formed from the hydrolysis and desilication of clay minerals (Freeze and Cherry, 1979; Sposito, 1989; Bird et al., 1994).

The clay minerals are important constituents of soil and they are excellent sorbents (e.g., gibbsite) for toxic metals and natural scavengers for cations and anions as water flows over soil or penetrates underground. The high specific surface area, high cation exchange capacity, layered structure, chemical and mechanical stability, Brönsted and Lewis acidity, and so on have made them excellent materials for adsorption (Tanabe, 1981; Dubbin et al., 2000; Gupta and Bhattacharyya, 2006; 2008; Bhattacharyya and Gupta, 2008a; 2008b).

1.3 OCCURRENCE OF ALUMINUM OXIDES AND HYDROXIDES IN SURFACE WATER

Surface waters contain a wide range of total aluminum (dissolved and solid phase) concentrations, with the amount depending on watershed soil characteristics and chemistry of rainwater and runoff. Generally, more acidic waters contain the most aluminum. Based upon data obtained in the northeastern United States, Canada, Sweden, Norway, and Germany, Cronan and Schofield (1979) concluded that one of the primary effects of acid deposition is increased mobilization of Al from soils to surface waters (Sullivan and Cosby, 1998). Acid rain has caused the aluminum level in many freshwater sources to increase (Schecher and Driscoll, 1988; Swistock et al., 1989). Al mobilization is now widely believed to be the most important ecological effect of surface water acidification (Sullivan and Cosby, 1998). Aqueous Al concentrations in acidified drainage waters are often an order of magnitude higher than those in circumneutral waters. Concentrations of Al in surface waters correspond reasonably well with the equilibrium solubility of $Al(OH)_3(s)$, but at low pH are generally undersaturated with respect to gibbsite (Sullivan and Cosby, 1998; Gensemer and Playle, 1999).

In aquatic systems, the solubility of Al is often controlled by amorphous $Al(OH)_3(s)$ or by microcrystalline gibbsite (Zänker et al., 2006), the latter being almost three orders of magnitude less soluble than the former. These solids exhibit solubility minima in the pH range 6–7. However, equilibrium conditions often do not exist because the Al hydroxo mono- and polynuclear complexes react slowly, and metastable polymorphs that form recrystallize slowly (Wesolowski, 1992; Zänker et al., 2006). The metastable species may exist for months or even years. There is some evidence that the metastable species grow in size as a function of time and ultimately are converted to microcrystalline gibbsite (Berkowitz et al., 2005). Gibbsite may form via different routes without hindrance in a short period of time varying from days to few weeks (May et al., 1979; Sposito, 1996).

Aluminosilicate minerals are formed when polysilicic acid reacts with dissolved aluminum species. Thus, the aluminosilicate secondary mineral kaolinite controls the equilibrium solubility of aluminum in natural waters as soon as the Si concentration exceeds about 1–3 ppm (Langmuir, 1997; Zänker et al., 2006). This condition is fulfilled by many natural water compositions. The dependence of kaolinite precipitation rates, which are slow, on pH under environmental conditions (at about 22°C) has not yet been determined. The rate of kaolinite precipitation could influence or even control the overall rate at which chemical weathering occurs (Yang and Steefel, 2008).

High aluminum concentrations have been reported in surface waters receiving large inputs of acid sulfate solution, such as acid rain and acid mine drainage, basically due to the enhanced dissolution kinetics of gibbsite (and by analogy, of other aluminum-containing minerals) (Ridley et al., 1997). Speciation calculations for aluminum in water samples taken from a basin imported by acid mine drainage demonstrate that above pH 4.9 dissolved Al is consistent with the equilibrium solubility of microcrystalline gibbsite or amorphous aluminum hydroxide (Nordstrom and Ball, 1986). Some investigators have reported that the activity of Al^{3+} in high sulfate loading in groundwaters and soil solution appears to be regulated by the solubility of jurbanite ($Al(OH)SO_4{\cdot}5H_2O$) (Karathanasis et al., 1988; Alvarez et al., 1993; Driscoll and Postek, 1996).

Particles bearing aluminum are common in surface waters. Clay and silt particles from erosion that are suspended in surface waters are sources of particulate Al.

In addition, aluminum hydroxide that is precipitated *in situ* contributes to the Al suspended solids, either as a separate solid or as a coating on other particles. In surface water, $Al(OH)_3$ solids with dissolved substances sorbed to them are common.

1.4 USE OF ALUMINUM HYDROXIDE IN WATER TREATMENT

The use of aluminum-bearing compounds in drinking water treatment has been conducted since the late 1800s. Al-based coagulants such as aluminum sulfate (alum, $Al_2(SO_4)_3(s)$) or polyaluminum chloride (PACl) are commonly used in drinking water treatment to enhance the removal of particulate and colloidal substances via coagulation processes (Srinivasan et al., 1999). Addition of the aluminum salts results in precipitation of voluminous $Al(OH)_3(s)$ particles that settle and remove finer particles. They are widely used because they are effective, readily available, and relatively inexpensive. The parallel processes that take place after aluminum sulfate addition to water to precipitate $Al(OH)_3(s)$ and form particle flocs during water treatment are depicted in Figure 1.1.

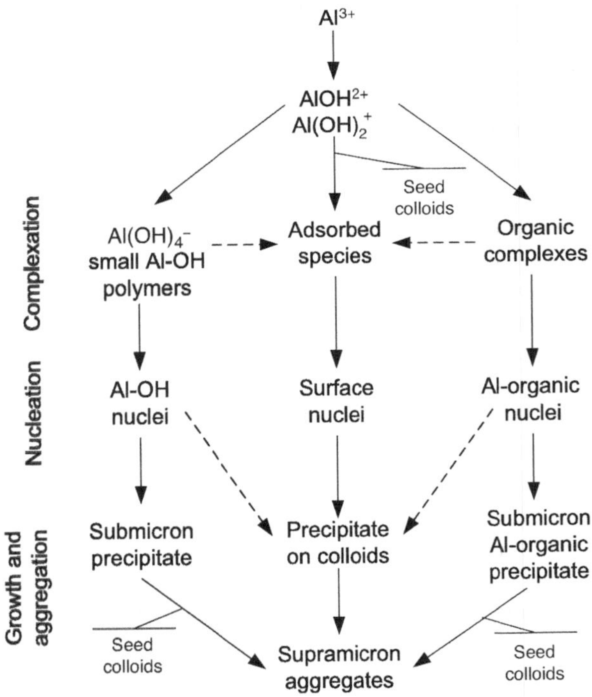

FIGURE 1.1 Parallel processes leading to incorporation of colloids into $Al(OH)_3$ flocs. Arrows indicate possible pathways; dashed lines are secondary pathways. (Adapted from Chowdhury et al., 1991.)

Precipitated $Al(OH)_3(s)$ also serves as an adsorbent in drinking water treatment, for removal of ionic contaminants such as arsenate (McNeill and Edwards, 1999) and dissolved natural organic matter (Huang and Shiu, 1996). The major mechanisms of organic acid removal by alum coagulation involve complexation, charge neutralization, precipitation, and adsorption entrapment (Huang and Shiu, 1996).

Typically, a portion of the alum added to the raw water is not removed during treatment and remains as residual aluminum in the treated water. The use of alum as a coagulant for water treatment often leads to higher concentrations of dissolved aluminum in the treated water than in the raw water itself. There is considerable concern throughout the world over the levels of aluminum found in drinking water sources (raw water) and treated drinking water. A high (3.6–6 mg/L) concentration of aluminum in treated water gives rise to turbidity, reduces disinfection efficiency, and may precipitate as $Al(OH)_3$ during the course of distribution (Srinivasan et al., 1999; Snoeyink et al., 2003).

Aluminum deposits can form in distribution systems because aluminum particles are not completely removed by sedimentation, and because the water is supersaturated with solids such as $Al(OH)_3(am)$ and $Al_2O_3(s)$, aluminosilicates, and aluminum phosphates. Supersaturation leading to Al precipitation may result because of (1) failure to reach equilibrium in the treatment plant during coagulation, flocculation, and sedimentation; (2) lowering of temperature during storage and transport; and (3) decreasing pH in the distribution system within the pH range of 6–10 (Snoeyink et al., 2003).

Granular aluminum oxide particles, typically granular activated alumina, are sometimes used in drinking water treatment and in treatment of industrial process water or wastewater via a fixed-bed configuration (Brattebo and Odegaard, 1986; Karthikeyan et al., 1997; Ghorai and Pant, 2004). Removal of fluoride has often been a target for such systems, but alumina has the ability to adsorb many other chemical contaminants, including oxyanions of arsenic and many dissolved metals (Manning and Goldberg, 1997; Paulson and Balistrieri, 1999; Lin and Wu, 2001; Goldberg, 2002; Singh and Pant, 2004; Ghorai and Pant, 2005; Violante et al., 2006).

1.5 SUMMARY

Gibbsite and amorphous aluminum hydroxide are the aluminum (hydr)oxide solids most commonly formed under environmental conditions in soils, sediments, surface waters, and shallow groundwaters. Gibbsite $Al(OH)_3(s)$ is the most common crystalline form of aluminum hydroxide in nature (Schoen and Robertson, 1970; Violante and Huang, 1993; Gale et al., 2001; Digne et al., 2002; Liu et al., 2004). The presence of gibbsite in soils is generally attributed to the action of weathering processes of high intensity and of long duration. It is especially abundant in highly weathered, acidic soils. It also occurs ubiquitously as part of the structure of common clay minerals, and can be liberated from weathering of clay minerals through hydrolysis and desilication. Gibbsite is very stable under most earth surface conditions, but may alter under special conditions to clay minerals, such as the kaolin minerals.

As is the case with most of the hydrous metal oxides, gibbsite has the ability to adsorb metal ions and anions as well as ligands on its surface. Ion binding on gibbsite in soils and sediments is well documented. Because of its sorptive role in nature and in water treatment, this book and the database it contains is focused on gibbsite.

2

FORMATION AND PROPERTIES OF GIBBSITE AND CLOSELY RELATED MINERALS

Aluminum oxides including gibbsite, boehmite, and diaspore are ubiquitous in soils in various crystalline and amorphous forms. Like the common oxides of iron, manganese, and silicon, aluminum oxides can sorb a host of chemical species and are important sorbents in natural systems.

Aluminum oxides with physical and surface properties engineered through thermal and other kinds of treatments are used in water treatment, catalysis, pharmaceuticals, separations, and other technology areas. Expansion of commercial applications has been driven by the increasing scientific understanding of the chemical, structural, and surface properties of aluminum hydroxides and oxides.

In this chapter, detailed descriptions of gibbsite and closely related aluminum oxide minerals and their properties under environmental conditions are provided. Nomenclature for gibbsite and its mineralogic neighbors is reviewed and clarified. Crystal properties that distinguish the various forms are explained and summarized.

2.1 AI POLYMERIZATION MODELS

The formation of aluminum hydroxides, such as gibbsite, is generally achieved by hydrolysis–polymerization of Al^{3+}. The hydrolysis–polymerization mechanisms and species conversions of Al^{3+} have been extensively explored for over a century. These mechanisms are explained by three widely accepted models: the "Core-links" model, the "Cage-like" Keggin-Al_{13} model, and the "Continuous" model (Bi et al., 2004). The three models are briefly described below.

Surface Complexation Modeling: Gibbsite By Athanasios K. Karamalidis and David A. Dzombak

2.1.1 The "Core-Links" Model

The model was initially introduced in 1954 by Brosset et al. who suggested a series of "Core-links" polymeric Al species whose form was $Al(Al_2(OH)_5)_n^{3+n}$ (Brosset et al., 1954). The Core-links model gives a distribution of the continuously changing transient state species of Al in the hydrolysis–polymerization process. It provides a basis for interpreting the various possible transient polymeric Al species and for explaining the experimental facts about how the monomeric and polymeric Al are converted into $Al(OH)_3$(am). However, it lacks direct and unequivocal evidence to prove the existence of these transient species, and it works under the condition of moderate rate of base injection in titrating Al^{3+} salts (Bi et al., 2004). The hydroxyl Al species change from monomeric to polymeric following the hexameric ring model (Fig. 2.1). The structure of OH–Al polymer in solution is the same as that of $Al(OH)_3$(s), whose basic units are either $Al_6(OH)_{12}(H_2O)_{12}^{6+}$ (single hexamer ring) or $Al_{10}(OH)_{22}(H_2O)_{16}^{8+}$ (double hexamer rings).

The Core-links model is the most commonly invoked model for Al polymerization, especially in the field of geochemistry because it follows the crystallographic law of gibbsite. Many scholars remain strong supporters of the model, so it has been coexisting with the "Cage-like" Keggin-Al_{13} model for more than 50 years (Bi et al., 2004).

2.1.2 The "Cage-Like" (Keggin-Al_{13} Structure) Model

The model was initially proposed by Johansson (1960, 1962) and it was based on sulfate precipitation from partially neutralized Al^{3+} solutions that were heated for 30 min at 80°C and aged for few days. The model considers that in Al solutions there are monomer, dimer, Keggin-Al_{13} polymer, and larger polymerized Al species. These species can be transformed from one to another directly (Akitt et al., 1972; Bottero et al., 1980; 1987; Bi et al., 2004). The metastable Keggin-Al_{13} polymer molecule is formed by the structural reordering of transient species after aging. Under the conditions of aging, heating, adding extra SO_4^{2-}, and slow addition of base, transient polymeric Al species transform from a state of disorder to a more ordered state (Bi et al., 2004). The Cage-like model is widely used in studies of the aqueous chemistry of aluminum because the polynuclear species of Al^{3+} invoked in the model are those identified experimentally.

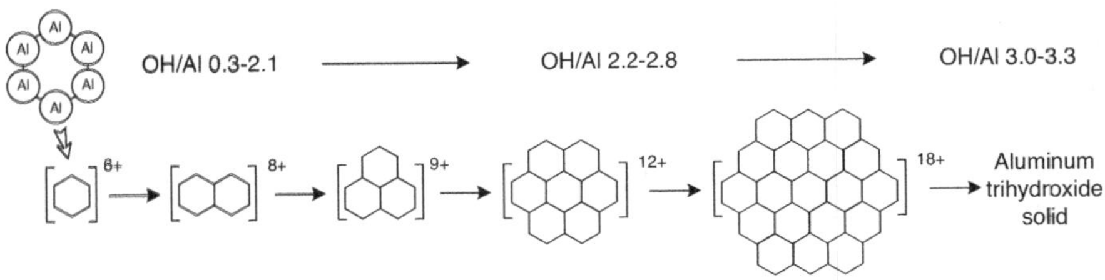

FIGURE 2.1 The polymerization of Al^{3+} via coalescence of the hexamer units. (*Source*: Bi et al., 2004.)

2.1.3 The "Continuous" Model

The model was introduced by Bi et al. (2004). Polynuclear Al species are considered to be a series of dynamic intermediates formed in the process of "hydrolysis–polymerization–flocculation–sedimentation." The model considers that in aged polymeric Al solutions, under a fixed molar ratio of OH/Al, if prolonging the aging time properly, only one polymeric Al species may exist, and this is Keggin-Al_{13} (Bi et al., 2004). The metastable Keggin-Al_{13} is formed by the structural reordering of transient species after aging. Aging is one prerequisite for Keggin-Al_{13} formation; elevating temperature and addition of extra SO_4^{2-} promote this conversion process. It is a combined model unifying the "Core-links" model and the "Cage-like" model. It can explain the entire course of hydrolysis–polymerization upon the addition of base to Al^{3+} solution (Bi et al., 2004). A conceptual representation of the "Continuous" model is shown in Figure 2.2.

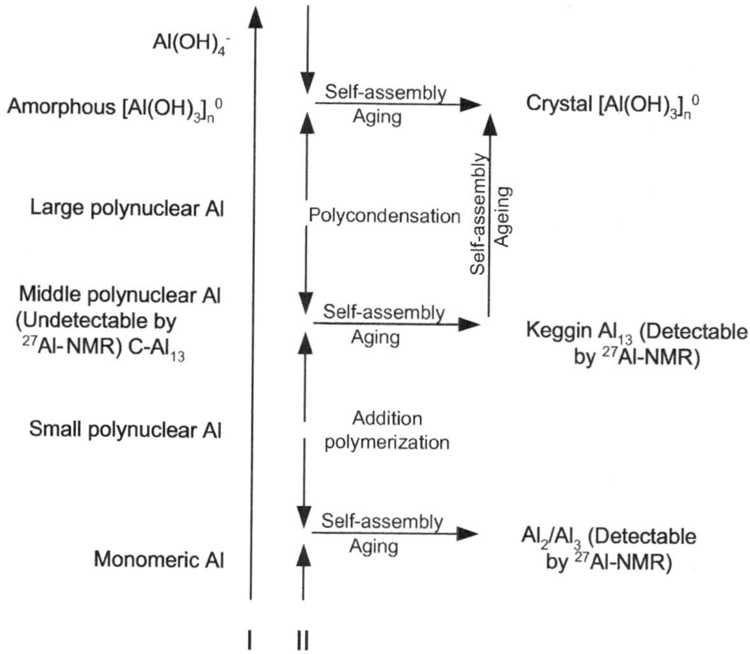

FIGURE 2.2 A conceptual representation of the "Continuous" model for Al^{3+} hydrolysis and polymerization, which builds on the "Core-links" model for evolution of transient polymeric Al species. Through self-assembly, the more stable species after aging are Al_2, K-Al_{13}, Al_{13}, and $[Al(OH)_3]_n(s)$. $[Al(OH)_3]_n(s)$ can be characterized by solid-state ^{27}Al NMR spectroscopy or X-ray diffraction. Other forms, including $Al(OH)_4^-$, can be measured by solution ^{27}Al NMR spectroscopy. Al_2: $Al_2(OH)_n(H_2O)_{10-n}^{(6-n)+}$ ($n = 2$–6); Al_3: $Al_3(OH)_n(H_2O)_{14-n}^{(9-n)+}$ ($n = 4$–9); Al_{13}: $Al_{13}O_4(OH)_n(H_2O)_{32-n}^{(31-n)+}$ ($n = 24$–31); K-Al_{13}: Keggin-Al_{13}; C-Al_{13}: Cage-Al_{13}. (*Source*: Bi et al., 2004.)

2.2 FORMATION OF GIBBSITE AND OTHER Al HYDROXIDES AND OXYHYDROXIDES

Gibbsite is one of the dominant forms of aluminum oxide resulting from precipitation and aging of aluminum hydroxide gels. Aluminum hydroxide gels are formed from solutions of aluminum salts by precipitation and aging. The relationship between the freshly precipitated amorphous aluminum hydroxide and the other forms has been summarized by several groups (Bye and Robinson, 1964; Schoen and Robertson, 1970; Alwitt, 1976). These groups document that in the environmental temperature range the amorphous aluminum hydroxide transforms first to pseudoboehmite, then to bayerite, and upon elevated pH to gibbsite (Alwitt, 1976). Norstrandite may also be present in the last two steps of the proposed transformation mechanism. Hsu (1966) proposed that the difference between the formation of the two polymorphs (gibbsite and bayerite) was in the source of the OH^- groups. When a strong alkaline solution is used, OH^- almost immediately couples with Al^{3+} to form bayerite. When acid solution is used, then the OH^- must come from the dissociation of water and gibbsite forms. Alwitt (1976) proposed that the transformation from pseudoboehmite can lead to any of the three forms of the aluminum trihydroxide, namely, gibbsite, bayerite, or norstrandite (Fig. 2.3), with the final product depending on aging conditions. In water, at temperatures up to 80°C and at a pH range of 7–11, the predominant product is bayerite (Alwitt, 1976). However, bayerite is considered a thermodynamically metastable phase and gibbsite is the stable phase below 90°C at 1 atm. It has been reported that upon aging bayerite can be transformed to gibbsite in several months (Pascal et al., 1963).

Bertsch and Parker (1996) proposed a mechanism for aluminum hydroxide solid formation similar to that in Figure 2.3, but which includes multiple reaction pathways as shown in Figure 2.4, many of which may occur simultaneously but in different relative proportions. Reaction pathway I occurs when solutions are brought to supersaturation through dilution or when samples are neutralized rapidly. Under these conditions, gibbsite crystals appear within days or weeks of aging. Reaction

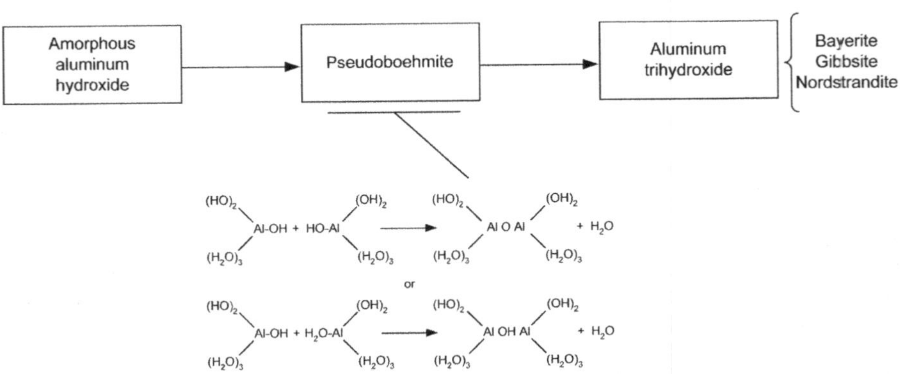

FIGURE 2.3 Sequence of transformations of amorphous to crystalline forms of aluminum hydroxide upon aging.

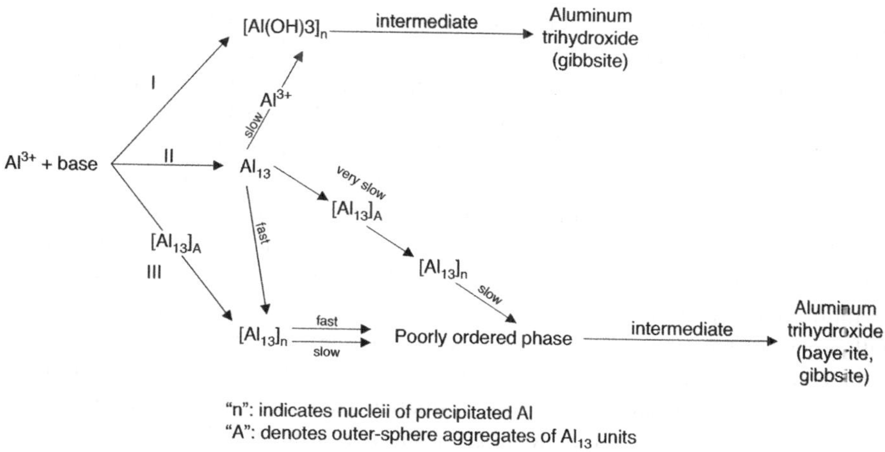

FIGURE 2.4 Reaction pathways for aluminum hydroxide formation from hydrolyzed aluminum solutions. The role of Al_{13} in controlling the reaction kinetics is depicted. (*Source*: Sposito, 1996, p. 148.)

pathways II and III both involve the formation of Al_{13} polymer through hydrolysis, with the resulting Al_{13} present either as soluble species or as colloidal aggregates (Bertsch and Parker, 1996). These mechanisms may require a longer time than the reaction pathway I. The time can vary from months to years if slow or weeks to months if fast. Again, the final product depends on the experimental conditions.

Under environmental conditions, gibbsite is usually formed through precipitation of amorphous aluminum hydroxide and extended aging. Precipitation occurs when aluminum salt solution is neutralized with a base and the precipitate is X-ray amorphous. The amorphous aluminum hydroxide is colloidal in nature, comprising discrete particles that are roughly spherical with diameter between 30 and 200 Å and that exhibit high solubility at room temperature and $pH < 7$ (Alwitt, 1976). With aging, this noncrystalline material decreases in abundance as $Al(OH)_3(s)$ polymorphs crystallize (Violante, 1989).

The rate of crystallization controls the final aluminum hydroxide products (Hsu, 1966, 1989; Violante and Violante, 1980; Violante and Jackson, 1981; Violante and Huang, 1993). A rapid crystallization yields bayerite and slow crystallization promotes gibbsite, whereas intermediate conditions favor the formation of nordstrandite. The abundance of gibbsite in the environment may thus be explained by the slow crystallization rates favored under environmental conditions (Cesteros et al., 2001; Liu et al., 2008). Faster formation rates can be obtained for synthetic gibbsite if (1) the aluminum gel hydroxide precursor is generated in a hot precipitation process, (2) the aluminum gel hydroxide is first aged in basic medium and then in acid medium, and (3) the ammonia concentration is low (Cesteros et al., 1999).

Organic and inorganic ligands delay or inhibit the crystallization rate of $Al(OH)_3(s)$ to varying degrees. The chemical composition, the molecular structure, and the nature of functional groups of each anion control the kinetics of crystallization (Violante and Violante, 1980; Violante and Huang, 1985). The nature of the precipitation products varies greatly with the ligands, the ligand/Al molar ratio, and the pH of systems. The effectiveness of each ligand in inhibiting aluminum hydroxide formation increases as its concentration increases and the pH decreases (Violante, 1989).

In an extensive investigation on the influence of inorganic and organic ligands on Al solid-phase formation, Violante and Huang (1985) demonstrated that various ligands promote and stabilize the formation of pseudoboehmite over the crystalline $Al(OH)_3(s)$ (Violante, 1989; Sposito, 1996). The ligands with a strong affinity for Al inhibit the crystallization of Al hydroxides or oxyhydroxides more than ligands with a poor or moderate affinity for Al. Also, polydentate and large ligands such as tannate, fulvate, and humate often favor stabilization of noncrystalline materials more than those with fewer functional groups or smaller size. This can be seen even if small ligands or those with fewer functional groups have a stronger affinity for Al than the polydentate or large ligands. The latter may promote structural changes as well as aggregation between the particles of the noncrystalline materials.

In an attempt to evaluate the influence of citric acid on the formation of $Al(OH)_3(s)$, Violante and Huang (1993) proposed the mechanism presented in Figure 2.3. Their experiments showed that the nature of the noncrystalline or poorly crystalline materials formed during the early stages of the aging process had an important effect on the Al transformation when the pH of the suspension was increased. The basic pathway of transformation of noncrystalline Al precipitation products to citrate-stabilized pseudoboehmite relies on the formation of various poorly ordered intermediate materials that increase in structural order and particle size and decrease in solubility as they age. When the pH of the suspension is increased to 10.0 without any preceding aging time, then fast solubilization and crystallization occur and the final product is bayerite (pathway A, Fig. 2.5). By increasing the pH of the original suspension to 10 after 1–6 days of aging, a mixture of gibbsite and nordstrandite is formed (pathway B, Fig. 2.5). A slower rate of crystallization promotes the formation of nordstrandite over bayerite (Violante and Violante, 1980; Violante et al., 1982). After aging of 11–43 days, pseudoboehmite starts to form along with other intermediate Al precipitation products, which in turn partially converts to gibbsite when the pH is around 10. After additional aging of 60 days or more, no gibbsite forms under these conditions (pH 10) (pathway C, Fig. 2.5). Crystallization of $Al(OH)_3$ polymorphs in slightly acidic, neutral, or alkaline media takes place via a dissolution–reprecipitation reaction (Calvet et al., 1953; Souza Santos et al., 1953; Papée et al., 1958; Bye and Robinson, 1964; Violante and Violante, 1980; Van Straten et al., 1984; Violante and Huang, 1985; Hsu, 1989; Violante, 1989). An alkaline medium favors the bayerite structure, whereas an acidic medium promotes the gibbsite structure. A mixture of nordstrandite with bayerite and/or gibbsite can be obtained in an intermediate pH value range (Jardine and Zelazny, 1996).

Synthetic alumina, especially γ-$Al_2O_3(s)$ that is used as a proxy for natural Al oxides, becomes supersaturated with gibbsite very rapidly (Goldberg and Glaubig, 1988), and can transform to gibbsite at pH ≥ 5. There are two mechanisms that can explain this transformation of γ-$Al_2O_3(s)$ to $Al(OH)_3(s)$ (gibbsite): (1) γ-$Al_2O_3(s)$

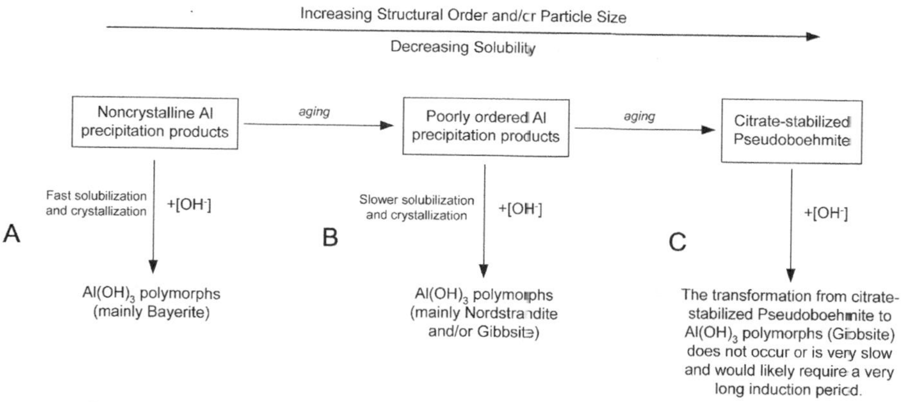

FIGURE 2.5 Proposed pathway of transformations of Al precipitation products in the presence of citrate as influenced by aging and change of pH in alkaline conditions. (*Source*: Violante and Huang, *Clays and Clay Minerals*, 1993, Figure 6, reproduced with permission of The Clay Minerals Society.)

surface hydration through hydrolysis of surface Al–O bonds and (2) dissolution of γ-Al_2O_3(s) and subsequent precipitation (Carrier et al., 2007). The composition of the final material is thus heterogeneous. In the acidic pH range, thermodynamically stable gibbsite particles are formed, while at near-neutral and higher pH values, bayerite becomes the predominant phase (Carrier et al., 2007).

2.3 ALUMINUM HYDROXIDE POLYMORPHS: STRUCTURE AND NOMENCLATURE

Aluminum hydroxide, $Al(OH)_3$(s), exhibits polymorphism, that is, when identical elemental compositions occur in more than one crystal structure (Hemingway and Sposito, 1996). The crystals comprising two or more compounds of identical composition are called *polymorphs*. The different structures of the polymorphic crystals result in differences in the physical properties and in the appearance of the crystals (Hemingway and Sposito, 1996).

Four known polymorphs of crystalline aluminum hydroxide exist: gibbsite, bayerite, nordstrandite, and doyleite. The pH, presence of anions or salt, and mineral surfaces in the parent solution play important roles in the formation of crystalline $Al(OH)_3$(s) polymorphs from Al interlayers and/or hydroxyl–Al polymers (Barnishel and Rich, 1965). All the polymorphs of aluminum hydroxide consist of layers of aluminum octahedra with hydroxyl groups on either side that hydrogen bond the layers together. The difference between the polymorphs is the stacking sequences of the layers, as shown in Figure 2.6 (Hsu, 1977; Digne et al., 2002). Gibbsite and bayerite represent the two extremes of stacking sequence, with nordstrandite and doyleite being intermediate structures. Gibbsite is the most thermodynamically stable form of aluminum hydroxide under environmental conditions (Gale et al., 2001; Digne et al., 2002; Veilly et al., 2008).

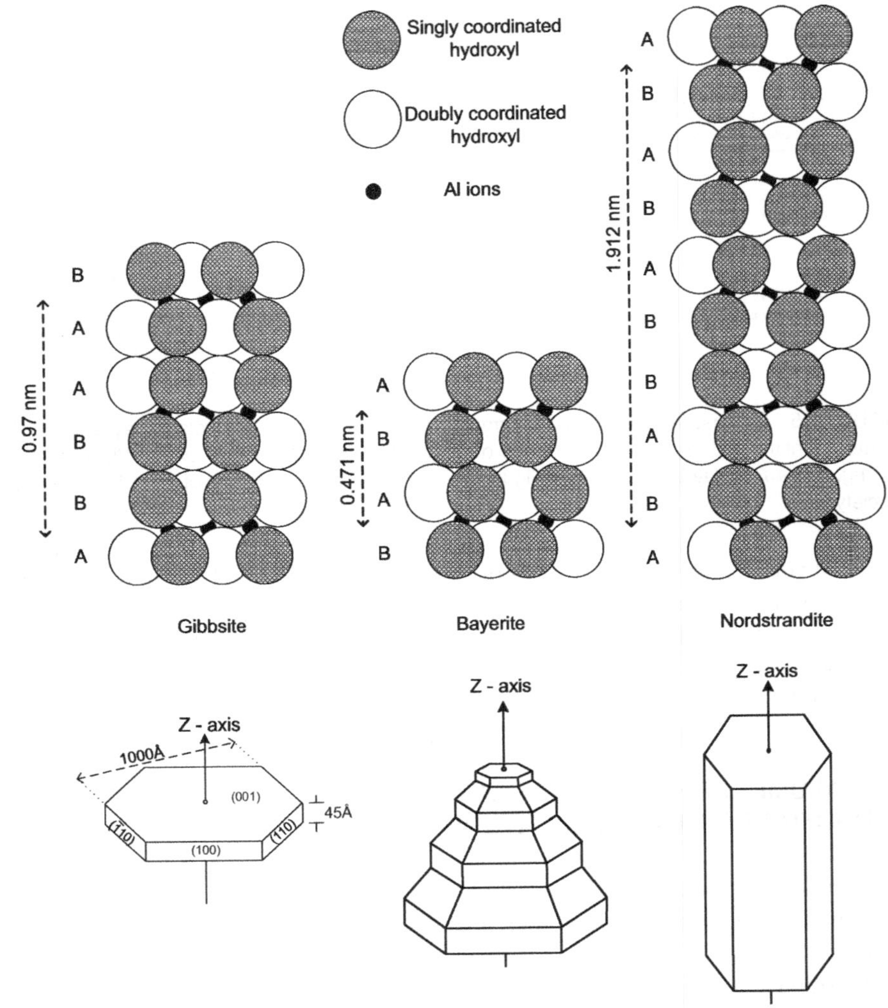

FIGURE 2.6 Schematic presentation of the layer stacking of $Al(OH)_3$ (above) and their respective common crystal morphologies (below) (after Schoen and Robertson, 1970; Wierenga et al., 1998; Hiemstra et al., 1999.)

There is much ambiguity in the various designations that have been given to the different forms of aluminum oxide and hydroxide. A comparison of nomenclatures for gibbsite (Table 2.1) and closely related minerals formed under environmental conditions illustrates the ambiguity.

Besides the mineral names, which are usually given by the name of the investigator who discovered them or analyzed their structure for the first time, there is also a one-letter designation accompanying the chemical formula of the minerals. Initially, aluminum hydroxides were assigned to two series distinguished with the Greek letters alpha (α) and gamma (γ), respectively (Lehl, 1934; Wefers and Misra, 1987). The

TABLE 2.1 Comparison of nomenclatures for gibbsite and closely related minerals formed under environmental conditions

Mineral Name	Chemical Composition	Common Crystallographic Designation[a]	Past Accepted Crystallographic Designation (1957)[b,c]	Alcoa (1930)[c]
Gibbsite (hydrargillite)[d]	Aluminum trihydroxide	α-$Al(OH)_3$	γ-$Al(OH)_3$	Alpha alumina trihydrate
Bayerite	Aluminum trihydroxide	β-$Al(OH)_3$	α-$Al(OH)_3$	Beta alumina trihydrate
Nordstrandite	Aluminum trihydroxide	$Al(OH)_3$	$Al(OH)_3$	
Doyleite	Aluminum trihydroxide	$Al(OH)_3$		
Boehmite	Aluminum oxyhydroxide	γ-AlOOH	γ-AlOOH	Alpha alumina monohydrate
Diaspore	Aluminum oxyhydroxide	α-AlOOH	α-AlOOH	Beta alumina monohydrate
Corundum (α-alumina)	Aluminum oxide	α-Al_2O_3	α-Al_2O_3	Alpha alumina

[a] There is no official universal standard nomenclature for aluminum oxides and hydroxides. The categorization was made according to their crystallographic structures found under environmental conditions and most cited in the literature. The terms in parenthesis refer to possible forms. The α-prefix is generally applied to hexagonal close-packed and related structures. The β-prefix is generally applied to designate instances of polymorphism and cases of alteration or dehydration. The γ-phase has cubic close-packed lattices or related structures.

[b] An attempt in the direction of a universal standard nomenclature was made in 1957 (Germany) at a symposium on alumina nomenclature (Ginsberg et al., 1957).

[c] *Source:* Wefers and Misra (1987). Alpha and beta were used to differentiate forms as the most abundant in nature or not. Alpha form was considered the most abundant.

[d] Hydrargillite is a mineral that was named after the Greek hyder (water) and argylles (clay). The name hydrargillite was mistakenly given to describe aluminum hydroxide, but later was proven to be aluminum phosphate. However, both names are still used to describe aluminum hydroxide: gibbsite is preferred in the United States and hydrargillite is used more often in Europe.

designation alpha was employed for minerals abundant in nature, while the designation gamma was given originally to all other aluminum hydroxides and hydrolyzed aluminas. Lehl (1934) noted that according to Haber, the aluminum hydroxides that, by elimination of water, have a stable cubic structure below 1000°C are termed γ-oxides, and this group included gibbsite along with boehmite and bayerite. Later, the γ-designation was used for all the alumina transition forms encountered in the low-temperature calcination of aluminum oxide compounds. Stumpf et al. (1950) restricted the prefix gamma to the product obtained in the dehydration sequence of boehmite at >800 K (Wefers and Misra, 1987). Lehl (1934) further reported that Tosterud gave the α- and β-prefix to gibbsite and bayerite, respectively.

Following the convention identified by Lehl (1934), and based on the crystallographic data for naturally occurring gibbsite that forms under lower temperatures in the range of those of the natural environment, the Greek letter alpha (α) is here considered to be the most appropriate designation reflecting its structure. Usually, gibbsite forms pseudohexagonal crystals and the prefix alpha is generally applied to hexagonal close-packed (hcp) and related structures. However, different morphologies of gibbsite are possible and it has been reported that it can form ultrathin or thicker oval-like crystals, prisms, and rod-like crystals grown from sodium aluminate solutions under various conditions (Sweegers et al., 1999, 2004). The industrially or laboratory produced gibbsite, where the temperature is usually higher than that exists in soils or surface waters, has sometimes been designated as gamma (γ-$Al(OH)_3$) (e.g., Sweegers et al., 1999; Vermöhlen et al., 2000; Kloprogge et al., 2002; Yang et al., 2007a). Nevertheless, the prefix alpha used for the gibbsite formed under environmental conditions is most common, α-$Al(OH)_3(s)$, and our approach is consistent with other investigators (Goldberg et al., 1993a, 1993b; Sposito, 1996; Hiemstra et al., 1999; Rosenqvist et al., 2002; Rosenqvist and Casey, 2004; Goldberg, 2005; Tabor and Yapp, 2005; Wang et al., 2007; Huittinen et al., 2009).

According to Wefers and Misra (1987), bayerite was initially labeled with the letter alpha (Table 2.1) after the Haber series, by which in 1925 aluminum and ferric hydroxides were divided into two series, alpha and gamma. From this legacy, many investigators still use the α-prefix for bayerite, because as with gibbsite, bayerite can have a pillared, double-layered structure with different stacking sequence compared to gibbsite. Others, though, prefer labeling bayerite with the Greek letter beta (β), because bayerite is considered a metastable form that is subject to alternation or dehydration (e.g., Galas et al., 1993; Laiti et al., 1998; Lefèvre et al., 2002; Kim et al., 2004a, 2004b; Carrier et al., 2007). In crystallography, the β-prefix is generally applied to designate instances of polymorphism or transformation. Although a metastable form can be stable for long periods of time, we prefer to label bayerite with the beta prefix, because under specific environmental conditions (e.g., pH, ionic strength, etc.) bayerite can be transformed to gibbsite. This transformation is irreversible (Wefers, 1967).

Other aluminum hydroxides, such as nordstrandite and doyleite, still do not have any specific crystallographic designation. Although nordstrandite has been labeled with the γ (gamma) prefix (Kodama and Schnitzer, 1980), it is usually referred to with its common name. Nordstrandite and doyleite are scarce in nature (Violante et al., 1982; Chao et al., 1985).

2.4 GIBBSITE

A number of crystalline aluminum hydroxides, oxyhydroxides, and oxides are found in nature or synthesized in the laboratory. One of the most common aluminum hydroxide polymorphs is gibbsite, which is the most important alumina chemical used as a precursor in the manufacture of advanced ceramics, catalysts, adsorbents, and aluminum (Souza Santos et al., 2000). Gibbsite occurs widely in soils and in a concentrated manner in bauxite deposits (Hsu, 1977).

In this section, we discuss the properties of gibbsite and how these properties relate to the method of synthesis. The average properties of gibbsite described here will be used as the basis for evaluating candidate sorption data sets to determine whether they were actually obtained with gibbsite as a sorbent. Sorption data for gibbsite have been obtained with laboratory synthesized gibbsite and with commercially synthesized gibbsite. The methods of synthesis and the resulting gibbsite properties are discussed.

2.4.1 Kinetics of Precipitation and Crystal Growth

In the Bayer process, which is industrially used to extract aluminum oxide from bauxite, gibbsite is the mineral recovered from the crystallization of aluminum hydroxide ($Al(OH)_3$) from alkaline, heated sodium aluminate solutions. Gibbsite growth proceeds very slowly as the solution is cooled, typically 1–2 μm/h. Despite a lot of research, the effects of crystallization conditions and impurities on the final gibbsite properties are still not well established (Baroni et al., 2000; Sweegers et al., 2001, 2002b). Both natural and synthetic gibbsite crystals exhibit a large variation in their morphological appearance (e.g., twinning). The reason for this high diversity is still unclear (Sweegers et al., 1999). The possibility of predicting the gibbsite growth behavior under various conditions will be very useful, especially for practical purposes (Sweegers et al., 2004).

Two crystalline polymorphs, gibbsite and bayerite, may nucleate from sodium aluminate solutions spanning a concentration range of 1–6 M NaOH ([NaOH]/[Al(III)] = 1.20–1.37) at moderate temperatures (<80°C). When the solution concentration is in the range of 4–6 M NaOH ([NaOH]/[Al(III)] = 1.20–1.37) at 65–80°C, the predominant crystalline precipitate is $Al(OH)_3(s)$, identified as the thermodynamically stable gibbsite phase. The concentration of bayerite coprecipitated depends on the specific solution conditions. Low crystallization temperatures (e.g., <40°C) favor the formation of bayerite (Li et al., 2005). However, gibbsite has been observed to crystallize from a solution containing 0.2 M Al(III) in the presence of acid (HCl or HNO_3, pH ≈ 7) at 23°C as hexagonal plates and prisms (100–150 nm) (Lee et al., 1999; Li et al., 2005). Bayerite transforms into gibbsite upon aging (Li et al., 2005).

In solutions of intermediate concentration (1–2.5 M NaOH, 0.82–2.05 M Al(III)), particle enlargement to an average diameter <300 nm occurs mainly by a single particle growth mechanism (Fig. 2.7). At larger sizes, particle aggregation is the dominant mechanism. Bayerite crystallization occurs with dilute sodium aluminate solutions. Solutions with an intermediate NaOH concentration (e.g.,

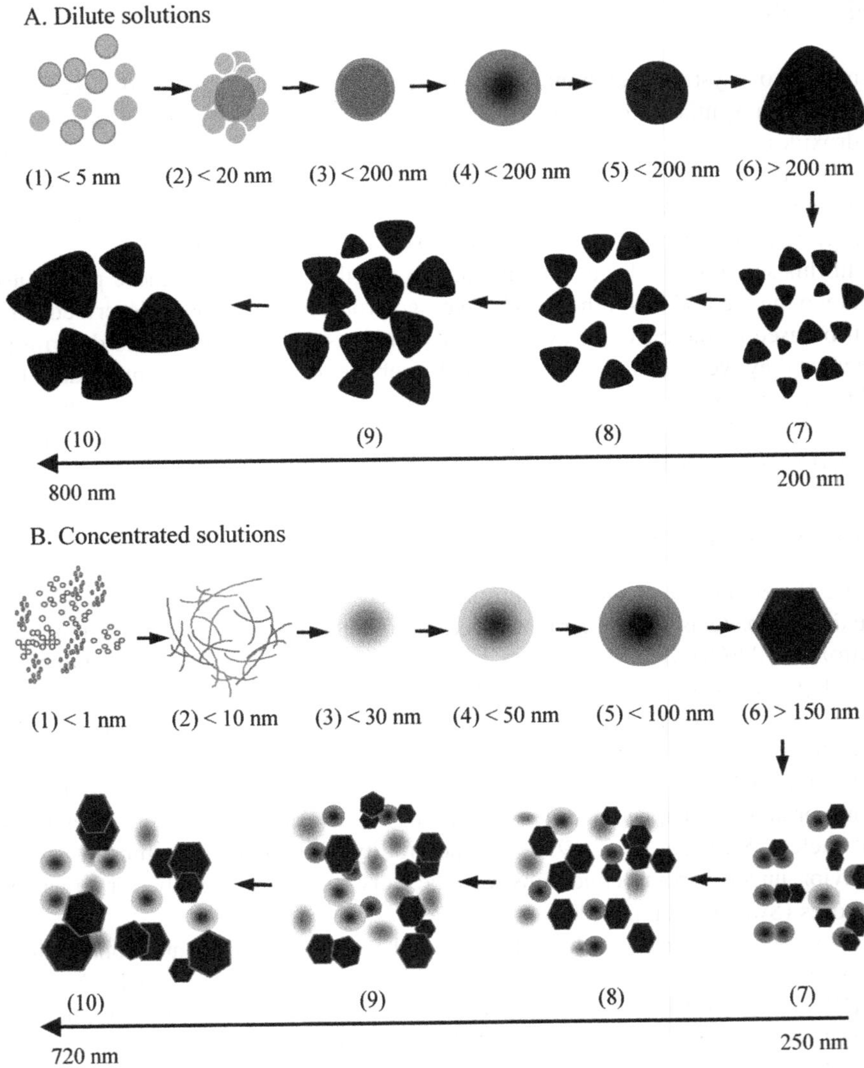

FIGURE 2.7 A schematic representation of the formation and enlargement of crystalline aluminum hydroxide from sodium aluminate solutions: (a) Dilute solutions; (b) concentrated solutions. The size ranges indicated are speculative and are based on literature sources and the work described herein. (*Source*: Li et al., 2005.)

1.5 M NaOH) may result in a two-phase mixture of bayerite and gibbsite. For concentrated solutions ([NaOH] > 2.5 M, [Al(III)] > 2.05 M), particle size enlargement occurs mainly through aggregation (Fig. 2.7). Gibbsite is the major component of the $Al(OH)_3$ phases crystallized from concentrated solutions (Li et al., 2005).

2.4.2 Structure

Gibbsite is monoclinic, with mostly a tabular pseudohexagonal crystalline form (Saalfeld and Wedde, 1974). The cell parameters of the structure of gibbsite (Table 2.2) were defined by Saalfeld and Wedde (1974), and verified later by Mercury et al. (2006). The two crystallographically independent Al atoms are located at the general positions (0.1679, 0.5295, −0.0023) and (0.3344, 0.0236, −0.0024). It is described by close-packed sheets of OH^- groups with Al^{3+} ions occupying two-thirds of the octahedral holes in alternate layers.

Figure 2.8 shows the structure of gibbsite, with sheets of hexagonal close-packed layers (vertical stacking of layers can be seen in Fig. 2.6) and open packing between successive sheets. In the lateral extension of the hexagonal close-packed sheets, each Al cation is octahedrally coordinated by six OH groups and each hydroxyl group is coordinated by two Al cations with one octahedral site vacant. The structure can also be viewed as double layers of OH groups with Al cations occupying two-thirds of the interstices within the layers. Each double layer is positioned in such a way that the adjacent upper and lower layers have their hydroxyl groups directly opposite to each other. This type of layer structure explains the perfect cleavage of gibbsite parallel to the basal plane (001). Due to the proximity of the hydroxyl groups, the O–H bonds are partially polarized causing displacement of the proton toward the vacant sites (Kloprogge et al., 2006).

So, for the purposes of the analyses of sorption data sets conducted in this study, sorbent material will be considered gibbsite if it exhibits pseudohexagonal tabular crystalline form, if it is naturally occurring, and/or if it is formed under the conditions of crystallization for synthetic gibbsite. Elongated prisms may form if gibbsite is grown from sodium aluminate solutions under conditions of low supersaturation and high temperature (>330 K), which is not the case for the data sets examined in this study.

In Table 2.2, the mineralogical and structural properties of hydrous aluminum oxides are presented. Along with the data presented for gibbsite, we present the respective data for bayerite and phases that might be mistaken or accidentally chosen as being gibbsite. We present these data for direct comparison with those of gibbsite. The space groups are expressed using the centrosymmetric system and the axis lengths are referenced to the unit cell of the crystalline gibbsite and its hydrous aluminum oxide polymorphs.

2.4.3 Common Techniques of Synthesis

Many procedures for the synthesis of crystalline aluminum hydroxides have been described, and techniques specific to the synthesis of gibbsite have been developed. Lippens (1961) observed that when aluminum salt solutions were used at pH values below 7.4 the precipitates obtained were amorphous to X-rays but noted that the precipitates invariably contained a large amount of anions, removal of which led to crystallization. Marboe and Bentur (1961) found that the crystallization of metal hydroxides in general and of aluminum hydroxides in particular is significantly

TABLE 2.2 Mineralogical and structural properties of gibbsite and other hydrous aluminum oxide polymorphs

Mineralogical Properties

	Index of Refraction, n_D								
Phase	α	β	γ	Average	Cleavage	Brittleness	Mohs Hardness	Luster	Reference
Gibbsite	1.568	1.568	1.587		(001) Perfect	Tough	2.5–3.5	Pearly vitreous	(1)
Bayerite				1.583					(1)
Nordstrandite	1.580	1.583	1.596–1.613		(110) Perfect		3	Vitreous, pearly	(2)
Doyleite	1.545	1.553	1.566		(010) Perfect	Flexible	2.5–3.5	Vitreous to pearly	(2)

Structural Properties

				Unit Axis Length (nm)					
Phase	Crystal System	Space Group[a]	Molecules per Unit Cell	*a*	*b*	*c*	Angle	Density	References
Gibbsite	Monoclinic	$P2_1/n^b$ (C_{2h}^5)	4	0.8684	0.5078	0.9736	94° 54′	2.42[b]	(3)
Gibbsite	Triclinic	—	16	1.733	1.008	0.9730	94° 10′, 92° 08′, 90° 0′		(4)
Bayerite	Monoclinic	$P2_1/a^b$ (C_{2h}^5)	2	0.5062	0.8671	0.9425[c]	90° 26′	2.53[b]	(5,6)
Nordstrandite	Triclinic	$P1^b$ (C_1^1)	2	0.5114	0.5082	0.5127	70° 16′, 74° 0′, 58° 28′	2.41–2.43[b]	(7)
Doyleite	Triclinic	$P\bar{1}^{\,b,d}$		0.5002	0.5175	0.4982	97° 50′, 118° 60′, 104° 74′	2.48[b]	(8, 9)

References: (1) Wefers and Misra (1987); (2) Anthony et al. (1997); (3) Saalfeld and Wedde (1974); (4) Saalfeld (1960); (5) Rothbauer et al. (1967); (6) Yamaguchi and Sakamoto (1958); (7) Bosmans (1970); (8) Chao et al. (1985); (9) Clark et al. (1998).

[a] Space group is given in international notation (Hermann–Mauguin notation). In parenthesis, space groups are given according to Schönflies notation.

[b] Densities are given in measured values (calculated values can be found in Anthony et al., 1997). Demichelis et al. (2009) proposed that doyleite and nordstrandite space group designation should be $P\bar{1}$ and $P1$, respectively.

[c] Anthony et al. (1997).

[d] Jampor et al. (1998).

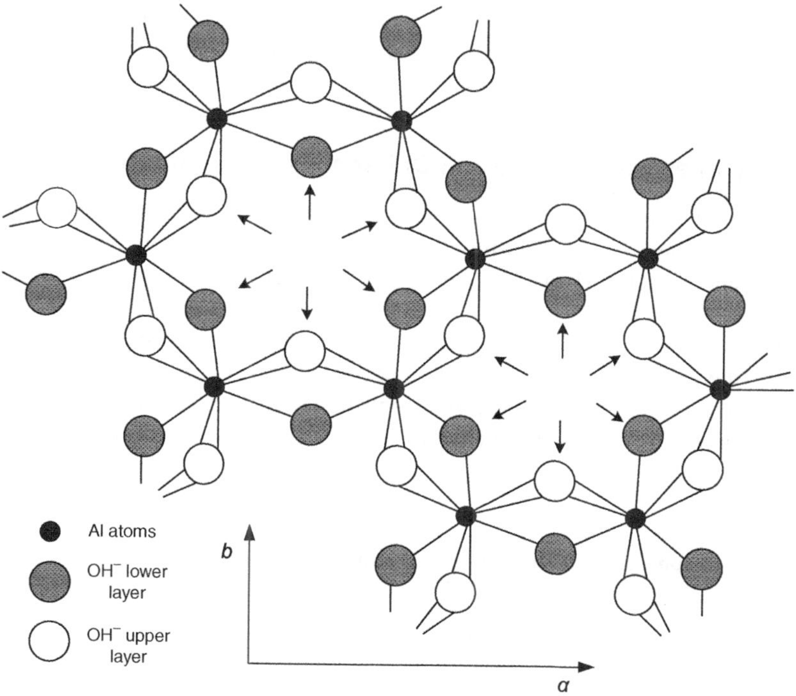

FIGURE 2.8 A schematic representation of gibbsite hexagonal close-packed layer. Each Al cation is octahedrally coordinated by six OH groups and each OH group is coordinated by two Al cations with one octahedral site vacant (arrows). (*Source*: Wefers and Misra, 1987; Sweegers et al., 2002a.)

affected by the presence of impurities. Gastuche and Herbillon (1962) studied the aging of gels in deionized media and concluded that the elimination of impurity ions by dialysis was more important in determining the rate of crystallization than the pH value. McHardy and Thomson (1971) found that introducing acetic acid in the hydrolysis of aluminum and precipitation of aluminum hydroxide yielded gibbsite as the main product after a month of aging at 23°C and depending on the conditions. At low concentration of acetic acid (10^{-4} to 10^{-3} M) and pH < 5.3, a mixture of gibbsite and bayerite was produced. At medium concentrations of acetic acid in the solution (about 10^{-3} M) and pH < 5.1, gibbsite was the main product, whereas at high concentrations (10^{-3} to 10^{-2} M) and pH < 5.1 a mixture of amorphous gel and gibbsite was produced (McHardy and Thomson, 1971). Gibbsite along with bayerite and/or pseudoboehmite can also be produced at ambient temperature and at a pH range of 9–10.5 or 8–8.7, when $Al_2(SO_4)_3$(s) reacts with either NaOH or NH_4OH (Prodromou and Pavlatou-Ve, 1995).

In 1962, Gastuche and Herbillon proposed one of the most commonly used techniques for gibbsite synthesis (Gastuche and Herbillon, 1962). Later, a modified synthesis technique was introduced by Kyle et al. (1975) and became commonly used.

Other techniques have been developed more recently, such as formation of gibbsite by laser ablation (Lee et al., 1999), but the methods of Gastuche and Herbillon (1962) and Kyle et al. (1975) are still commonly used.

2.4.4 Synthesized Gibbsite and Differences from Natural Gibbsite

Gibbsite is an important product of the Bayer process in industrial extraction of alumina from bauxite ores. During this process, gibbsite is extracted along with boehmite and diaspore (Huneke et al., 1980; Ruan et al., 2001; Kloprogge et al., 2006).

Although the Bayer process is the most applied industrial way to produce gibbsite as an intermediate, many commercially synthesized gibbsite solids are manufactured with different properties. Manufactured gibbsite is designated with more than 120 different names. Properties of some commercial grade gibbsite materials are given in Table 2.3.

TABLE 2.3 Properties of commercial grade aluminum hydroxides

Property	Normal Coarse Grade[a]	Normal White Grade[b]	Ground[c]	Fine Precipitated[d]
Al_2O_3 (wt%)	65.0	65.0	65.0	64.7
SiO_2 (wt%)	0.012	0.01	0.02	0.04
Fe_2O_3 (wt%)	0.015	0.004	0.03	0.01
Na_2O (total) (wt%)	0.40	0.15	0.30	0.45
Na_2O (soluble) (wt%)	0.05	0.05	0.05	0.1–0.25
LOI at 1200°C (wt%)[e]	34.5	34.5	34.5	34.5
Moisture at 100°C (wt%)	0.1	0.1	0.4	0.3–1.0
Specific gravity	2.42	2.42	2.42	2.42
Bulk density (loose) (g/cm^3)	1.2–1.4	1.0–1.1	0.7–1.25	0.13–0.22
Surface area (m^2/g)	0.1	0.15	2–4	6–8
Color	Off-white	White		White
Refractive index	1.57	1.57	1.57	1.57
Mohs hardness	2.5–3.5	2.5–3.5	2.5–3.5	2.5–3.5
Particle size, cumulative (wt%)				
Retained 100 mesh = 149 mm	5–20	0–1		
Retained 200 mesh = 74 mm	65–90	5–15		
Retained 325 mesh = 44 mm	90–98	30–65	1–2	0.1–0.2
Passing 325 mesh = 44 mm	2–10	35–70	98–99	99.8
Median particle size (mm)			6.5–9.5	0.6

Source: Adapted from Misra (2003).

[a] Alcoa C-30.

[b] Alcoa C-31.

[c] Alcoa C-330.

[d] Alcoa Aydral 710.

[e] Loss on ignition.

Naturally occurring gibbsite may differ from its manufactured counterpart. The size distribution of gibbsite particles is one aspect of the difference. Gibbsite particles obtained from Bayer liquors range from a few to several hundred micrometers, with diameters between 40 and 200 μm, but each tabular or prismatic crystal unit is between 5 and 10 μm in diameter (Souza Santos et al., 2000). Efforts for synthesis of nanosized gibbsite with dimensions of particles up to several hundred nanometers have been made. Successful synthesis of hexagonal plate-like particles of nanosized gibbsite was achieved when long aging time ranging up to several days was applied (Liu et al., 2008 and the references therein). Solubility is a second important aspect of the difference between natural and manufactured gibbsite. Natural gibbsite is relatively more soluble than the synthetic one (May et al., 1979; Wesolowski, 1992).

The crystallization time of natural versus synthetic gibbsite may differ dramatically. The aggregation step is an important size enlargement process during crystallization since the growth rates of gibbsite are generally slow (Cesteros et al., 2001). The synthesis of gibbsite of small particle size can be accomplished in 2 days (Cesteros et al., 2001), whereas in nature this process can vary dramatically from few days to years (Sposito, 1996).

Finally, manufactured gibbsite particles tend to have higher degrees of crystallinity in contrast to gibbsite found in nature. Naturally occurring gibbsite solids often contain impurities of other aluminum hydroxide polymorphs, such as bayerite, or other ions incorporated in their structure. Solution-phase constituents existing under environmental conditions, such as natural organic acids, hinder crystallization of gibbsite precipitates (Kodama and Schnitzer, 1980; Violante and Violante, 1980; Violante and Huang, 1985; Colombo et al., 2004; Yu et al., 2007). These constituents can also be incorporated into the gibbsite structure resulting in a solid with different structural and surface properties than pure-phase gibbsite.

2.5 BAYERITE

Bayerite is closely related in structure to gibbsite and is formed under similar conditions. It is used commercially to prepare aluminum oxide catalyst supports (Saalfeld and Wedde, 1974). Bayerite can be formed in various stages of the Bayer process (Gale et al., 2001), but it is an undesirable by-product, since the process is principally used for refining bauxite to produce alumina.

2.5.1 Kinetics of Precipitation and Crystal Growth

Like gibbsite, bayerite forms upon the aging of aluminum hydroxide gels precipitated from aluminum salt solutions in water. The abundance of bayerite depends on the concentration of aluminum in water (McHardy and Thomson, 1971), as indicated in Table 2.4, and on the electrolyte composition of the water. This polymorph is stable only in the absence of alkali ions, particularly Na^+ and K^+. In their presence, bayerite converts irreversibly into gibbsite (Wefers, 1967).

TABLE 2.4 Aging of suspensions prepared by hydrolysis of aluminum in water

Aging	Hydrolysis Products (by moles Al Hydrolyzed/L H_2O)		
Time (h)	0.017	0.05	0.20
1	Pseudoboehmite	Pseudoboehmite	Pseudoboehmite
3	Pseudoboehmite	Pseudoboehmite	Pseudoboehmite
5	Pseudoboehmite	Pseudoboehmite	Pseudoboehmite + bayerite
9	–	Pseudoboehmite + bayerite	—
24	Predominantly bayerite with some gibbsite	Bayerite with few crystals of gibbsite	Bayerite

Reproduced from McHardy and Thomson (1971), with the permission of the Mineralogical Society.

2.5.2 Structure

The crystal structure of bayerite is monoclinic, with the properties given in Table 2.2 (Zigan et al., 1978). The Al atoms are located at the general positions (0.5114, 0.1642, 0.0004) and (0.0326, 0.3315, −0.0015). The structure is similar to that of gibbsite. The Al atoms also form hexagonal close-packed layers resulting in higher density of bayerite compared to that of gibbsite. The Al atoms deviate a maximum of 0.015 Å from the (002) plane, so that they are closer to the structure of graphite layers. The Al–Al distances range from 2.82 to 3.01 Å (mean value = 2.90 Å), also close to the distances observed in the parent metal (Ramos-Gallardo and Vegas, 1996).

The most commonly observed forms of bayerite are spindle or hourglass shapes formed by stacking of $Al(OH)_3$ layers in a direction perpendicular to the basal plane. The bayerite lattice is also composed of double layers of OH, but hydroxyl groups of one layer lie in the depressions between the OH positions of the second (Misra, 2003).

2.5.3 Differences from Gibbsite

The four polymorphs, gibbsite, bayerite, nordstrandite, and doyleite, differ in the relative disposition of the double layers of OH groups and in the hydrogen bonding between and within the layers. The stacking of the double layers of OH groups can be summarized as follows (see Table 2.2 for more details):

- In bayerite, each layer is stacked directly above the layer below in an -AB-AB-AB-sequence (where A represents the close-packed oxygen layer above the Al atoms, B represents the close-packed oxygen layer below the Al atoms, and the combination AB represents one oxygen double layer) (Zigan et al., 1978; Clark et al., 1998).
- In gibbsite, each alternate layer inverted in an –AB–BA–AB–sequence (Saalfeld and Wedde, 1974; Clark et al., 1998).

- In nordstrandite, the stacking is –AB–AB–AB– as in bayerite, but with an accompanying shift of 1.82 Å along a direction inclined approximately at an angle of 35° to the bayerite *a*-axis (Bosmans, 1970; Clark et al., 1998).
- In doyleite, the stacking is again –AB–AB–AB– but the lateral displacement is approximately at right angles to that of nordstrandite, being at 25° to the *b*-axis of bayerite (Clark et al., 1998).

Doyleite appears to lack well-defined intralayer hydrogen bonds, which makes it the sole $Al(OH)_3$ polymorph to do so. The crystal structure determinations of bayerite and gibbsite allow all hydrogen atoms to be reliably located and to establish the presence of both inter- and intralayer hydrogen bonding. In nordstrandite, there is no exact deduction about the position of the hydrogen bonds, but the existence of these bonds is reported.

2.5.4 Synthesized Bayerite and Transformation to Gibbsite

Bayerite is synthesized and is rarely found in nature. Bayerite is a commercially available technical product that is produced in small quantities mainly for alumina catalyst manufacture. High-purity aluminum metal has been converted to bayerite to produce very high-purity aluminum oxides (Misra, 2003).

Bayerite has been synthesized by several methods. A pure product is prepared by the Schmah method in which dissolved aluminum hydrolyzes at room temperature (Misra, 2003). Other methods include rapid precipitation from sodium aluminate solution, aging of gels produced by neutralization of aluminum salts with NH_4OH, and rehydration of transition ρ-alumina (Misra, 2003). Metastable γ-alumina can be progressively transformed to bayerite (β-$Al(OH)_3$, hydration reaction) mainly due to its high specific surface area and the presence of aluminol groups on its surface (Dyer et al., 1993; Galas et al., 1993; Lefèvre et al., 2002):

$$\gamma\text{–}Al_2O_3(s) + 3H_2O \rightarrow 2\beta\text{–}Al(OH)_3(s) \tag{2.1}$$

Bayerite forms in significant amounts only in the absence of alkali ions (Na and K) and there is evidence that bayerite converts irreversibly to gibbsite in the presence of alkali metal ions (Wefers, 1967; Ramos-Gallardo and Vegas, 1996).

2.6 NORDSTRANDITE

Van Nordstrand et al. (1956) identified a third crystalline form of $Al(OH)_3(s)$ as a synthetic product mixed with gibbsite and bayerite (Violante et al., 1982). The name of nordstrandite was established by Papée et al. (1958) when they obtained the new material in essentially pure form by aging Al gels at high pH. Nordstrandite has been synthesized in a pure state by reacting slightly hydrolyzed Al foil or noncrystalline, precipitated aluminum hydroxide gel in an ethylenediamine solution (Hauschild, 1963; Violante et al., 1982) or by aging hydrolytic precipitation products of aluminum with aqueous ethylene glycol (Aldcroft and Bye, 1967; Violante et al., 1982). Organic chelating anions and clay mineral surfaces, particularly at pH > 8, have been shown to help induce the crystallization of pure nordstrandite (Violante and Violante, 1980; Violante and Jackson, 1981; Violante et al., 1982).

Naturally occurring crystals of nordstrandite have also been identified at various places. Natural norstrandite has been found on the island of Guam, in microscopic solution cavities in upper Miocene limestone, Borneo, in Montenegro, in the Dinaric Alps of Croatia, in Hungary, in Jamaica, in Sydney Basin, Australia, and in northwestern Colorado (as reported by Violante et al., 1982).

Nordstrandite has a triclinic crystal structure, with the properties given in Table 2.2 (Bosmans, 1970). The Al atoms are located at the general position (0.337, 0.333, 0.010). The structure of nordstrandite is intermediate to those of gibbsite and bayerite. Double layers of hydroxyl groups of the gibbsite structure alternate with those of the bayerite structure. Also, in the structure of norstrandite the Al atoms form hexagonal layers that are somewhat more regular and stacked in the ...ABBAAB... sequence (Fig. 2.4), though slightly displaced with respect to each other (Ramos-Gallardo and Vegas, 1996).

2.7 DOYLEITE

Doyleite was named after E.J. Doyle of Ottawa, Ontario, who identified this distinct $Al(OH)_3(s)$ polymorph. The mineral and its name have been approved by the Commission on New Minerals and Mineral Names of International Mineral Association, and its relation to the other aluminum hydroxide polymorphs has been established (Chao et al., 1985; Clark et al., 1998). It appears to be rare in nature, with only a few reports mentioning its occurrence in Quebec, Canada, and at some locations in China (Chao et al., 1985; Tarassoff et al., 2006). Doyleite is unstable at ambient temperature and pressure, which might be attributed to the absence of intralayer hydrogen bonds (Clark et al., 1998).

Doyleite has a similar bilayer structure to the other modifications of $Al(OH)_3(s)$, comprising double layers of hexagonal close-packed O atoms with Al atoms occupying two-thirds of the octahedral interstices, but is distinct from the other forms in its lateral displacement of adjacent layers and in aspects of the hydrogen bonding pattern (Clark et al., 1998). The structure of doyleite has been characterized as a triclinic, $P\bar{1}$, with unit cell dimensions as given in Table 2.2, unit cell volume $V = 104.39(3)$ Å^3, and space group $Z = 2$. Intralayer interactions between hydroxyl groups appear to involve limited, if any, bonding between a hydrogen and the next-nearest-neighbor oxygen, in contrast to other $Al(OH)_3(s)$ polymorphs (Clark et al., 1998).

2.8 OTHER FORMS OF ALUMINUM OXIDES AND OXYHYDROXIDES

2.8.1 Corundum (α-Al_2O_3)

Corundum (or α-alumina) occurs naturally as an igneous and metamorphic mineral and is the only thermodynamically stable oxide of aluminum (Wefers and Misra,

1987). It is also synthesized by various high-temperature techniques. The crystal structure of corundum is described by Wefers and Misra (1987) and others as a hexagonal close packing of oxygen ions with Al^{3+} ions occupying two-thirds of the octahedral interstices. The structure comprises alternating layers of O^{2-} and Al^{3+} ions. Thus, interlayer bonding between aluminum and oxygen is well established (as presented by Goldberg et al., 1996a). Wefers and Misra (1987) believe that the actual aluminum–oxygen bond angles show significant departure from ideal values for regular octahedra. The surface properties of different crystal faces of corundum are expected to be similar. However, the mode of formation may affect the crystal structure and the distribution of crystal defects (Goldberg et al., 1996a). Smit and Holten (1980) estimated from crystal lattice considerations the surface hydroxyl density of α-Al_2O_3(s) to be six groups per square nanometer (Goldberg et al., 1996a).

2.8.2 Boehmite (γ-AlOOH)

Boehmite is a relatively common mineral with the composition AlOOH. It is a major constituent in many bauxites, especially those of the Mediterranean type (Wefers and Misra, 1987). The processes by which boehmite is produced in bauxite deposits in preference to gibbsite are not well understood. The structure of boehmite has been described as composed of double layers of oxygen and hydroxide in cubic packing (Goldberg et al., 1996a). As in gibbsite, the double layers are joined by hydrogen bonds, and crystals exhibit preferential cleavage in the direction perpendicular to the general direction of the hydrogen bonding. Aluminum oxyhydroxide structures have some similarities with those of iron oxyhydroxides, and a significant amount of Fe^{3+} can replace Al^{3+} in the boehmite crystal lattice (Goldberg et al., 1996a). According to Goldberg et al. (1996a), De Boer et al. (1963) determined the hydroxyl group density on γ-AlOOH(s) from infrared spectroscopy studies as 16.5 groups per square nanometer.

Boehmite in bauxites is an important raw material for the production of aluminum oxide. Fine crystalline, synthetic boehmite is produced as a precursor for activated aluminas used in the manufacture of catalysts and absorbents (Wefers and Misra, 1987).

2.8.3 Diaspore (α-AlOOH)

Diaspore is a naturally occurring mineral that is a major constituent in some bauxite ores and in certain aluminum-rich clays and metamorphic rocks (Goldberg et al., 1996a). Use of diaspore bauxites for the production of alumina via the Bayer or the sintering process is possible, but the use of gibbsite bauxites for alumina production is more common due to lower requirements in processing temperatures (Hind et al., 1999; Authier-Martin et al., 2001).

Studies of the crystal structure of diaspore quoted by Wefers and Misra (1987) indicate that the aluminum ions are in a somewhat distorted octahedral coordination with the O^{2-} and OH^- ions in the crystal. Diaspore has a similar structure to goethite (α-FeOOH). Its oxygen sheets are in hexagonal close packing with two-thirds of the octahedral sites filled with Al ions (Kloprogge et al., 2002). Diaspore differs from boehmite in the coordination of the oxygen atoms, all of which are hydrogen bonded to another oxygen. The hydrogen atoms are closest to the oxygen farthest from the Al cation. Because of this compact arrangement, diaspore has greater density than boehmite (Kloprogge et al., 2002).

2.9 OTHER FORMS MANUFACTURED UNDER HIGH TEMPERATURE AND PRESSURE

There are other aluminum oxide forms that are manufactured and used in a variety of applications. One of the most important manufactured aluminum oxide forms is alumina, which comprises anhydrous aluminum oxides of the chemical formula Al_2O_3 produced upon heating and dehydroxylation of precursor $Al(OH)_3$ solids.

Corundum (α-Al_2O_3(s)) is the thermodynamically stable form of alumina that occurs in nature, but alumina exists in many metastable polymorphs as well, which are not formed under environmental conditions. The most common naturally occurring aluminum oxides and hydroxides utilized for the formation of different metastable alumina structures upon heating are depicted in Figure 2.9.

Transition aluminas are characterized by their fine particle size, high surface area, and catalytic activity of their surfaces. These properties have made aluminas ideal in many industrial applications including use as adsorbents, catalysts, catalyst supports, coatings, and soft abrasives (Levin and Brandon, 1998a). Understanding of the metastable intermediate polymorphic structures and the transformation mechanisms is essential for controlling the properties of manufactured aluminas (Wefers and Misra, 1987).

According to Levin and Brandon (1998b), the manufactured transition aluminas can be divided into two broad groups: those with a face-centered cubic arrangement of oxygen anions and those with hexagonal close-packed arrangement. In the first subgroup (Fig. 2.7), the polymorphs included are the γ (gamma) and η (eta) with cubic structure, θ (theta) with monoclinic structure, and δ (delta) with either tetragonal or

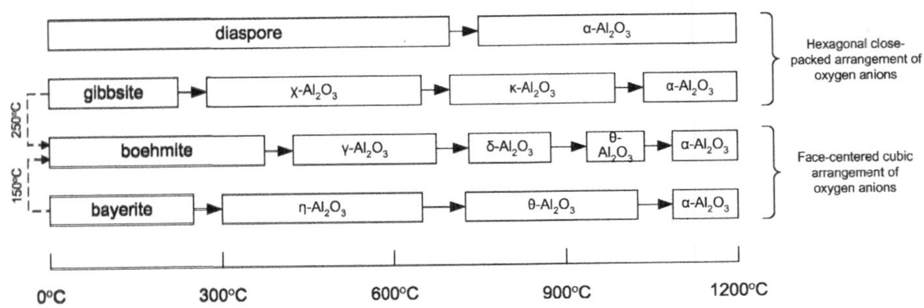

FIGURE 2.9 Transformation sequences of hydroxides from transition aluminas (Al_2O_3) to α-Al_2O_3 (corundum) upon heating (modification of Franks and Gan, 2007, after Levin and Brandon, 1998a; Digne et al., 2002.)

orthorhombic structure. In the second subgroup (Fig. 2.7), the polymorphs included are the α (alpha) with trigonal structure, κ (kappa) with orthorhombic structure, and χ (chi) with hexagonal structure. Also, Levin and colleagues report the existence of three other polymorphs of aluminum oxide: θ′, θ″ (monoclinic) (Levin et al., 1997), and λ (monoclinic) (Levin and Brandon, 1998b).

Among the precursors of corundum, gibbsite is of major importance because of its wide occurrence in nature and its industrial production in the Bayer process. During heating, dehydroxylation occurs, and gibbsite transforms to one or more transition aluminas with partially disordered structures. Transformation sequences of gibbsite through transition aluminas to α-Al_2O_3 (corundum) upon heating are depicted in Figure 2.10.

Considering alterations during heating under vacuum, *in vacuo*, gibbsite transforms via ρ-alumina (100–400°C), η-alumina (270–500°C), and θ-alumina (870–1150°C) to α-Al_2O_3. In air, the transformation follows two paths: (1) via boehmite (60–300°C), γ-alumina (500–850°C), δ-alumina (850–1050°C), and θ-alumina (1050–1150°C); and (2) via χ-alumina (300–500°C) and κ-alumina (800–1150°C) to the more thermodynamically stable form α-Al_2O_3 (Slade et al., 1991). The relative proportions following these two paths depend on a number of factors, including gibbsite particle size, moisture, alkalinity, pressure, bed depth, and heating rate (Slade et al., 1991). Characteristics of the sequences of the transition aluminas shown in Figure 2.8 are described by Slade et al. (1991). Other pathways for gibbsite dehydration to corundum at conditions relevant to Bayer refineries have also been proposed, but in some cases there is lack of detailed information about the experimental procedures followed. Whittington and Ilievski (2004) studied dehydration of gibbsite at 970°C and found that the majority (~70 percent) of gibbsite dehydrates via χ-alumina and that significant concentrations of χ-alumina react to form γ-alumina,

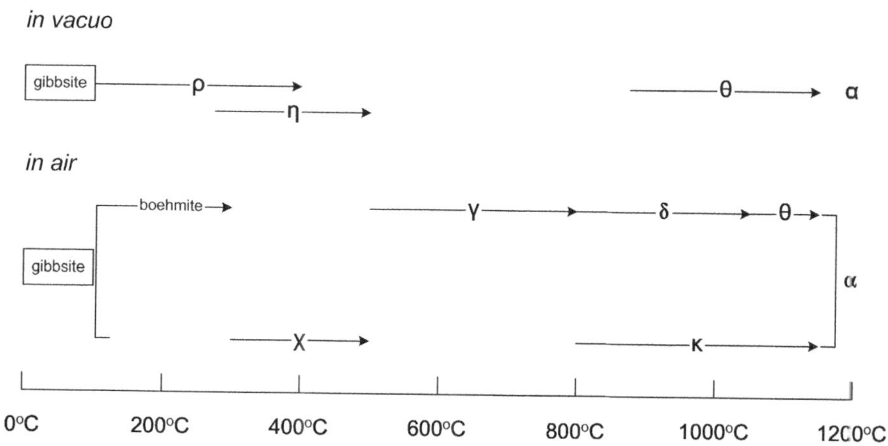

FIGURE 2.10 Transformation sequences of gibbsite, in air and under vacuum, from transition aluminas to α-Al_2O_3 (corundum) upon heating (after Slade et al., 1991, reproduced with permission of the Royal Society of Chemistry.)

and then to form θ- and α-aluminas (Whittington and Ilievski, 2004). Their proposed pathway is similar to that proposed by Ingram-Jones et al. (1996) and displays some similarities with the pathway proposed by Yamada et al. (1984) that χ-alumina transforms to γ- (pseudo-gamma) alumina.

3

TYPES OF AVAILABLE DATA

In this book, the generalized two-layer sorption model is used for modeling the sorption of cations and anions on hydrous aluminum oxide. The model presented in detail by Dzombak and Morel (1990), has been used to fit cation and anion sorption data for hydrous ferric oxide (Dzombak and Morel, 1990), goethite (Mathur and Dzombak, 2006), and manganese oxide (Tonkin et al., 2004). In this chapter, we examine the types of available data and identify the data considered as quantitative model constraints, with particular attention to the unique characteristics of gibbsite. Additional information on oxide-sorption data and modeling of such data with the generalized two-layer model is given in Dzombak and Morel (1990).

3.1 GIBBSITE STRUCTURE VERIFICATION

Gibbsite (α-$Al(OH)_3(s)$) is a sheet structure, usually forming hexagonal-like platelets with monoclinic symmetry (Saalfeld and Wedde, 1974; Mercury et al., 2006). As described in Chapter 2, the crystal structure is composed of layers of aluminum atoms, arranged between two layers of hydroxyl groups. Each of these hydroxyl groups is bonded to two aluminum atoms, resulting in electrically neutral sheets that are kept together by hydrogen bonds (Frenzel et al., 2005).

Gibbsite was successfully characterized with respect to its X-ray diffraction properties by Saalfeld and Wedde (1974). Their X-ray diffraction patterns are commonly regarded as the standard for the identification of gibbsite. The XRD technique is used to verify the end product of the synthesis of gibbsite or the degree of

Surface Complexation Modeling: Gibbsite By Athanasios K. Karamalidis and David A. Dzombak

crystallinity of the adsorbent. In this study, sorption data were collected only when the X-ray patterns of the gibbsite used in the experiments presented similarity to the X-ray patterns reported by Saalfeld and Wedde (1974).

3.2 PHYSICAL–CHEMICAL PROPERTIES

The physical–chemical properties of hydrous oxides affect the sorption of aqueous solutes. These properties include chemical composition, structural morphology, particle size distribution, surface area and porosity, and the nature and quantities of surface sites. The most important of these properties in terms of experimental data that support key modeling concepts and parameters are discussed below.

3.2.1 Specific Surface Area

When the area of the interface between two phases is proportional to the mass of one of the phases (e.g., for a solid adsorbent, for an emulsion, or for an aerosol), the specific surface area is defined as the surface area divided by the mass of the relevant phase (IUPAC, 1997). The ability of gibbsite to sorb chemical species from either liquids or gases is strongly dependent on the specific surface area. The exposed solid surface contains the functional groups and/or lattice ions and atoms that interact with the liquid or the gas-phase species to produce sorption reactions. Many techniques have been employed to measure specific surface area from various sorbents.

The most popular of these techniques is the gas sorption and data analysis method proposed by Brunauer, Emmett, and Teller (1938). This technique is commonly recognized as the BET method and involves the measurement of nitrogen ($N_2(g)$) gas sorption on the dried solid at constant temperature and variable gas pressure. The resulting gas-sorption isotherm is then analyzed to determine the surface area of the sample. Details about the advantages and disadvantages of using the BET technique are presented by Dzombak and Morel (1990).

Another method that has been used to estimate surface areas of solids in aqueous suspension involves the use of microscopic images in addition to gas-adsorption isotherms (e.g., Tournassat et al., 2003; Gan and Franks, 2006). The measurement of the specific surface area using gas-adsorption isotherms alone, similar to the BET method, provides no information about the distribution of geometrical specific surface area of individual particles.

Imaging flat surfaces with atomic resolution is feasible with atomic force microscopy, which yields information about the surface structure, strength, and geometry of chemical bonds between single atoms on surfaces (Lloyd et al., 1998; Giessibl, 2003; Gan and Franks, 2006). The use of a three-dimensional AFM can characterize the distribution of geometrical features of hexagonal plate-like gibbsite particles without microporosity (Jodin et al., 2004). This approach can provide information about the distribution of specific surface area (basal, lateral, and total) in gibbsite and other minerals (Hiemstra et al., 1999; Rosenqvist et al., 2002). Microscopic interpretations based on geometrical parameters derived from averages on a number of particles can lead to the specific surface area determination. The value of the specific surface area is obtained by a method initially implemented by Bickmore et al. (2002).

In order to avoid the assumption of a sample of particles of a granular solid being homogeneous, as is done with BET analysis, more sophisticated experimental and theoretical approaches have been developed. Infrared spectrum analysis coupled with Kr-adsorption isotherm analysis has been applied to oxides such as gibbsite (Phambu et al., 2000). However, this technique uses a description of the energetic heterogeneity of the participating surfaces which depends on prior knowledge of the shape and the type of crystal faces of the particles (Jodin et al., 2004). In the case of well-crystallized hexagonal gibbsite particles, Kr-adsorption isotherms can display two well-separated energetic domains (Phambu et al., 2000), which can be evaluated by applying the derivative isotherm summation method (Villiéras et al., 1992, 1997). The intensity of the infrared component at 3460/cm of a well-crystallized gibbsite has been found to be linearly related to the lateral surface area measured by gas-adsorption isotherms (Phambu et al., 2000).

Although other techniques have been developed to measure the specific surface area of hydrous oxides (Dzombak and Morel, 1990), the BET method is still the most popular and frequently applied. Due to its popularity, a great amount of surface-area data has been gathered with the method. Therefore, in this study, we used specific surface area measurements produced with the BET method.

3.2.2 Surface Site Characterization

3.2.2.1 Hydroxyl Surface Sites The dissociative sorption of water molecules forms surface hydroxyl groups at the surface of a metal oxide when it is exposed to water or its vapor. At the surface layer of a dry oxide, vacant coordination sites are available which can be occupied by water molecules via chemisorption as surface metal atoms are usually not fully coordinated. The surface becomes hydroxylated by proton transfer from the bound water molecules to the neighboring oxide ions, a process that is energetically favored because better charge neutralization in the lattice is achieved (Dzombak and Morel, 1990). James and Park (1982) report that hydroxylation of many metal oxide surfaces is complete at humidities well below 1 percent.

The water/solid interface is well defined and the interaction of water with solid surfaces is well understood at a molecular level by making use of surface science techniques at low pressure and temperatures. However, processes in environmental systems take place at ambient or higher pressures and elevated temperatures (Yamamoto et al., 2008). It is known that water adsorption (wetting) on solid surfaces may be affected by surface hydroxyl groups. Surface OH groups act as anchoring sites for adsorbed water molecules through H_2O–OH H-bonding that is stronger than the H_2O–H_2O H-bonding. Different wettabilities of surfaces can result due to two reasons: the difference in the kinetic barrier for OH formation, and the number and nature of surface OH groups (Yamamoto et al., 2008).

Many spectroscopic techniques have been employed which verified the existence of surface hydroxyl groups on metal oxides exposed to water. Dzombak and Morel (1990) provide an overview of the kinds of data and evidence that has been developed.

3.2.2.2 Surface Site Density The number of reactive OH sites per unit surface area is the surface site density of a hydrous oxide. Various methods have been employed to determine the density of surface hydroxyl sites.

The maximum sorption density in an ion-sorption isotherm provides one measure of surface site density, assuming 1 : 1 surface-ion binding. The maximum sorption density depends on the pH at which the experiment is conducted. The site density is generally deduced from the maximum value of the sorption density obtained for the highest (cation sorption) or lowest (anion sorption) pH values (Lefèvre et al., 2004).

Determination of the proton-active site density via acid–base titration is another method used to measure site density. Site-density determination is based on the fact that the maximum proton uptake and release occurs at extreme pH values at which estimation of the concentration of ionizable hydroxyl groups can be performed. Proton sorption is maximum at low pH, and proton desorption is greatest at high pH. However, most acid–base titrations are not performed at extreme pHs or high ionic strengths, where ionization of all sites is expected, and this may consequently underestimate site density (Sahai and Sverjensky, 1997).

Crystallographic data can be used for the determination of site densities on oxide surfaces. Pivovarov (1997) pointed out that the calculated site densities of metal oxide faces are numerically equal to the corresponding lattice spacings, with the reticular density of crystallographic plane directly proportional to the lattice spacing. Thus, to determine the site density it is enough to know the stoichiometry of the adsorption site (e.g., $\equiv$AlOH or $\equiv Al_2OH$), the density of the mineral, and the lattice spacing. In the general case,

$$\text{site density} = \frac{d}{2nV} \tag{3.1}$$

where d is the lattice spacing (or average lattice spacing), V the volume per cation (the volume of elementary cell, divided on the number of cations), 2 the coefficient in which the doubling of the surface at the dividing of crystallographic plane on the two faces is taken into account, and n the number of cations per site. The method has been compared to standard site-density determination techniques for various minerals (e.g., goethite) or to spectroscopic techniques and found to produce equivalent results (Randall et al., 1999; Villalobos et al., 2003; Chvedov and Logan, 2004; Pokrovsky et al., 2006; Huang et al., 2009).

Another widely used technique for measuring site densities on oxide surfaces is the isotopic-exchange technique, usually using tritium or deuterium (e.g., Yates, 1975). The extent of H_2O–D_2O exchange between water and D_2O-labeled oxide or the extent of ^1H–^3H exchange between water and ^3H labeled oxide is determined (Sahai and Sverjensky, 1997). The amount of exchangeable H on the surface is then calculated based on this result. The method uses the adsorbate of interest (i.e., H^+) for determining site density in terms of the number of exchangeable protons per unit surface area and yields estimates for N_s that do not depend on pH (Sahai and Sverjensky,

1997). The isotopic-exchange techniques, as well as theoretical methods for counting sites based on crystal structures, likely give maximum values for site density.

Other techniques that have been used for the determination of site density include chemical reactions with various adsorbates (e.g., fluoride-sorption isotherms), weight-loss methods upon heating, infrared and H_2O-adsorption/desorption studies (Sahai and Sverjensky, 1997). Although numerous experimental methods have been used to estimate the site densities for oxides and hydroxides, estimates from different techniques should be handled with caution, as they can be quite different for the same mineral.

In this work, most of the site-density values extracted and considered from the literature are based on the crystallographic determination. These values along with their respective references are presented and discussed in Chapter 5.

3.3 ACID–BASE TITRATION DATA

Oxides can obtain surface charge as a result of proton-transfer reactions at their surface hydroxyl groups. It is usually assumed that oxides surface hydroxyls are truly amphoteric and they are modeled as diprotic acids. The amphoteric surface reactions can be described as:

$$\equiv AlOH_2{}^+ = \equiv XOH^0 + H^+ \quad K_{a1}^{app} \tag{3.2}$$

$$\equiv AlOH^0 = \equiv XO^- + H^+ \quad K_{a2}^{app} \tag{3.3}$$

where $\equiv AlOH_2{}^+$, $\equiv AlOH^0$, and $\equiv AlO^-$ represent positively charged, neutral, and negatively charged surface sites, respectively, and K_{a1}^{app} and K_{a2}^{app} are apparent equilibrium acidity constants. As reactions 3.2 and 3.3 indicate, excess of protons at the surface (positive charge) occurs by the addition of acid to an oxide suspension and a proton deficit (negative charge) by the addition of base. If H^+ is the only specifically sorbing ion present, the proton surface-charge density σ is given by

$$\sigma = (F/AS)[(\equiv AlOH_2{}^+) - (\equiv AlO^-)] = F[\Gamma_H - \Gamma_{OH}] \tag{3.4}$$

where F is the Faraday constant (96,485 C/mol), A the specific surface area (m^2/g), S the solid concentration (g/L), and Γ_H and Γ_{OH} the experimental sorption densities (mol/m^2) of H^+ and OH^-. In this case, the surface charge is given by the proton excess or proton deficiency at the surface. In the presence of other specifically sorbing cations and anions the surface charge resulting from surface protolysis is only one of the components of the net surface charge.

There are two commonly used methods to measure proton excess or deficiency. The first one is the potentiometric acid–base titration and the second one is the comparison of the oxide titration curve with that of the medium alone by subtracting the titration curve for the background electrolyte in the absence of oxide from that of the oxide suspension.

In the first case, acid or base is added to a gibbsite suspension and the pH is monitored potentiometrically. The potentiometric acid–base titration is usually conducted in the absence of $CO_2(g)$, in closed reaction vessels and using an inert electrolyte. Equation (3.5) can be used for the calculation of H^+ or OH^- consumption:

$$\Gamma_H - \Gamma_{OH} = \left(\frac{1}{AS}\right)[C_A - C_B - (H^+) + (OH^-)] \tag{3.5}$$

where, C_A is the molar (or molal) concentration of added acid and C_B of added base. The concentrations of (H^+) and (OH^-) are calculated from the pH measurements. The specific equation is valid only for low-solubility hydrous metal oxides where the uptake or release of OH^- by dissolved metal species is negligible.

As gibbsite has relatively high solubility compared to the iron oxides and hydroxides, Equation 3.5 in principle must be modified to account for the hydrolysis reactions with released metal. If the equation does not account for the uptake or release of OH^- by dissolved metal species, then an artificially high- or low-surface-charge estimate may result. In the case of gibbsite where several important hydrolysis species are formed, Equation 3.5 (TOTH) would be modified to consider the dissolution of aluminum hydroxide as follows:

$$\begin{aligned}(\equiv AlOH_2{}^+) - (\equiv AlO^-) = C_A - C_B - (H^+) + (OH^-) + (AlOH^{2+}) \\ + 2(Al(OH)_2{}^+) + 3(Al(OH)_3{}^0) + 4(Al(OH)_4{}^-)\end{aligned} \tag{3.6}$$

Table 3.1 presents the mass law and mole balance equations used to develop the tableau from which Equation 3.6 was derived. The corresponding tableau is given in Table 3.2.

To determine surface-charge values corrected for oxide solubility, the total dissolved metal concentration at each titration point must be known (or calculated assuming equilibrium) and the concentrations of dissolved species calculated. The soluble aluminum concentration changes in the suspension with pH. Thus operational determination of surface charge by the usual subtraction of a blank titration curve from a suspension titration curve (Dzombak and Morel, 1990) may not be appropriate in this case, if the dissolved aluminum species contribute significantly to the acid–base balance.

The traditional way of describing the protonation and deprotonation reactions has been the use of the two pK_a model, where each site is considered to have a noncharged ground state and that is able to adsorb/desorb protons. New models for describing oxide surface acid–base chemistry have been presented in the last two decades. Bolt and Van Riemsdijk (1982) considered the crystal arrangement around a singly coordinated $\equiv AlOH$ group in the gibbsite and concluded that such a site should have only two possible states, $\equiv AlOH_2{}^{0.5+}$ or $\equiv AlOH^{0.5-}$. With this approach, only one pK_a value is needed to describe the protonation/deprotonation. Doubly coordinated $\equiv Al_2OH$ groups are nearly completely neutralized and are therefore considered

TABLE 3.1 Mass law and mole balance equations for surface acid–base chemistry of a soluble hydrous aluminum oxide

Species

H^+, OH^-, $\equiv AlOH^0$, $\equiv AlOH_2^+$, $\equiv AlO^-$, Al^{3+}, $AlOH^{2+}$, $Al(OH)_2^+$, $Al(OH)_3^0$, $Al(OH)_4^-$

Mass Law Equations

$(OH^-) = (H^+)^{-1}$				K_w
$(AlOH^{2+}) = (H^+)^{-1}$	(Al^{3+})			K_1
$(Al(OH)_2^+) = (H^+)^{-2}$	(Al^{3+})			K_2
$(Al(OH)_3^0) = (H^+)^{-3}$	(Al^{3+})			K_3
$(Al(OH)_4^-) = (H^+)^{-4}$	(Al^{3+})			K_4
$(\equiv AlOH_2^+) = (H^+)$		$(\equiv AlOH^0)$	P	$1/K_{a1}^{int}$
$(\equiv AlO^-) = (H^+)^{-1}$		$(\equiv AlOH^0)$	P^{-1}	K_{a2}^{int}

where $P = \exp(-F\Psi/RT)$

Mole Balance Equations

TOTH $= (H^+) - (OH^-) - (AlOH^{2+}) - 2(Al(OH)_2^+) - 3(Al(OH)_3^0) - 4(Al(OH)_4^-) + (\equiv AlOH_2^+) - (\equiv AlO^-) = C_A - C_B$

TOTAl $= (Al^{3+}) + (AlOH^{2+}) + (Al(OH)_2^+) + (Al(OH)_3^0) + (Al(OH)_4^-) =$ total dissolved Al

TOT($\equiv$AlOH) $= (\equiv AlOH_2^+) + (\equiv AlOH^0) + (\equiv AlO^-)$

TOTP $= (\equiv AlOH_2^+) - (\equiv AlO^-) = \sigma(AS/F)$

Charge–Potential Relationship

$$\sigma = 0.1174c^{1/2}\sinh(Z\Psi \times 19.46)$$

TABLE 3.2 Tableau for surface acid–base chemistry of a soluble hydrous aluminum oxide

	H^+	Al^{3+}	$\equiv AlOH^0$	$P = \exp(-F\Psi/RT)$
H^+	1			
OH^-	−1			
Al^{3+}		1		
$AlOH^{2+}$	−1	1		
$Al(OH)_2^+$	−2	1		
$Al(OH)_3^0$	−3	1		
$Al(OH)_4^-$	−4	1		
$\equiv AlOH_2^+$	1		1	1
$\equiv AlOH^0$			1	
$\equiv AlO^-$	−1		1	−1
	TOTH	TOTAl	TOT($\equiv$AlOH)	TOTP $= \sigma(AS/F)$

less reactive. However, if doubly coordinated $\equiv Al_2OH$ groups are reactive, two pK_a values are needed to describe the reactions (Rosenqvist et al., 2002).

Macroscopic information about the surface-charge characteristics of oxide surfaces can be obtained by acid–base titration data. They reveal the manner in which proton surface charge varies with pH for a particular oxide, including the pH at which the net proton charge is zero. The *point of zero net proton charge* (PZNPC) indicates in what pH regions there is proton excess or deficiency at the surface. If H^+ is the only specifically sorbing ion in the system, proton-transfer reactions are solely responsible for surface-charge development and the PZNPC is equal to the *point of zero charge* (PZC). In this special case, the PZC is also called the *pristine point of zero charge* (PPZC). When other specifically sorbing cations and anions are present, however, they contribute to the overall surface charge and hence PZNPC and PZC differ.

3.4 CATION AND ANION SORPTION DATA

The sorption of cations and anions on a hydrous oxide surface and other solids is commonly determined in two ways. The first is by measuring the change in solute concentration after addition of the solid phase, and the second is by measuring the amount of solute that becomes associated with the solid phase. Desirable sorption data are those for which no solid–liquid separation technique is needed. Separation techniques, which usually are needed, increase the uncertainty of the results due to losses experienced in the separation.

The limited experimental data for cation and anion sorption on gibbsite reported in the literature before 1974 were not used because of variations in the methods for identification of the crystal structure of gibbsite. In 1974, the crystal structure of gibbsite was identified by Saalfeld and Wedde (1974) and subsequently more stringent criteria were developed to distinguish gibbsite from among the many aluminum oxides reported in the literature (see Chapter 2).

The mechanism of metal-ion association with hydrous oxide surfaces involves an ion-replacement process in which the adsorbed metal ions replace bound protons. In contrast, anions can sorb on the surface of aluminum oxides by ion-pair formation with positively charged surface sites or by ligand exchange with surface hydroxyls (Goldberg et al., 1996a).

Sorption of both cations and anions onto hydrous oxide surface is strongly dependent on pH. There is a pH region in which sorption changes significantly. This pH region depends on cation concentration. For cations, sorption increases at high pH and the fraction sorbed increases from zero to one at a narrow range. On the contrary, sorption of anions is greatest at low pH and decreases gradually as the solution becomes more alkaline.

The sorption of anions that are weakly bound is affected by the concentration of the background electrolyte, while the sorption of strongly bound anions (e.g., PO_4^{3-}, AsO_4^{3-}, SeO_3^{2-}) is less affected by the ionic strength (Goldberg et al., 1996a).

Surface precipitation can be involved in the sorption of cations and anions at high sorbate concentrations. Surface precipitation is the formation of precipitates at the surface prior to precipitation in bulk solution (Dzombak and Morel, 1990).

For example, in the sorption of phosphate (PO_4^{3-}) on gibbsite, surface precipitation of aluminum phosphate can occur at high phosphate concentrations (Johnson et al., 2002). In contrast to the unlimited sorption densities that can result from surface precipitation on cation hydroxide species (OH^- ions are effectively in unlimited supply in water), surface precipitation of anions ultimately results in maximum sorption densities due to the limited supply of the participating cations (Dzombak and Morel, 1990).

3.5 SPECTROSCOPIC DATA FOR SORPTION ON GIBBSITE

Surface-complexation models were developed in the 1970s and 1980s based on macroscopic partitioning data for sorption on hydrous oxides. Since that time, there has been substantial development of spectroscopic techniques to study metal-ion sorption at the oxide/water interface. Spectroscopic measurements can provide valuable molecular-scale information with respect to the mechanisms of ion adsorption. Spectroscopic techniques that have been used to study sorption of cations and anions on gibbsite include Fourier Transform Infrared (FTIR), Raman, and X-ray photoelectron spectroscopy (XPS), and extended X-ray absorption fine structure (EXAFS). Maurice (2009) provides an overview of the most common techniques used for the spectroscopic investigation of mineral surfaces and reactions on such surfaces. Each of these techniques has provided information about the mechanism of ion retention by gibbsite, but data interpretation has been found to be constrained by certain limitations.

Infrared spectroscopy (IR) has been applied extensively in sorption studies, mainly because of its ability to probe chemical bonds. However, interpretation of IR data is challenging when water is present because of its strong IR absorption, and therefore FTIR is typically used for the study of bonding in aqueous systems (Lefèvre, 2004). The strong IR absorption by water can be addressed by using internal reflection spectroscopy, that is, by probing a short depth of sample with an evanescent wave existing in a lower index refraction medium (water) in contact with a denser medium in which IR radiation is propagating (Lefèvre, 2004). This spectroscopic technique is called attenuated total reflection (ATR). ATR results have been little used to refine sorption modeling (Hug, 1997) of inorganic ions onto oxyhydroxides for several reasons. Interpretation of spectra to obtain information on inner or outer-sphere complexes is difficult, especially for ions that can be sorbed with various geometries (Lefèvre, 2004).

Another technique similar to IR which yields complementary information for gibbsite or sorption of different species on gibbsite is Raman spectroscopy (Veilly et al., 2008). Raman spectroscopy provides information about the chemical bonds in molecules or crystallographic orientation of a sample by measuring vibrational, rotational, and other low-frequency modes in a system and relies on inelastic scattering, or Raman scattering of monochromatic light, usually from a laser in the visible, near-infrared, or near-ultraviolet range. The Raman spectrum shows characteristic wave numbers of OH-stretching vibrations between 3300 and 3700/cm (Ruan et al., 2001). The four dominant peaks at 3620, 3526, 3436, and 3364/cm and two shoulders at 3510 and 3372/cm are characteristic of gibbsite crystals (Phambu et al., 2000; Wang and Johnston, 2000). The Raman spectrum for bayerite differs from that of gibbsite at two peaks, 3658 and 3546/cm. In the case of ion sorption on gibbsite, a normalized subtraction of the initial sample Raman spectrum by the sorbed sample Raman spectrum can give valuable information about the sorbed ion on the surface of gibbsite (Veilly et al., 2008).

The oxides, hydroxides, and oxyhydroxides of aluminum cannot be distinguished by their characteristic X-ray photoelectron (XPS) regions, but have significantly different valence-band regions. Limited studies have been conducted using valence-band XPS (Thomas and Sherwood, 1992). XPS is performed inside a high-vacuum chamber, and dehydration may modify the structure of surface complexes. Attempts have been made to distinguish between the major aluminum oxide, hydroxides, or oxyhydroxide minerals, using high-resolution XPS (Kloprogge et al., 2006). From this work it has been concluded that very small chemical shifts in binding energies among the oxides, hydroxides, and oxyhydroxides lead to nonconclusive interpretation of XPS peaks among minerals with the same chemical composition such as gibbsite and bayerite. XPS data were not used in this study, due to data interpretation ambiguity for most of the elements.

X-ray absorption spectroscopy (XANES, EXAFS) is the most widespread technique applied to investigate the type and geometry of surface complexes on hydrous oxides. EXAFS determines the coordination numbers of the sorbed ion and the radial distances of the shells surrounding it (Fendorf and Sparks, 1996). Although EXAFS is a commonly employed technique it has not been extensively used in the characterization of ion surface species on gibbsite or natural samples (Oliveira et al., 2006), as it was used for other metal (hydr)oxides (e.g., hydrous ferric oxide). However, it is increasingly gaining attention. Specific insights provided by available EXAFS information for cation and anion surface species on gibbsite are summarized in Chapters 6 and 7, respectively.

Tools that have been used to interpret experimental data obtained by EXAFS include Density Functional Theory (DFT), and cluster models for the study of the sorption of As(V) complexes and of protons (as reported by Frenzel et al., 2005). Computational chemistry methods, such as DFT, are used to interpret spectroscopic measurements, predict adsorption rate constants, activation barriers, and to estimate the energies of adsorption reactions (Paul et al., 2007). In general, they are used to give insight into the sorption process at the atomic level and to contribute to the interpretation of the experimental results. There is limited information, however, on the electronic properties of hydrous aluminum oxides available in the literature (Thomas and Sherwood, 1992).

DFT calculations have been performed with edge-sharing dioctahedral Al^{3+} or Fe^{3+} cluster models to simulate the surfaces of Al and Fe (hydr)oxide minerals, and they have been shown to reproduce accurately the main geometrical features of metal oxides and clay minerals (Sainz-Diaz et al., 2000; Paul et al., 2007). However, this approach cannot predict the binding geometries (e.g., interatomic distances) of anions (e.g., oxyanions) on every surface of a soil mineral. An alternative approach is to model a soil mineral surface as a two-dimensional periodic surface constrained by boundary conditions (e.g., Paul et al., 2007). Usually, DFT with one or the other approach is used to interpret experimental data obtained by EXAFS or FTIR. DFT calculations have been used (Ladeira et al., 2001; Oliveira et al., 2006) to interpret EXAFS data for the sorption of cations and anions on gibbsite, but the available data span limited conditions and thus are not yet broadly useful as constraints for surface-complexation modeling.

3.6 PROTON RELEASE/UPTAKE DATA

The stoichiometry of sorption can be studied by measuring the shift in the acid–base titration curve that occurs in the presence of specifically sorbing cations and anions. When cations are sorbed, protons release and the titration curve shifts to lower pH values, whereas when anions are sorbed, protons uptake and the titration curve shifts to higher pH and the PZNPC changes accordingly (Dzombak and Morel, 1990). The extent of proton release or uptake is determined by comparing acid–base titration curves obtained with or without the specific ion being present.

Proton release and uptake associated with ion sorption can also be assessed by backtitration to constant pH of an oxide suspension in which the specifically sorbing ion has been added. Knowledge of the amount of acid or base needed to maintain the pH after the addition of a specific amount of sorbing cation or anion gives information about H^+ release/uptake per mole of cation/anion sorbed. This stoichiometric information can help guide the formulation of surface-complexation reactions to describe ion sorption. While proton release and uptake data can be useful in guiding and constraining surface-complexation modeling of ion sorption on hydrous oxides such as gibbsite, such data are time consuming to obtain and are not commonly available. Thus, in this work, proton release and uptake data were not used to constrain formulation of the surface-complexation reactions used to fit sorption data for gibbsite.

3.7 ELECTROKINETIC DATA

Electrokinetic methods for estimating the electrical potential difference between the metal oxide surface and bulk solution for oxide suspensions in water rely on the relative movement of a charged surface and an electrolyte solution. The potential difference estimated from the velocity of particle in water under an applied electric field—electrophoresis—is referred to as the zeta (ζ) potential. Zeta potential is the electric potential in the interfacial double layer at the shear or slipping plane versus a point in the bulk solution away from the interface. Zeta potential is calculated using theoretical models from experimentally determined electrophoretic mobility, the most common electrophoretic measurement for oxide suspensions.

Zeta potential is related to the magnitude of the electrical charge on an oxide surface, but is not necessarily equal to the Stern potential or electric surface potential in the double layer. The location of the slipping plane will be variable and is not known with accuracy. Thus, zeta-potential measurements are generally not useful as quantitative constraints for surface-complexation modeling. Limitations about the use of the electrophoretic mobility data have been discussed by Dzombak and Morel (1990).

Zeta-potential data are useful as quantitative constraints only for when the net surface charge is zero. In this case, the potential at the edge of the diffuse layer (Ψ) and the potential at the plane of shear (ζ) must both be equal to zero. Zeta-potential data can thus be used to identify the PZC, the pH value of the solution where total net particle charge is zero (Dzombak and Morel, 1990). When the PZC is measured using electrokinetic experiments it is called isoelectric point, IEP.

3.8 SUMMARY

X-ray diffraction is a well-established technique which is commonly used to verify the structure of synthesized and natural gibbsite used in most sorption experiments. The X-ray diffraction patterns reported by Saalfeld and Wedde (1974) are the standard used for gibbsite confirmation.

The chemical data that are most commonly available to constrain surface-complexation modeling of gibbsite are acid–base titrations and ion-sorption data as a function of sorbent and sorbate concentrations, pH, and ionic strength. Measurements of the proton release and uptake which accompany cation and anion sorption are potentially useful, but such data are scarce. With the exception of confirming predictions of zero surface-charge conditions, zeta potentials determined by electrophoresis are not useful as surface-complexation model constraints as they do not provide an accurate measure of the surface potential.

Spectroscopic data can provide information about the nature of surface complexes associated with the sorption of cations and anions on the surface of gibbsite. The spectroscopic techniques employed most often for the study of surface complexation on gibbsite and other hydrous oxides are FTIR, Raman, XPS, and EXAFS spectroscopy. The surface-complexation modeling conducted in this study was constrained, in some cases, by surface-complexation species reported from the application of spectroscopic techniques. In general, however, spectroscopic data for ion sorption are limited in scope and uncertainty in data interpretation is common. Thus, spectroscopic data were not sufficient in extent or robustness to constrain surface-complexation modeling in a consistent manner. A variety of spectroscopic techniques are being applied to natural samples of gibbsite, which will constrain future models to a greater extent. Although advanced techniques are being used to identify surface complexes there is much debate regarding the mechanisms by which cations and anions are bound as inner-sphere complexes or as outer-sphere complexes to the surface of gibbsite.

4

DATA COMPILATION AND TREATMENT METHODS

In this chapter, we describe in detail the data compilation and the evaluation parameter extraction methodology. We acknowledge at the outset that subjective analysis of the extracted data sets was required to a certain extent, as we found the experimental results reported in literature for sorption on gibbsite to have various ambiguities and inconsistencies. Nonetheless, by using a systematic and reproducible parameter extraction procedure we endeavored to minimize our personal biases and provide anyone with the means to update our results as new data sets become available.

4.1 COLLECTION OF DATA

An extensive search of the English language scientific literature (including translations and limited literature in French and German) was conducted to indentify and collect all refereed papers containing information on structural properties, surface properties, and sorption capacities of gibbsite.

The experimental details for each paper containing experimental data are given in Appendix A. Details are provided for each data set cited in the study, including those data sets considered inappropriate for use. Data accepted for parameter extraction, if not tabulated, were obtained from the published graphs with the aid of an electronic digitizer (DegitizeIt 1.5.8, ShareIt Inc., Köln, Germany).

Surface Complexation Modeling: Gibbsite By Athanasios K. Karamalidis and David A. Dzombak

4.2 ASSESSMENT OF DATA QUALITY

The data from which the model parameters were extracted were first screened to ensure that appropriate experimental methods had been employed. The criteria for acceptance of data set were similar to those used by Dzombak and Morel (1990): (1) use of the standard oxide preparation method in the laboratory or acquisition from a widely accepted source; (2) experimental temperature $25 \pm 5°C$; (3) absence of competing ions; (4) absence of CO_2; (5) use of an appropriate equilibration time; and (6) use of a reasonable method of solid–liquid separation in sorption experiments.

4.2.1 Solid-Preparation Method

Gibbsite-sorption data sets were evaluated with respect to solid-preparation methods. Close examination of method of synthesis, of reported surface properties, and reported X-ray diffraction data was performed in order to verify that gibbsite was the solid used in the study. Differences in conditions during synthesis of aluminum oxides can lead to large variations in its crystal formation and consequently to structural and surface properties of the oxide (see Chapter 2).

As documented in Chapter 2, there is inconsistency in the literature regarding the nomenclature and distinguishing features of aluminum oxides and hydroxides. Many sorption experiments have been performed with aluminum hydroxides being wrongly identified as gibbsite.

To ensure fitting of sorption data obtained in experiments with gibbsite having consistent structural and surface properties, it is important to ensure that the solids used in the experiments were synthesized with methods known to yield gibbsite, or were obtained from sources known to provide material documented to be gibbsite. In this section, the widely accepted methods of synthesizing gibbsite are described. The most widely applied methods for the synthesis of gibbsite, other than the industrial Bayer method, are those of Gastuche and Herbillon (1962) and Kyle et al. (1975) and these methods were considered as standard.

The nature of aluminum hydroxide precipitated from solution by raising the pH depends on the pH of the solution. Gastuche and Herbillon (1962) identified two possible routes for the formation of gibbsite crystals depending on the pH range in which precipitation is conducted. The first route considers that starting with a solution of aluminum bicarbonate and HCl, gibbsite crystals are obtained by continuous agitation at a final pH of 8. In the second route, starting with a solution of aluminum chloride, aluminum hydroxide gels are precipitated at either pH 4.5 or pH 6.5 using NaOH. In this method, the fresh precipitates are aged at room temperature in the parent solution for a curing time between 1 and 3 months, and/or in a no-background electrolyte solution for 1 month. The procedure involves pH control of the parent solution, separation of the precipitated solid by sedimentation, rinsing with water followed by acetone, and drying at 100°C for 24 hours. When the no-background electrolyte approach is used, then the freshly formed crystals are put into cellulose bags and suspended in distilled water that is renewed daily for one month.

The solutions are then evaporated using silica gel or centrifuged, and the solids produced are dried at 105°C for 24 hours.

Kyle et al. (1975) proposed a method similar to that of Gastuche and Herbillon (1962) for the preparation of gibbsite crystals which is also widely used. The method involves precipitation of aluminum hydroxide crystals from aluminum chloride solution using NaOH at pH 4.6. The difference is that they precipitate using heat at 40°C for 2 hours before being put in cellulose tubing. Nevertheless, both methods are well-established and result in the formation of crystalline gibbsite.

Experimental data sets produced with gibbsite reported to be prepared by the two aforementioned methods were used for modeling, along with some data sets obtained using gibbsite material from various sources. Sources that were considered to provide reliable experimental data were those that showed proof of gibbsite crystallinity (e.g., via spectroscopic evidence), stability of synthesized product, and a natural solid of known origin and properties. Although in this study we tried to consistently use experimental data produced with alpha-form gibbsite, data reported using γ-$Al(OH)_3$(s) were sometimes used due to limited available sorption data. For the purpose of the study, the γ-$Al(OH)_3$(s) was considered as the closest to the naturally occurring α-$Al(OH)_3$(s).

4.2.2 Type of Reaction Vessel

Reaction vessel material is important in sorption experiments for various reasons. Leaching of contaminants (e.g., silicon from glass reaction vessels) has been previously reported by Anderson and Benjamin (1985). The importance of this was mentioned by Dzombak and Morel (1990) regarding experiments on hydrous ferric oxide. Struempler (1973) demonstrated that glass and polypropylene reaction vessels sorb cations and the phenomena are more pronounced as pH and equilibration time increases. In the case of sorption experiments with gibbsite suspensions, the most frequently used material for the reaction vessel is polypropylene. Other materials that have been used for reaction vessels in sorption experiments with gibbsite include pyrex, teflon, plastic, polyethylene, polysterene, and polycarbonate. Studies conducted with polypropylene vessels have shown minimal adsorption of cations on the walls, at least for the equilibration time used in the experiments (e.g., Zhang et al., 2006). Sorption data from experiments employing other materials for reaction vessels were used in this study. We tried to utilize data for which there was no indication of cation or anion loss on the reaction vessel walls or were proven to have minimal measurable loss of sorbates (e.g., Weerasooriya et al., 2006). In the analysis of the gibbsite-sorption data presented in Chapters 5 to 7, we assume that reported data have been appropriately corrected for sorbate loss to reaction vessel walls and that the sorption of the sorbates to walls is minimal where the type of reaction vessel is not specified.

4.2.3 Nature of Background Electrolyte

The composition of the background electrolyte used in the acid–base titration and ion-sorption experiments with gibbsite was checked for the presence of competing ions.

Only data for sorption of single sorbates on gibbsite were considered in this study. Therefore, if competing ions were present the data set was excluded. When the background electrolyte used in the experiment to fix the ionic strength of the solution includes salts with specifically sorbing ions (e.g., Ca^{2+}, Mg^{2+}, SO_4^{2-}) this can influence the results. In experiments conducted in solutions with several competing ions (e.g., sea water), the nature of the competition needs to be understood a priori in order to determine sorption behavior for particular ions and this is not possible. Data sets in which monovalent ions were used to fix ionic strength (e.g., K^+, Na^+, Cl^-, NO_3^-), were accepted, as these ions exhibit negligible specific sorption on oxide surfaces.

4.2.4 Sorption Kinetics

Evaluation of the adequacy of reaction time is important for both potentiometric titration and ion-sorption data for oxides in the development of a thermodynamic database. Sorption of inorganic ions on hydrous oxides is a two-step process. The first step consists of a rapid initial uptake; the second step is slower and the equilibrium sorption density is approached asymptotically (Yates, 1975; Dzombak and Morel, 1990). Equilibration times for proton-exchange reactions are not necessarily the same as specific sorption reactions for other ions.

4.2.4.1 Proton Sorption Kinetics The traditional way of describing the proton uptake on metal oxide surfaces is by distinguishing two consecutive steps, a fast and a slow step. Potentiometric titration experiments with iron oxides (Parks and DeBruyn, 1962; Schindler and Kamber, 1968; Dzombak and Morel, 1990) have demonstrated that the fast step is complete within a few minutes, whereas the slow step can last for weeks or months. In the fast step, surface charge is developed by ionization of the surface hydroxyl groups, while in the slow step structural rearrangement of the surface occurs (Yates, 1975; Dzombak and Morel, 1990).

We have determined surface acidity constants from the analysis of relatively slow titration data. The reason was that slow titration allows the surfaces to equilibrate. Contrary to rapid acid–base titration data that was used earlier for HFO (Dzombak and Morel, 1990), titration step times for gibbsite show great variability ranging from 2 minutes (Hiemstra et al., 1999) to many hours (e.g., $>$ 14 hours, Rosenqvist et al., 2002). Rapid acid–base titration data were chosen in the past because it is believed they suppress the slow proton uptake that is related to structural changes (Yates, 1975).

Gibbsite has different types of proton reactive surface sites located on its basal or edge planes (Rosenqvist et al., 2002; Jodin et al., 2005), and the nature and distribution of those sites apparently change with time after formation. Rosenqvist et al. (2002) demonstrated that due to smoothing of the surface of gibbsite with time there was a difference in surface proton reactivity between aged and nonaged gibbsite suspensions. They also found that the formation of ion pairs at the basal planes is a comparatively slow process. Alteration of surface proton reactivity probably occurs on the continuously changing surface of freshly changed gibbsite suspensions until they reach more stable forms. Thus, in this study, experimental potentiometric titration data sets produced on freshly precipitated or nonaged gibbsite were not considered.

The surface of aluminum hydroxide is amphoteric and the behavior of its reactive groups is of importance with respect to sorption studies of many chemical species. At high proton sorption levels at low pH, the degree of surface saturation is strongly determined by the reactive site density. This site density is an important parameter in the available proton sorption models (Hiemstra et al., 1987). Here we consider the whole gibbsite crystal surface as proton active and we do not distinguish different sorption kinetics between the different surface planes. We found that distinguishing between the surface planes by introducing two types of surface sites in surface-complexation modeling of gibbsite titration and sorption data does not significantly improve the ability to fit the data.

4.2.4.2 Cation and Anion Sorption Kinetics Inorganic ion sorption on hydrous oxides has been studied using various equilibration times, for example, hours to weeks for anions on gibbsite (Helyar et al., 1976a; Farrah et al., 1987). Although most of these studies had similar experimental designs and objectives, a wide range of equilibration times has been used. The two-step character of the kinetics of cation and anion sorption on hydrous oxides at constant pH has revealed equilibration-time discrepancies as found in the literature (Yates, 1975; Adu-Wusu and Wilcox, 1991a). In many sorption studies, it has been assumed that the contribution of the slow-sorption step is small, and short equilibration times have been selected (Dzombak and Morel, 1990). It has been demonstrated, however, that significant sorption may occur in the slow-sorption step, that the relative importance of the second sorption step increases at high ionic strength and high sorbate/sorbent ratio, and that sorption kinetics are influenced by pH. Cation sorption is faster at higher pH and anion sorption is faster at lower pH. Additional information can be found in Dzombak and Morel (1990).

In ion-sorption experiments appropriate reaction times are critical to ensure the achievement of equilibrium. In this study, most of the sorption data sets for cations (other than protons) and anions used in parameter extraction employed equilibration periods ranging from 24 to 48 hours. Sorption data for gibbsite resulting from experiments in which short (e.g., 0.5 to 2 hours) reaction times were employed often do not represent equilibrium and thus were not considered. However, equilibration varies from system to system and depends on various factors, for example, sorbate/sorbent ratio. In cases where sufficient evidence of equilibration was presented even though the reaction period allowed was less than 24 hours, data sets were included.

4.2.5 Method of Solid–Liquid Separation

The method of solid–liquid separation after the reaction period in a sorption experiment is another important experimental detail. Metal oxide suspensions may contain a significant fraction of colloids or very small particles. To ensure the complete separation of the solid from the solution after the desired equilibration time, colloids should be removed along with the rest of the solids. Usually, large centrifugal acceleration (>1500*g*) or filtering with a small pore size filter (<0.45 μm) is adequate to separate solids from the supernatant. In this study most of the data sets used were extracted from published experiments with clear indication of use of centrifugal separation technique. In rare cases where relatively large pore size filters (e.g., >0.45 μm) were used instead of centrifugation we still accepted those data sets as long as the rest of the criteria for acceptance were fulfilled. Considering the uncertainties regarding particle size under the various experimental conditions investigated, we have rejected no data solely on the basis of the method of solid–liquid separation.

4.2.6 CO_2 Exclusion

Exclusion of $CO_2(g)$ while conducting acid–base titration experiments is important, as it is for sorption experiments with cations that form carbonate precipitates. The presence of $CO_2(g)$ in the solution may cause various complications in analyzing data to determine fundamental surface-complexation constants. The first one is the potential for drift of pH with time, and thus nonestablishment of equilibrium conditions, due to the buffering capacity of dissolved $CO_2(g)$ in the solution. The second one is the hydrolysis of CO_2 which produces CO_3^{2-} and HCO_3^- in the solution. These ions may complex with the sorbate ion in solution or sorb weakly on the surface and thus influence the sorption of the sorbate ions (Su and Suarez, 1997). Previous studies with aluminum oxide (Schulthess and McCarthy, 1990), amorphous iron hydroxide (Zachara et al., 1987), and goethite (Van Geen et al., 1994) found that carbonate ions can sorb competitively, if weakly, under some conditions on Al or Fe sites, and can also form discrete solid phases with Al.

Several studies have focused on carbonate species sorption on gibbsite, including sorption experiments at different partial pressures of CO_2 (e.g., Kavanagh et al., 1975; Helyar et al.,1976a, 1976b; Tochyama et al., 1996; Jodin et al., 2005). Study of the effect of different concentrations of the Cl salts of Na, K, Mg, and Ca on the adsorption of phosphate by gibbsite with continuous injection of CO_2 of known partial pressure revealed that HCO_3^- had little effect on phosphate sorption. Studies of Eu(II) and Co (II) sorption on amorphous aluminum hydroxide in the presence of CO_2 indicated that carbonate species have no significant impact on their sorption (Tochyama et al., 1996). Gibbsite, in general, sorbs much less carbonate than amorphous $Al(OH)_3$ due to limited functional groups on the surface (Su and Suarez, 1997).

Due to the fact that only limited results have been published and based on the aforementioned reasons, experimental data sets of acid–base titration or ion-sorption edges which did not involve exclusion of CO_2 were not considered. In this book if there was an indication of interference due to the presence of carbonate for acid–base or ion-sorption data, the data set was not used. While HCO_3^- and CO_3^{2-} were considered to be weakly sorbing species based on available data, and no surface-complexation constants were determined for these ions, we decided to analyze only data for which no CO_2 was present in order to avoid a potential source of experimental interference and a source of uncertainty in the extracted surface-complexation constants.

4.2.7 Experimental Temperature

In this study of ion sorption on gibbsite, only those data sets were accepted in which the experimental temperature during sorption was maintained at $25 \pm 5°C$. This temperature constraint for accepting various data was done for internal consistency of the database. Temperature variation may affect both reaction constants and crystalline structure of gibbsite. Weerasoriya et al. (2006) demonstrated that the sorption capacity of mercury on gibbsite increased as the temperature increased from 25 to 50°C. For sorption of anions, such as borate, there are contradictory results in the literature on whether temperature increases sorption capacity or not. Goldberg et al. (1993a, 1993b) demonstrated that a temperature variation of 10–40°C was enough to cause changes in sorption capacity not only for gibbsite but also for other clay-like minerals that possibly contain gibbsite. Adu-Wusu and Wilcox (1991a, 1991b) studied silicate sorption on gibbsite in the temperature range of 30–60°C and they showed that sorption increases as the concentration and temperature increase.

4.3 COMPILATION OF SURFACE PROPERTIES

In Chapter 5, we present compilations of measured surface properties of gibbsite: surface area, site density, and pristine point of zero charge. Surface acidity constants determined in this study are also presented. Best estimates for surface properties are given and explained.

Specific surface area is an important surface property that can be measured by various methods, as discusses in Chapter 3. The BET method, which involves measurement of gas (typically N_2) adsorption on dried solid, is the most popular method for specific surface area determination. This method is relatively easy and precise, but the applicability of the result to a hydrated solid is obviously an issue, though less so for crystalline solids such as gibbsite compared to freshly precipitated, amorphous, highly hydrated solids such as hydrous ferric oxide. In this study, as presented in Chapter 5, the specific surface area for gibbsite was determined using reported BET values due to the predominance of use of this technique and the abundance of BET measurements for gibbsite in the literature.

4.4 EXTRACTION OF EQUILIBRIUM SORPTION CONSTANTS

It is important for database development that the methods and procedures followed for data analysis be reproducible, uniform, and consistent. This enables other researchers to rederive the same equilibrium constants under the same conditions of data sets and model, and to update the best estimates for the derived constants as new data become available and are analyzed.

TABLE 4.1 Activity coefficients for dissolved ionic species

	$-\log \gamma_z$ for				
I (mol/L)	$Z=1$	$Z=2$	$Z=3$	$Z=4$	$Z=5$
0.0001	0.005	0.02	0.05	0.08	0.13
0.0005	0.01	0.04	0.10	0.18	0.28
0.001	0.02	0.06	0.14	0.25	0.39
0.005	0.03	0.13	0.30	0.53	0.83
0.01	0.05	0.18	0.40	0.72	1.13
0.05	0.09	0.35	0.78	1.39	2.19
0.1	0.11	0.44	0.99	1.76	2.77
0.3	0.13	0.52	1.17	2.08	3.61
0.5	0.15	0.60	1.35	2.40	3.77

Note: Z is the charge number. For $I \leq 0.1$ M (for Z from 1 to 4), and for $I \leq 0.5$ M (for $Z=5$), $\log \gamma_Z$ values are calculated with the Davies equation; for $I > 0.1$ M, $\log \gamma_Z$ values are taken from Morel and Hering (1993).

4.4.1 Solution Activity Coefficients

Chemical equilibrium calculations in this study were performed using the Davies equation for ionic strength correction of equilibrium constants for solution phase reactions. We corrected equilibrium constants for ionic strength using the single-ion activity coefficients given in Table 4.1. The $\log \gamma_Z$ values presented in Table 4.1 were calculated using the Davies equation:

$$\log \gamma_z = -AZ_i^2 \left[\frac{\sqrt{I}}{1+\sqrt{I}} - bI \right] \tag{4.1}$$

where Z_i is the charge number, I the ionic strength (mol/L), and A and b are constants with values of $A=0.512$, and $b=0.24$. The Davies equation is accurate for aqueous systems with $I \leq 0.5$ M (Morel and Hering, 1993).

4.4.2 FITEQL

FITEQL (Westall, 1982; Herbelin and Westall, 1999) is a computer program for the determination of chemical equilibrium constants from experimental data. It uses nonlinear least-square optimization for model calculations and optimization of model constants. Technical details of the FITEQL algorithm can be found in both Herbelin and Westall (1999) and Dzombak and Morel (1990). The FITEQL version 4.0 (Herbelin and Westall, 1999) used in this study has been improved over its precursor by the addition of new features, including the choice of preprocessors for different operating systems, the choice of entering serial data in table or column format, and various other improvements regarding the error estimates, and initial guess management.

The program includes a multicomponent chemical equilibrium model and an algorithm for repeated application of this model to a chemical equilibrium data set.

First, the user specifies the appropriate equilibrium model for the system of interest in FITEQL. Second, the equilibrium problem is solved at each of the conditions corresponding to the data points provided. Third, specified parameters in the model are adjusted until the model results coincide with the data. Parameter adjustment is done to minimize the difference between values calculated from the model and the values observed experimentally.

The goodness-of-fit measure in the FITEQL program is based on the overall variance V_Y, which is the weighted sum of squares of residuals divided by the degrees of freedom (WSOS/DF). Values of V_Y between 0.1 and 20 indicate a reasonably good fit, whereas values around 1 indicate an excellent fit. Values of V_Y much higher than 20 indicate insufficient data for the equilibrium model specified or that the model is not entirely appropriate for the data, whereas values significantly less than 0.1 indicate that the model has too many adjustable parameters.

The value of V_Y is dependent on the experimental error estimates specified in the input, as well as on the selection of data for the construction of the input file. FITEQL has an inherent tendency of heavily weighing the data points representing very high sorption values (low solution concentration of the sorbates ion), leading to good fits for data in this range and, sometimes, poorer fits (typically underprediction of sorption) for low-sorption data. This is attributable to the weighting algorithm incorporated in the program. To improve the quality of fits, even at low sorption values, and to obtain fits consistent with the entire range of experimental data, pH edge data points that represented sorption of 98 percent or more were not considered in the input files for the model.

The error estimates used for the model are shown in Table 4.2. These were fixed uniformly in the parameter extraction process. For consistency with the previously developed databases for hydrous ferric oxide (Dzombak and Morel, 1990), goethite (Mathur and Dzombak, 2006) and hydrous manganese oxide (Tonkin et al., 2004) performed with FITEQL and the generalized two layer model, we used the error estimates defined by Dzombak and Morel (1990). All experimental measurements have a degree of uncertainty. Although it would be desirable to use individual error

TABLE 4.2 Error estimates for gibbsite acid–base titration data and sorption data

Measurement[a]	S[rel][b]	S[abs][c]	Remarks
X_H	0.05	0.0	±0.02 pH units
X_H	0.10	0.0	±0.04 pH units
T_H	0.01	$0.01 \times \min T_H$	
X_M	0.05	0.0	±0.02 pM unit
T_M	0.01	$0.01 \times \min T_M$	
X_A	0.05	0.0	±0.02 pA unit
T_A	0.01	$0.01 \times \min T_A$	

[a] H, H^+; M, cation; A, anion.

[b] S_H[rel] = 0.05 used for pH measurements in sorption experiments; S_H[rel] = 0.1 used for pH measurements in titration experiments. [rel], relative.

[c] For $b = \log a$, S_b[abs] = $0.434 S_a$[ref]; see Westall (1982) or Skoog and West (1982). [abs], absolute.

estimates for each data point or at least each data set, the lack of reported error analysis for most of the sorption data sets compelled the fixation of error estimates for the sake of consistency. Thus, error estimates given in Table 4.2 were applied uniformly in the parameter extraction. Fits for some data sets yielded high values of V_Y yet the parameters extracted enabled reasonably close simulation of the data. Using larger values for error estimates in most of these cases could yield almost the same parameter values with lower values of V_Y for the fits, but for the purpose of consistency the final data fitting was done using the error estimates of Table 4.2.

For some data sets, convergence with FITEQL could not be achieved due to some combination of the nature of the data, the nature of the chemical model, and the error estimates. In such cases, which occurred mostly with anions, some of the surface-complexation constants in the model were fixed manually after trial fitting with FITEQL and/or data set simulation with Visual MINTEQ. If FITEQL convergence still could not be achieved for a particular data set, manual fitting was performed through trial simulations with Visual MINTEQ (Gustafsson, 2009).

All derived or fixed constants were adjusted to zero ionic strength by the application of appropriate solution-phase activity coefficients. Thus, all reported values of acidity or surface-complexation constants in this study are for zero ionic strength.

4.4.3 Data Grouping

An important issue regarding the way that sorption data sets should be treated in parameter optimization is whether to consider all the data sets simultaneously or to determine optimum parameter values for one data set at a time. In this study, we used the traditional approach of data set by data set, which is also consistent with the development of the HFO, goethite, and HMO databases.

Dzombak and Morel (1990) demonstrated that the traditional data set by data set approach has some advantages over the simultaneous consideration of all data. The data set by data set approach avoids the problems associated with nonidentical oxides, different experimental techniques, and different experimental systems requiring different chemical models. Also the influence of the individual data sets determining parameter estimates is apparent and can be weighted by their variance or standard deviations when best overall estimates are determined. Another significant advantage of the data set by data set approach is that the database can be easily updated when new parameter estimates are derived from new sources. Otherwise, considering all data sets simultaneously will require manipulation of all raw data previously examined. The latter increases complexity and encumbers the update of the database. Additional discussion of the data set by data set versus all data sets simultaneously approach can be found in Dzombak and Morel (1990).

4.4.4 Selection of Surface Species

Selecting the appropriate surface species for the extraction of surface-complexation constants from a particular data set is the critical part of the formulation of the surface-complexation chemical equilibrium model. In the case of a chemical model for inorganic- data three types of reactions are included in the model: solution complexation reactions, surface hydrolysis reactions, and surface-complexation reactions for specifically sorbing cations and anions in addition to H^+. Selection of surface reactions and species in this study followed the rules and the generalized two-layer model constraints as described by Dzombak and Morel (1990). The general approach was that surface reactions were formulated with the simplest possible surface species (e.g., $\equiv AlOM^+$ for divalent cations, and $\equiv AlA^-$ and $\equiv AlHA^0$ for divalent anions). Past experience with surface complexation (Dzombak and Morel, 1990; Turner et al., 1996; Fendorf et al., 1997; Tonkin et al., 2004; Mathur and Dzombak, 2006) has shown that two or three surface species, in addition to surface hydrolysis species, may be required to fit cation or anion data.

Another potentially important type of reaction for the formulation of the surface-complexation chemical model is surface precipitation. At high concentrations, both cations and anions may sorb via surface precipitation, and the generalized two-layer model can be extended to include this sorption mechanism. Surface precipitation is only important for very large sorbate concentrations and hence need not be considered in most applications. The number of data sets with high sorbate concentrations was very small in this study, surface precipitation was not invoked, and therefore no surface precipitation species were needed.

4.4.5 Selection of Best Estimates

The application of FITEQL to all available data sets for titration data or a particular sorbate–sorbent system produces a family of "optimum" values for the equilibrium surface acidity or sorption constants involved. The question of how to select a best estimate among the list of optimum *K* values must then be addressed.

For each sorbate ion studied, the best overall estimates for the relevant surface-complexation constants were calculated according to the procedure developed by Dzombak and Morel (1990):

$$\overline{\log K} = \sum w_i(\log K)_i \tag{4.2}$$

where w_i is a weighting factor calculated from

$$w_i = \frac{(1/\sigma_{\log K})_i}{\sum (1/\sigma_{\log K})_i} \tag{4.3}$$

The parameter $(1/\sigma_{\log K})_i$ is the standard deviation in log *K* calculated by FITEQL for the *i*th data set modeled for the sorbate ion of interest.

A set of confidence intervals was also calculated for the weighted average of each surface-complexation constant to indicate the uncertainty associated with the best estimate. To derive a confidence interval corresponding to a specific level of significance (α), the sample variance must first be calculated. The latter is calculated using the sum of squares. If $x_i = \log K_i$ and $\bar{x} = \overline{\log K}$, the sample variance s_x^2

is given by

$$s_x^2 = \frac{\sum_{i=1}^{n} w_i(x_i - \bar{x})^2}{n-1} \tag{4.4}$$

The sample variance is then used to calculate the standard average error for the weighted average $\bar{x}$:

$$s_{\bar{x}} = \frac{s_x}{\sqrt{n}} \tag{4.5}$$

Knowing $s_{\bar{x}}$, the $1-a$ confidence limits for x can be calculated as:

$$x_- = \bar{x} - t_{a/2} s_{\bar{x}} \tag{4.6}$$

$$x_+ = \bar{x} + t_{a/2} s_{\bar{x}} \tag{4.7}$$

where $t_{\alpha/2}$ denotes the t deviate corresponding to the single-tailed area $\alpha/2$ in Student's t-distribution. Equations 4.6 and 4.7 were implemented in a computer program to calculate the weighted-average best-estimate surface acidity and ion-sorption constants.

4.5 OPTIMAL-FIT SIMULATIONS

Visual MINTEQ (Gustafsson, 2009) was used to calculate the fitting curves for the individual data sets with the individual data set optimum surface-complexation constants, and also with the overall best-estimate surface-complexation constants. Equilibrium surface acidity and surface-complexation constants calculated with FITEQL were used in Visual MINTEQ for a forward calculation. For each of these calculations, the parameters used were for the specific experiment on which the individual constants were calculated. In rare cases, when Visual MINTEQ produced fitting curves that did not fit the data sets well, possible outliers from the individual data set or some data points related to maximum sorption were removed and FITEQL was rerun. High sorption density data are weighted heavily by FITEQL, as discussed previously. The problem set up in Visual MINTEQ was same as that in FITEQL. The calculated "optimal" fitting curve for each individual data set is plotted along with the overall best-estimate acidity or surface-complexation constant presented in the graphs of acid–base titration data, and of cation and anion sorption data for gibbsite in Chapters 5, 6 and 7.

4.6 PRESENTATION OF RESULTS

The following three chapters summarize the results of the sorption data fitting for gibbsite. Intrinsic surface-complexation constants obtained for acid–base titration

TABLE 4.3 Identification code for data sets

First character: solid identifier[a]
4 = Gibbsite, α-$Al(OH)_3$

Second character: data type
T = Acid–base titration
E = Sorption pH edge
I = Constant pH isotherm

Third and fourth characters (titration data): data set number used for cross-reference between tables and figures

Third and fourth characters (sorption data): sorbate identifier

CD = Cadmium	AS = Arsenate
CA = Calcium	AT = Arsenite
CO = Cobalt	BO = Borate
CU = Copper	CH = Chromate
FE = Iron (II)	F = Fluoride
PB = Lead	MO = Molybdate
MN = Manganese	PO = Phosphate
HG = Mercury	SE = Selenate
TH = Thorium	SO = Sulfate
UO = Uranium	SI = Silicate
ZN = Zinc	

Fifth and sixth characters (sorption data): data set number used for cross-reference between tables and figures

[a]Previously assigned solid identifier codes: 1, Hydrous ferric oxide (Dzombak and Morel, 1990); 2, Goethite (Mathur, 1995; Mathur and Dzombak, 2006); 3,Hydrous manganese oxide (Tonkin et al., 2004).

data (Chapter 5), cation-sorption data (Chapter 6), and anion sorption data (Chapter 7) are tabulated for each data set. Experimental data for every data set used for parameter extraction are plotted together with the individual best fit (— lines) determined by FITEQL and the fit corresponding to the best-estimate log K values (- - - lines) for the particular sorbate–sorbent system.

All the data points are labeled with an identifier code of six characters (or less) and important experimental variables are listed in the figure. The meaning of the code is given in Table 4.3. For each sorbate studied, a table has been provided containing the source of each data set and identifier code assignments.

5

SURFACE PROPERTIES OF GIBBSITE

This chapter presents a synthesis of the available information on the major physical–chemical properties of gibbsite, α-$Al(OH)_3(s)$. Supplementary to Chapter 2, in which the basic structural characteristics of gibbsite are discussed, this chapter summarizes and analyzes measurements of surface area, site densities, and surface acid–base properties for gibbsite. Best estimates for the surface properties of gibbsite are provided.

5.1 SURFACE AREA

A list of gibbsite-specific surface areas measured by various techniques is given in Table 5.1. A discussion of surface-area-measurement techniques in use is provided in Chapter 3. The most widely used method remains that of nitrogen gas adsorption coupled with BET analysis of the resulting isotherm, considering the gibbsite surface as homogeneous. The inherent limitations of the BET method are that the validity of the monolayer gas-sorption capacity is questionable, the monolayer structure differs for each solid, and the potential for localized monolayer coverage and/or primary micropore filling resulting from the strong adsorption at very low equilibrium/saturation pressure ratios (Dollimore et al., 1976; Sing, 2001). Despite these limitations, the BET method remains widely used because it can be conducted relatively rapidly and it is a reproducible technique.

Surface Complexation Modeling: Gibbsite By Athanasios K. Karamalidis and David A. Dzombak

TABLE 5.1 Specific surface areas for gibbsite

A (m^2/g)	Method	Sources
13	BET, $N_2(g)$	Weerasooriya et al. (2000, 2002)
29.5	BET, $N_2(g)$	Rosenqvist et al. (2002)
25	BET, $N_2(g)$	Scheinost and Sparks (2000)
40.3 (H)[a]	BET, $N_2(g)$	Hiemstra et al. (1999)
40.8 (L)[b]	BET, $N_2(g)$	Hiemstra et al. (1999)
13.5	BET, $N_2(g)$	Ladeira et al. (2001)
9.97	BET, $N_2(g)$	Yang et al. (2007a)
19	BET, $N_2(g)$	Kaiser and Guggenberger (2003)
19.8 (3.4 edge face surface area)	BET, $N_2(g)$	Hiemstra et al. (1989)
1.7	BET, $N_2(g)$	Dietzel and Böhme (2005)
26–39	BET, $N_2(g)$	Van Riemsdijk and Lylkema (1980)
71	BET, $N_2(g)$	Molis et al. (2000)
37.7 (H)	BET, $N_2(g)$	Mitchell (2005)
45.9 (L)	BET, $N_2(g)$	Mitchell (2005)
76	BET, $N_2(g)$, and N_2 isotherms	Wierenga et al. (1998)
7.67	BET, $N_2(g)$	Bloom and Erich (1987)
1.04	BET, $N_2(g)$	Bénézeth et al. (2008)
7.2	BET, $N_2(g)$	Addai-Mensah et al. (1998); Prestidge et al. (1999)
7.51	BET, $N_2(g)$	Bloom (1983)
19.6	BET, $N_2(g)$	Rosenqvist and Casey (2004)
0.217	BET, $N_2(g)$	Gorman-Lewis and Fein (2004)
94	BET, $N_2(g)$	Dubbin and Sposito (2005)
66.5	BET, $N_2(g)$	Goldberg et al. (1993a)
49.56	BET, $N_2(g)$	Guan et al. (2007)
25	BET, $N_2(g)$	Scheckel and Sparks (2000)
0.37 ± 0.04	BET, $N_2(g)$	Ganor et al. (1999)
1.48	BET, $N_2(g)$	Hongxia et al. (2006)
1.5	BET, $N_2(g)$	Baumann et al. (2005)
14 ± 1	BET, $N_2(g)$	Foster et al. (1998)
47 (L)	BET, $N_2(g)$	Hingston et al. (1972)
58 (L)	BET, $N_2(g)$	Hingston et al. (1972)
48	BET, $N_2(g)$	Muljadi et al. (1966)
31	BET, $N_2(g)$	Hingston et al. (1971)
32.9	BET, $N_2(g)$	Goldberg et al. (1993b)
5.9	BET, $N_2(g)$	McBride (1982)
32.5	BET, $N_2(g)$	McBride and Wessellnk (1988)
7.67	BET, $N_2(g)$	Bloom and Erich (1987)
45	BET, $N_2(g)$	Manning and Goldberg (1996)
47	BET, $N_2(g)$	Kyle et al. (1975)
48	BET, $N_2(g)$	Kavanagh et al. (1975)

TABLE 5.1 (*Continued*)

A (m^2/g)	Method	Sources
45	BET, N_2(g)	Parfitt et al. (1977)
96	BET, N_2(g)	Roberts et al. (2003)
56.5	BET, N_2(g)	Goldberg et al. (1996)
37	BET, N_2(g)	Hattori et al. (2009)
	Values Not Used	
8.3 ± 0.4	Ethylene glycol adsorption (26 ± 0.5°C)	Helyar et al. (1976a, 1976b)
100	Estimated by electron microphotographs	McBride and Wessellnk (1988)
68.5	Ethylene glycol adsorption	Okazaki et al. (1986)
119	Ethylene glycol adsorption	Okazaki et al. (1989)
192 (probably amorphous)	Ethylene glycol adsorption	Okazaki et al. (1989)
9.7	Unspecified	Alvarez et al. (1976b)
0.1	Unspecified (C-31, Alcoa)	Adu-Wusu and Wilcox (1991a, 1991b)
11	Unspecified	Papelis et al. (1995)
1.82	BET, N_2(g) (0.2–20 μm fraction/Fisher Chemicals)	Bloom (1983)
40.8 43.6	Estimated with AFM and computerized image analysis routines that include mass-normalization and corrections for microtopography (Bickmore et al., 2002)	Mitchell (2005)
14	Unspecified	Beattie et al. (1996)
6.4	Unspecified	Tsuchida and Ichikawa (1989)
4.5	Estimated with AFM and computerized image analysis routines that include mass-normalization and corrections for microtopography (Bickmore et al., 2002)	Jodin et al. (2004)
86.7 (H) 51.2 (L)	Estimated using TEM and platinum shadowing, and mass-normalization	Hiemstra et al. (1999)

(*Continued*)

TABLE 5.1 *(Continued)*

A (m^2/g)	Method	Sources
4 (1.5 of lateral and 2.5 of basal)	Estimated using Kr-adsorption and infrared spectroscopy	Jodin et al. (2004)
1.5 ± 0.2 4 ± 0.3 6 ± 0.5 70 ± 5	Estimated from krypton adsorption-desorption isotherms at 77 K	Phambu et al. (2000)
66.5	BET, $N_2(g)$ (gibbsite and bayerite)	Xu et al. (2008)
0.6	BET, $N_2(g)$ (pretreatment may have caused greater particle sizes due to sedimentation and smoothing of the mineral surface during predissolution)	Dietzel and Böhme (2005)
96	Estimation only on (001) face	McBride et al. (1984)
91	Estimated using AFM and number-averaged particle dimensions	Rosenqvist et al. (2002, 2003)

[a]H: High, refers to the temperature at which the gibbsite sample used for the SSA measurement was prepared (crystallized). For "high" 70°C was used.

[b]L: Low, refers to the temperature at which the gibbsite sample used for the SSA measurement was prepared (crystallized). For "low" 4°C was used.

Gibbsite is a crystalline solid and should have reasonably consistent measured surface areas. However, the published values for specific surface areas of gibbsite range between 0.37 and 94 m^2/g, reflecting different conditions of crystallization and synthesis techniques.

In order to maintain a consistent surface-area value throughout modeling in this study, a surface area of 32 m^2/g was used. This value represents the average of the reported BET data (32.03 m^2/g), adjusted to two significant figures. Reported surface-area measurements for gibbsite obtained with methods other than the BET method (see bottom of Table 5.1) were not considered in the calculation of the average value. Surface-area measurements for gibbsite made with other techniques are few in number.

5.2 SITE DENSITY

The site density (N_s) indicates the number of surface hydroxyl sites available per unit amount of solid, expressed as sites/nm^2 or mol/mol Al, and determines the site concentration associated with a particular solid concentration and the maximum concentration of sorbed ions. This is a critical parameter for surface-complexation modeling. The total concentration of sites, N_T (mol/L), for a particular solid concentration, C_s (g/L), is related to the site density N_s (sites/nm^2) through

$$N_T = N_s \times A \times C_s \times \frac{1}{N_A} \times \frac{10^{18}\text{nm}^2}{\text{m}^2} \quad (5.1)$$

where A is the specific surface area (m^2/g) of the solid, and N_A the Avogadro's number (6.02×10^{23} sites/mol).

There have been attempts to estimate site density theoretically based on crystal structure (James and Park, 1982, and references therein; Sposito, 1984; Hiemstra et al., 1989; Barron and Torrent, 1996; Sahai and Sverjensky, 1997; Koretsky et al., 1998). The usefulness of theoretical estimates has been hindered, however, by lack of information regarding crystallinity, grain size, and morphology of the mineral samples used in most adsorption studies (Sahai and Sverjensky, 1997).

In most applications of surface-complexation modeling, a constant value of N_s has been employed and determined by fitting sorption data for the solid of interest for a range of loading conditions. For example, Dzombak and Morel (1990) used 2.31 sites/nm^2 for hydrous ferric oxide, Hayes et al. (1991) suggested 10 sites/nm^2 for α-FeOOH, α-Al_2O_3, and TiO_2, and Mathur and Dzombak (2006) used 2 sites/nm^2 for goethite. While oxide minerals will have some variability in hydroxyl site density due to differences in formation conditions, consideration of a constant value of N_s provides consistency when modeling natural systems (Sahai and Sverjensky, 1997).

In the case of gibbsite, which crystallizes in hexagonal shape, the proton-reactive surface sites are often considered to be the singly coordinated lateral surface groups (Hiemstra et al., 1999). Rosenqvist et al. (2002) suggested that basal surface sites can sorb protons at slow rates, and Gan and Franks (2004) that the basal plane is proton active at pH less than 7 based on atomic force microscopy (AFM) observations. Surface hydroxyl sites on gibbsite have been distinguished for basal and edge face surfaces and different site-density values have been reported for each type of these surfaces (Bickmore et al., 2004). The differences in reactivity of the basal and edge surface sites are not clear, however. Therefore, in this study the site density was considered to be the total sites on the gibbsite crystal surface.

Surface hydroxyl site density can be estimated from crystallographic data for gibbsite reported in the literature (e.g., Hiemstra and Van Riemsdijk, 1991; Rosenqvist et al., 2002), and also by fitting the density of the total reactive sites available to sorb protons in modeling of acid–base titration data. The reported values of site densities for gibbsite range between 8.0 and 8.8 sites/nm^2, as shown in Table 5.2. In modeling trials with gibbsite acid–base titration data conducted in this study, assuming the best-estimate surface-area value of 32 m^2/g, most of the data sets tested seemed to be well fitted using high site-density values (e.g., ~8 sites/nm^2). Using lower values for site density (e.g., 2 sites/nm^2) did not improve data fitting. The chosen value of 8 sites/nm^2 enabled good fits of most of the data sets, and also falls within the range of site-density values for gibbsite specified in the literature.

TABLE 5.2 Site densities (N_s) for gibbsite

N_s (sites/nm^2)	Method	Sources
8.5	Not specified	Hiemstra et al. (1999)
8.0	Crystallographic determination	Hiemstra et al. (1991)
8.0	Not specified	James and Park (1982)
8.0	Not specified	Meng and Letterman (1993)
8.11	Not specified	Weerasooriya et al. (2007a, 2007b)
8.8	Crystallographic determination	Rosenqvist et al. (2002)

5.3 POINT OF ZERO CHARGE

The pristine point of zero charge (PPZC) is the point of zero charge (PZC) in the absence of specifically sorbing ions (Sposito, 1996). PPZC values for gibbsite have been determined by various techniques, including acid–base titration, potentiometric titration, microelectrophoresis, and salt titration. As shown in Table 5.3, PPZCs reported in literature for gibbsite are in the range of 7.8–10.0 with an average value of 9.0. PZC values measured in solutions with 1 : 1 electrolytes of nonsorbing monovalent ions such as NaCl, $NaNO_3$, and KCl were considered to be PPZC measurements. PPZC values that were predicted using models or reported without mention of the measurement technique were excluded.

TABLE 5.3 Pristine points of zero charge (PPZC) for gibbsite

PPZC	Method	Sources
7.8	Potentiometric titration	Kaiser and Guggenberger (2003)
10.4	Potentiometric titration	Addai-Mensah et al. (1998)
9.2	Potentiometric titration	Addai-Mensah et al. (1998)
9.6	Microelectrophoresis (NaCl)	Goldberg et al. (1993b)
9.8	Potentiometric titration/$NaNO_3$	Kavanagh et al. (1975)
8.1–8.3	Potentiometric titration/KNO_3	Yang et al. (2007a)
8.7	Potentiometric titration/$NaNO_3$	Weerasooriya et al. (2001a, 2001b)
8.35	Potentiometric titration/NaCl	Weerasooriya et al. (2007a, 2007b)
8.7	Potentiometric titration/$NaClO_4$	Weerasooriya et al. (2007a, 2007b)
8.9	Potentiometric titration/$NaNO_3$	Weerasooriya et al. (2007a, 2007b)
8.7	Potentiometric titration/NaCl	Weerasooriya et al. (2007a, 2007b)
9.8	Potentiometric titration/NaCl	Manning and Goldberg (1996)
9	Potentiometric titration/NaCl	Rosenqvist et al. (2002)
10	Potentiometric titration/$NaNO_3$	Hiemstra et al. (1987)
9.41	Electrophoretic mobility/NaCl	Goldberg et al. (1996a, 1996b)
9.9	Predicted with MUSIC model	Hiemstra et al. (1999)
9.1	Not specified	Jepson et al. (1976)
7.7–9.4	Not specified	Yopps and Fuerstenau (1964)
9.5	Not specified	Hingston et al. (1972)
7.8	Not specified	Hingston et al. (1972)

PPZC values reported for amorphous aluminum hydroxide suspension are very close to the average PPZC value found for gibbsite. The average experimental PPZC value obtained for amorphous aluminum hydroxide or freshly prepared aluminum gel is 9.0 (Feldkamp et al., 1981; Beyrouty et al., 1984; Micera et al., 1986; Rakotonarivo et al., 1988; Manning and Goldberg, 1997; Gu and Karthikeyan, 2005; Fukushi et al., 2006), while for alumina the PPZC generally is lower, ranging between 6.4 and 8.8 (Stumm and Morgan, 1981; Brady, 1994; Yang et al., 2007b; Su et al., 2008) and rarely above 9.0 (Brady et al., 1994). Worth noting is the fact that both amorphous hydrous ferric oxide and crystalline goethite have an average PPZC at pH of 8.0, and derived values from fitting acid–base titration data sets with generalized two-layer model were 8.1 (Dzombak and Morel, 1990) and 8.3 (Mathur and Dzombak, 2006). Accordingly, both amorphous hydrous aluminum oxide and crystalline gibbsite have an average PPZC at pH of 9.0, higher than iron hydroxides.

5.4 SURFACE ACID–BASE CHEMISTRY

The surface protonation and deprotonation characteristics of gibbsite were determined by fitting the diprotic acid (2-pK_a) model (see Chapter 3) to available acid–base titration data for gibbsite. Effects of gibbsite dissolution on the acid–base balance were not considered, as discussed and justified in Section 5.6. A literature search was conducted to identify available data sets for rapid acid–base titration of gibbsite suspensions, and these data sets were evaluated for their suitability for modeling by applying the criteria described in Chapter 4. Only data from experiments that employed gibbsite synthesized by one of the standard methods or obtained from a source known to provide certifiable gibbsite were used. In addition, only data obtained in experiments with inert electrolytes were used, that is, no sorbing ions other than H^+.

Due to the presence of impurities (e.g., bayerite) in some gibbsite samples, and a wide range in concentrations of suspension used (from 0.25 g/L (Li and Xu, 2008) to 37 g/L (Hiemstra et al., 1987)), acid–base titrations for gibbsite exhibit variability. The reported values for gibbsite surface charge at particular pH values as determined from acid–base titration data show considerable scatter (Rosenqvist et al., 2002). Nevertheless, the acid–base titration data available for gibbsite are sufficiently consistent for average surface acidity characteristics to be determined.

The available acid–base titration data sets for gibbsite used for surface acidity constant extraction are presented in Table 5.4. Other data available in the literature, but not considered for various reasons, are reported in Table 5.5. The data sets from Hiemstra and colleagues (Hiemstra et al., 1987, 1999) and Jodin et al. (2005) were not considered, because the suspension concentrations used in their experiments were very high (>30 g/L). Such concentrated suspensions favor agglomeration of the suspension and consequently partial coverage of the active sorption sites. Another data set that was not used was that of Kavanagh et al. (1975), where the ionic strength of the background electrolyte solution was high (1 M) and the Davies equation correction for activities used in this study can not be applied. Finally, one data set

TABLE 5.4 Gibbsite acid–base titration data

ID	Source	Medium	Data Set
4T1	Weerasooriya et al. (2000)	0.1 M $NaNO_3$	Fig. 1
4T2		0.01 M $NaNO_3$	
4T3		0.001 M $NaNO_3$	
4T4	Weerasooriya et al. (2001)	0.1 M NaCl	Fig. 5
4T5		0.01 M NaCl	
4T6		0.001 M NaCl	
4T7	Yang et al. (2007a)	0.1 M KNO_3	Fig. 6c
4T8		0.01 M KNO_3	
4T9		0.001 M KNO_3	
4T10	Rosenqvist et al. (2002)	0.02 M NaCl	Fig. 2
4T11		0.1 M NaCl	
4T12	Kavanagh et al. (1975)	0.1 M $NaNO_3$	Fig. 2
4T13		0.01 M $NaNO_3$	
4T14	Li and Xu (2008)	0.0001 M NaCl	Fig. 4

TABLE 5.5 Gibbsite acid–base titration data not used

Source	Data Set	Medium	Reason
Rosenqvist et al. (2002)	Fig. 3	0.1 M NaCl	Nonaged gibbsite
Kavanagh et al. (1975)	Fig. 2	1 M $NaNO_3$	High ionic strength. Davies equation cannot be applied
Hiemstra et al. (1999)	Fig. 3	0.1 M $NaNO_3$ 0.005 M $NaNO_3$	High suspension concentration. Possible agglomeration and partial coverage of the active surface sites. Also for data of Fig. 7 low (4°C) and high (70°C) temperatures were used
	Fig. 5	0.1 M $NaNO_3$ 0.005 M $NaNO_3$	
	Fig. 7	0.1 M $NaNO_3$ 0.005 M $NaNO_3$	
Hiemstra et al. (1987)	Fig. 4	0.5 M NaCl 0.05 M NaCl 0.005 M NaCl	High suspension concentration. Possible agglomeration and partial coverage of the active surface sites
Jodin et al. (2005)	Fig. 4	0.1 M NaCl 0.01 M NaCl 0.001 M NaCl	High suspension concentration. Possible agglomeration and partial coverage of the active surface sites. Also, reported hysteresis

extracted from Rosenqvist et al. (2002) was not used, because the gibbsite solid phase used in this potentiometric titration was not aged sufficiently.

Test fitting of data spanning a wide range of suspension concentrations revealed inconsistencies, and the need to exclude consideration of data obtained with high suspension concentrations. Suspension concentrations up to 30 mg/L were found to yield good fits using consistent surface properties of $A = 32\,m^2/g$ and $N_s = 8$ sites/nm^2 for all acid–base titration data sets. Higher suspension concentration led to progressively poor fits, as evidenced by lower goodness-of-fit (WSOS/DF) and different values for N_s were required to fit the data acceptably. A possible explanation is that high suspension concentration favors agglomeration and consequently partial coverage of the active surface sites. To achieve acceptable data fits, lower values of N_s had to be applied. Decreasing the N_s value to 2 sites/nm^2 improved fitting for the high suspension concentration data sets, but it did not have the same effect for all data sets. Most of them were well fitted with high N_s. The best-estimate value of N_s of 8 sites/nm^2 was determined by trial and error and was found to yield lower calculated WSOS/DF values with FITEQL, and good fits for most acid–base titration data sets.

An example of acid–base titration data obtained with a high concentration suspension (32 g/L) and fitting using higher and lower site-density values is given in Figure 5.1. In the example, the titration curve was fitted first with the optimal values of specific surface area (32 m^2/g) and site density ($N_s = 8$ sites/nm^2) determined in this study. The fit was poor and a lower site-density value (2 sites/nm^2) was then applied.

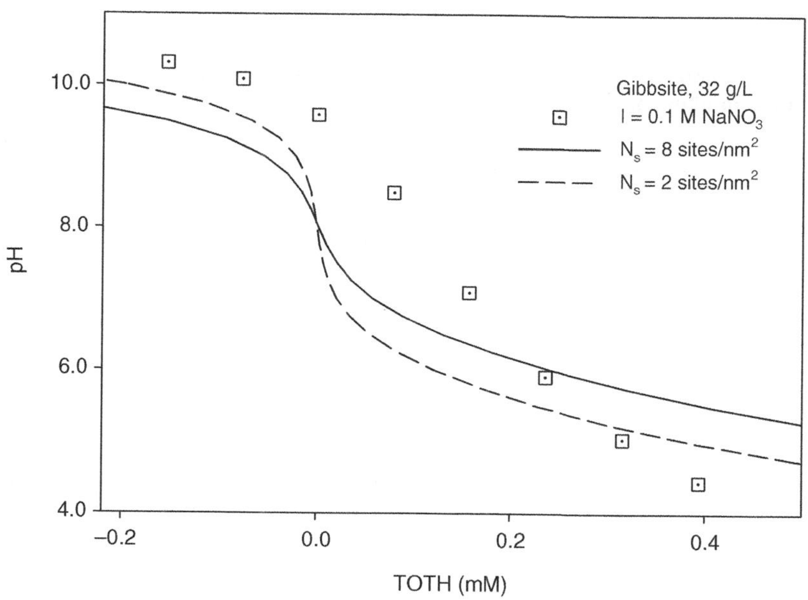

FIGURE 5.1 FITEQL fitting of acid–base titration data at ionic strength 0.1 *M* and high suspension concentration (data from Jodin et al., 2005). $A = 32\,m^2/g$. No dissolution of gibbsite.

The fit was better than that determined for the higher site density, but still poor. Attempts at fitting other acid–base titration data obtained with higher gibbsite suspension concentrations also resulted in poor fitting. Therefore, only titration data obtained for systems with suspension concentrations less than 30 g/L were considered for the extraction of the optimal surface acidity constants in this study.

The results of fitting the two-layer model to the gibbsite acid–base titration data sets are summarized in Table 5.6. The high residual sum of squares associated with some of the data sets in Table 5.6 is attributed primarily to an insufficient number of titration data at high pH, above the PPZC. The limited amount of such data meant that most of pK_{a2}^{int} values had to be manually adjusted to an optimized value.

Extracted experimental acid–base titration data and titration simulations with the fitted surface acidity constants in the two-layer model are presented in Figures 4T1 to 4T13. Solid lines present best fits of individual experimental data sets; dashed lines represent fits calculated based on the best-estimate acidity constants. The high residuals in most of the fits can be attributed in not adjusting the relative error estimate (s_x(rel)) used in FITEQL calculations. Increasing the relative error estimate of H^+ from 0.05 to 0.2, that is ±0.02 to ±0.08 pH, will decrease the residuals causing marginal change to the acidity constant extracted from each experimental data set. For consistency with related studies (Dzombak and Morel, 1990; Mathur and Dzombak, 2006) conducted previously, the same relative error estimate value of 0.05 for pH measurements was maintained in modeling the titration data.

The weighted averages for $\log K_{a1}^{int}$ and $\log K_{a2}^{int}$ are given in Table 5.7 along with confidence intervals for these estimates. At the confidence level of 95 percent, $pK_{a1}^{int} = 7.17 \pm 0.19$ and $pK_{a2}^{int} = 11.18 \pm 0.22$, indicating that available data are

TABLE 5.6 Gibbsite intrinsic surface acidity constants extracted from acid–base titration data

ID	pK_{a1}^{int}	$\sigma_{\log K}$	pK_{a2}^{int}	$\sigma_{\log K}$	WSOS/DF
4T1	5.59	0.015	−11.66[a]	0.15[b]	42.6
4T2	6.17	0.024	−11.66[a]	0.15[b]	6.1
4T3	8.76	0.028	−11.13	0.084	42.0
4T4	4.20	0.011	−11.66[a]	0.15[b]	113
4T5	4.39	0.015	−11.45	0.072	35.9
4T6	4.47	0.027	−11.53	0.15	5.4
4T7	6.20	0.013	−10.31	0.042	16.4
4T8	6.66	0.023	−10.30	0.097	58.7
4T9	8.01	0.014	−11.57[a]	0.15[b]	35.1
4T10	9.18	0.010	−11.61[a]	0.15[b]	1981
4T11	9.04	0.006	−11.66[a]	0.15[b]	381
4T12	7.71	0.010	−11.66[a]	0.15[b]	277
4T13	8.45	0.011	−11.60[a]	0.15[b]	266

[a] Convergence not possible unless one surface acidity constant fixed.
[b] Fixed at this value by convention.

TABLE 5.7 Best estimates for gibbsite intrinsic surface acidity constants

Best-Estimate K	Confidence Level	$pK_a^{int}(-)$	$pK_a^{int}(+)$
$\log \overline{K_{a1}^{int}} = -7.17$	0.99	6.91	7.43
	0.95	6.97	7.36
	0.90	7.00	7.33
	0.80	7.04	7.29
$\log \overline{K_{a2}^{int}} = -11.18$	0.99	11.47	10.89
	0.95	11.39	10.96
	0.90	11.36	10.99
	0.80	11.32	11.03

reasonably consistent. At a confidence level of 99 percent the deviations from the best estimates are 0.5 log units.

The pristine PZC is given by the average of the two acidity constants pK_a values:

$$\text{PPZC} = \tfrac{1}{2}(pK_{a1}^{int} + pK_{a2}^{int}) \tag{5.2}$$

PPZC refers to the pH at which the number of the positively and negatively charged sites is equal and the surface is uncharged. That pH is halfway between the two acidity constants. Equation (5.2) applies to the condition where there is no cation or anion sorption other than H^+ and OH^- sorption. Substituting the best estimates for pK_{a1}^{int} and pK_{a2}^{int} from Table 5.7 into Equation (5.2), the derived PPZC value is 9.17, which is within the reported range of measured values (see Table 5.5), and close to the average value of 9.0 for reported values.

The intrinsic constants in Table 5.7 are reported in terms of $I = 0$ and temperature of 25°C, and have to be corrected for ionic-strength effects when used in chemical-equilibrium calculations. In such calculations, which are performed on a concentration basis (usually, molar), intrinsic acidity constants should be corrected for solution activity coefficient for H^+. That is

$$K_{a1}^{int} = \frac{\{\equiv \text{AlOH}^0\}\{\text{H}^+\}}{\{\equiv \text{AlOH}_2^+\}} P = \frac{(\equiv \text{AlOH}^0)\gamma_\text{H}(\text{H}^+)}{(\equiv \text{AlOH}_2^+)} P \tag{5.3}$$

$$K_{a2}^{int} = \frac{\{\equiv \text{AlO}^-\}\{\text{H}^+\}}{\{\equiv \text{AlOH}^0\}} P = \frac{(\equiv \text{AlO}^-)\gamma_\text{H}(\text{H}^+)}{(\equiv \text{AlOH}^0)} P \tag{5.4}$$

where {} represent activities, () represent molar concentrations, γ_H is the activity coefficient for H^+, and $P = \exp(-F\Psi/RT)$. All the data fits shown in Figure 4T1 to 4T13 and throughout Chapters 6 and 7, were done with the intrinsic constants corrected with γ_H.

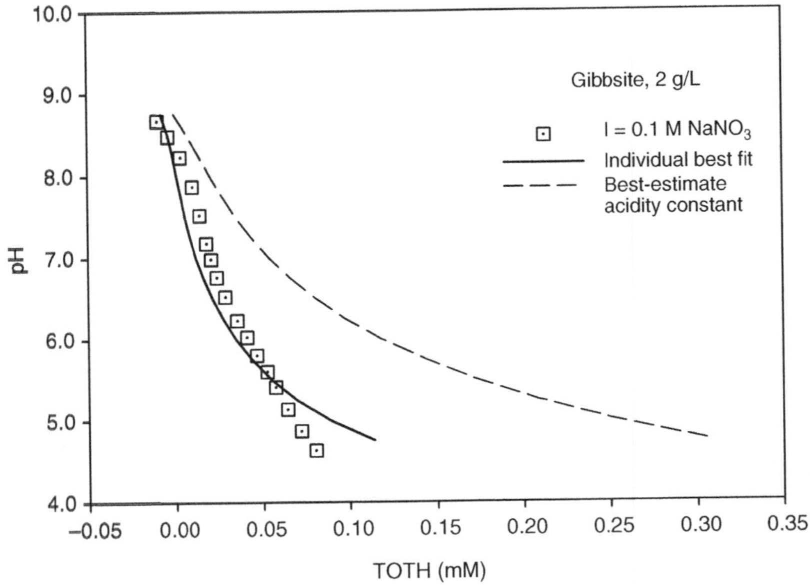

FIGURE 4T1

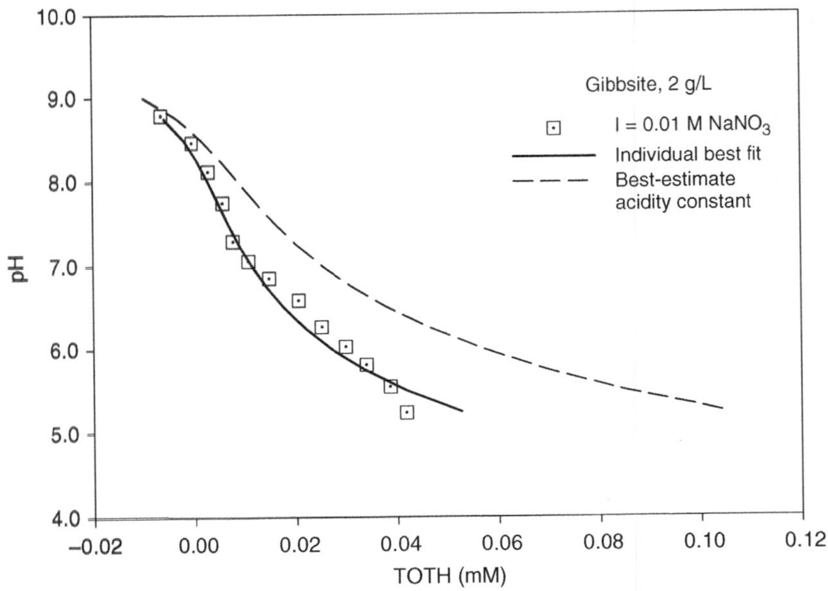

FIGURE 4T2

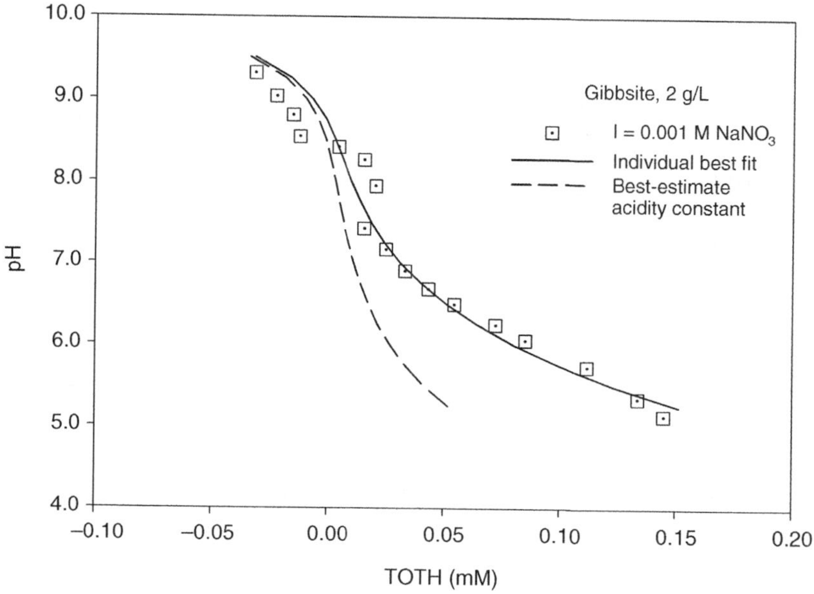

FIGURE 4T3

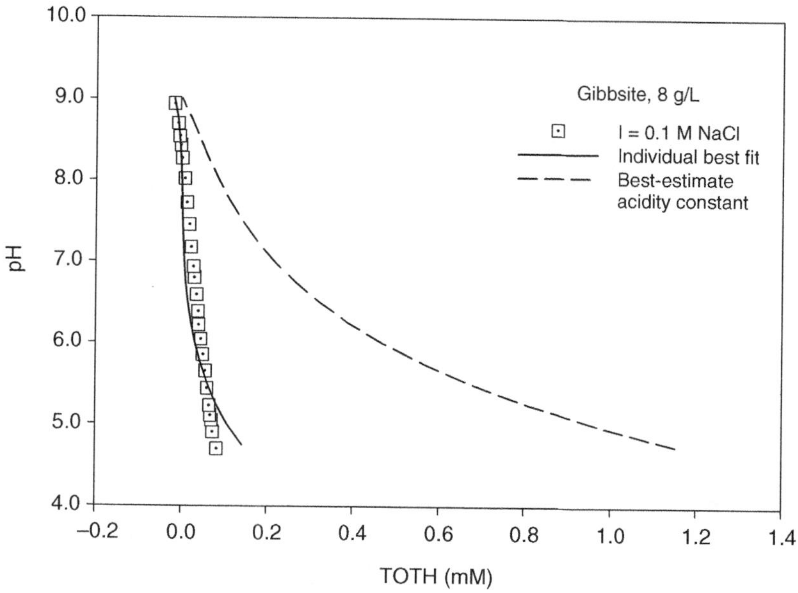

FIGURE 4T4

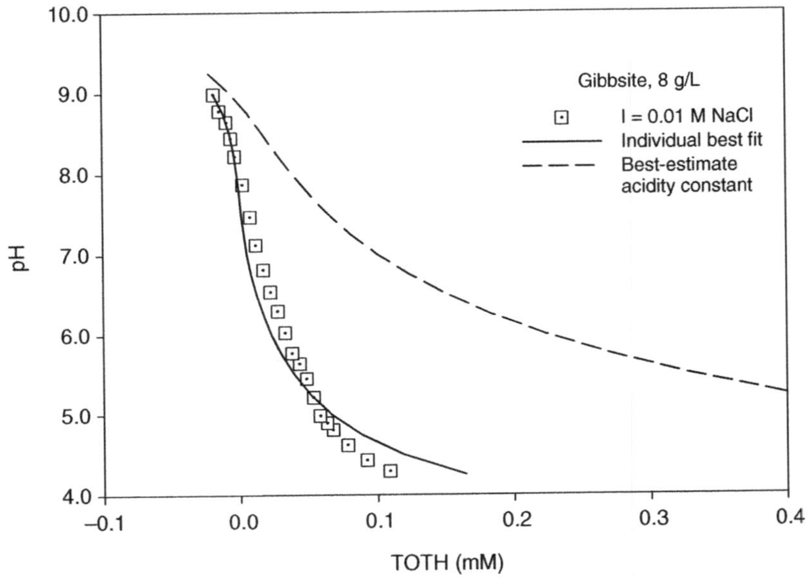

FIGURE 4T5

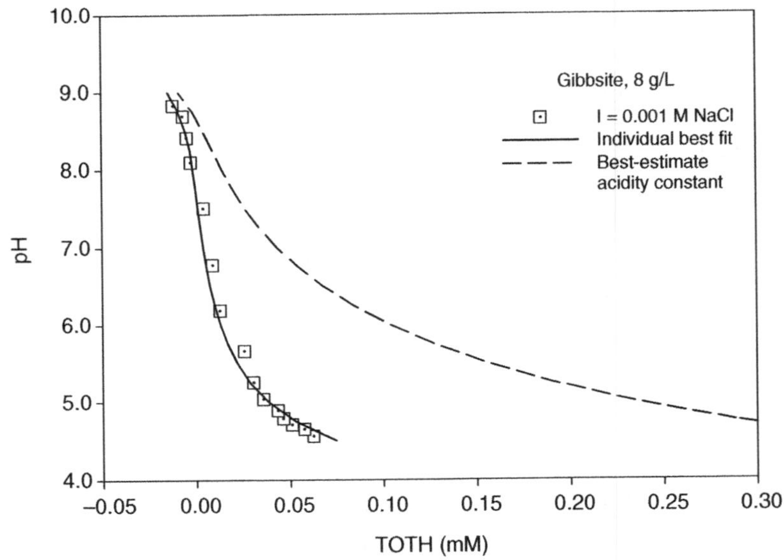

FIGURE 4T6

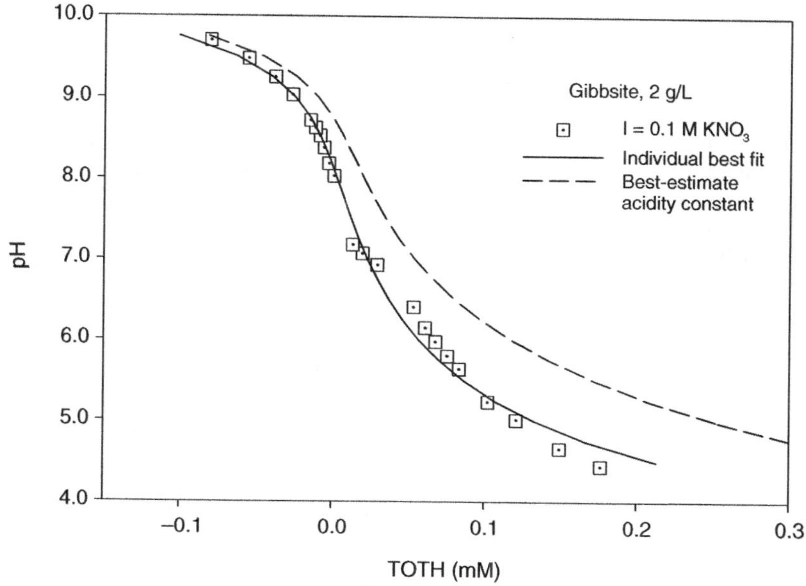

FIGURE 4T7

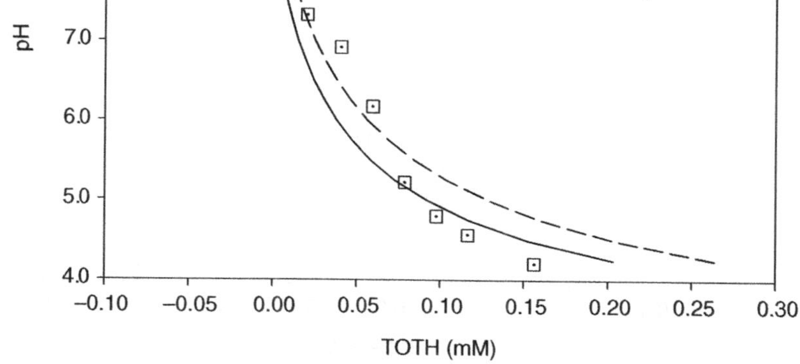

FIGURE 4T8

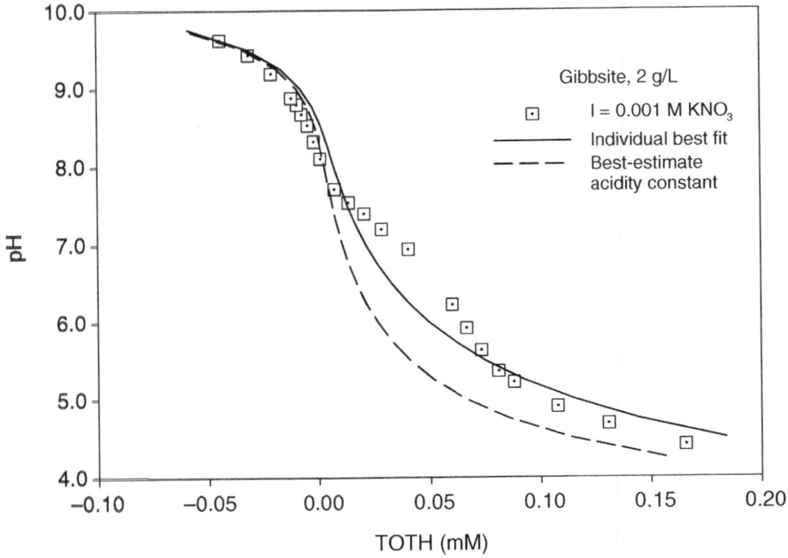

FIGURE 4T9

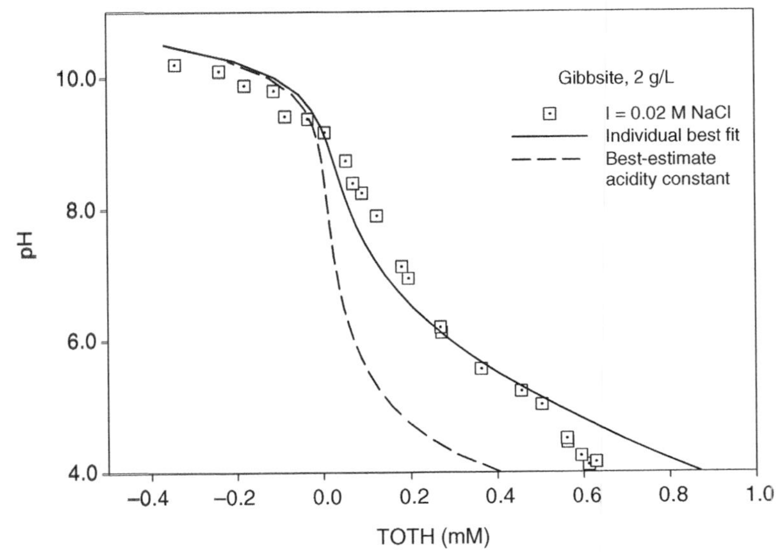

FIGURE 4T10

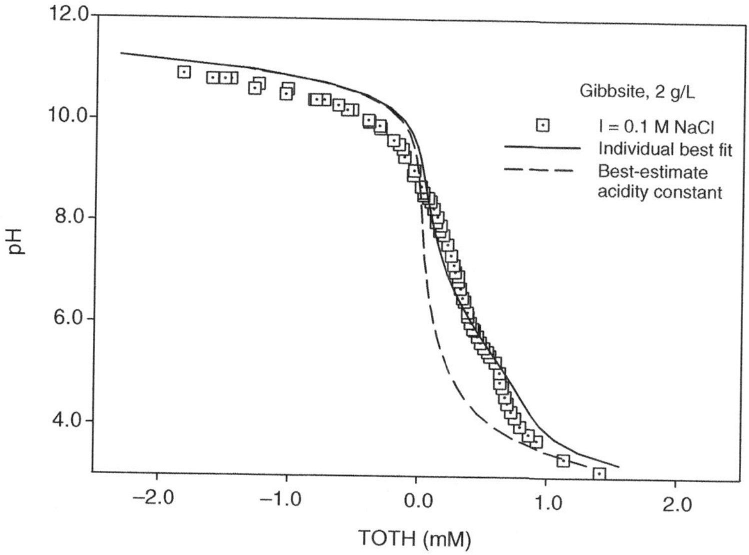

FIGURE 4T11

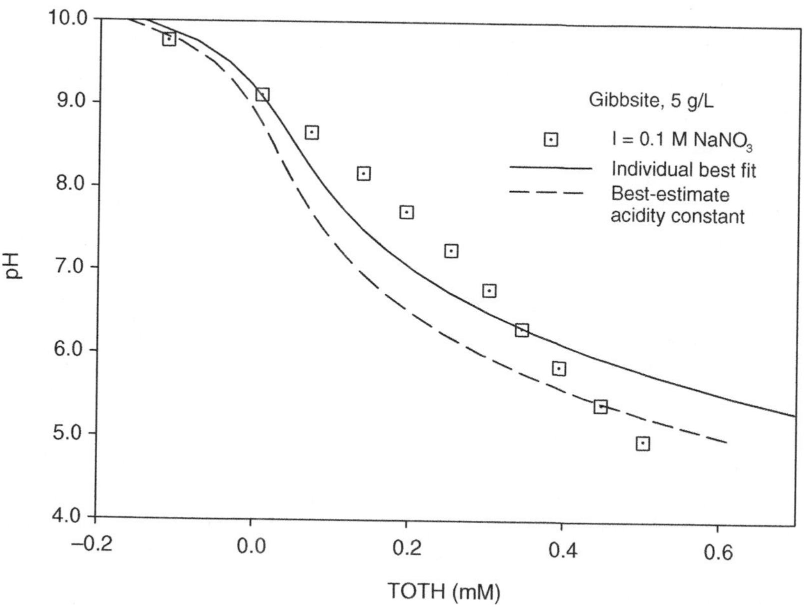

FIGURE 4T12

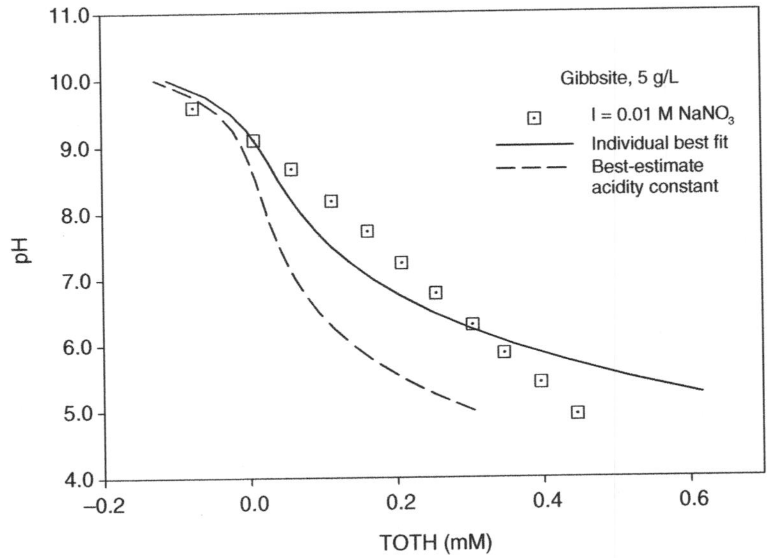

FIGURE 4T13

5.5 EFFECTS OF DISSOLUTION ON GIBBSITE SURFACE ACID–BASE CHEMISTRY

In fitting titration data sets 4T1–4T13 with the two-layer model, gibbsite dissolution and the contributions of Al^{3+} hydrolysis reactions to the acid–base balance in solution was ignored. Effectively, gibbsite was treated as an insoluble solid, and the surface protonation–deprotonation reactions were considered to dominate the acid–base balance of the solution. This same approach was used with hydrous ferric oxide (Dzombak and Morel, 1990) and goethite (Mathur and Dzombak, 2006), but those two iron oxides have considerably lower solubility than gibbsite and other aluminum oxides.

The dissolution products of gibbsite under environmental conditions in aqueous systems are monomeric or polynuclear aluminum hydroxide ions. The consideration of gibbsite as soluble in modeling means the inclusion of these aluminum dissolved species in the TOTH equation, as shown in Tables 3.1 and 3.2. Consequently, the effect that these concentrations of dissolved aluminum species may have on the proton balance in solution during the titration might be significant (Schulthess and Sparks, 1988). Applicable mass law and mole balance equations, including the hydrolysis species, are given in Table 5.8 and the corresponding tableau is presented in Table 5.9. The reactions in Table 5.8, and the corresponding entries in the tableau of Table 5.9, have been formulated in terms of $Al(OH)_3(s)$ for the aluminum species, rather than in terms of Al^{3+} as in Table 3.1 and 3.2. The alternative model formulation is presented

TABLE 5.8 Mass law and mole balance equations for surface acid–base chemistry of a soluble hydrous aluminum oxide

Species

H^+, OH^-, $\equiv AlOH^0$, $\equiv AlOH_2{}^+$, $\equiv AlO^-$, Al^{3+}, $AlOH^{2+}$, $Al(OH)_2{}^+$, $Al(OH)_3{}^0$, $Al(OH)_4{}^-$, $Al_2(OH)_2{}^{4+}$, $Al_3(OH)_4{}^{5+}$, $Al(OH)_3(s)$

Mass Law Equations

(OH^-)	$= (H^+)^{-1}$				K_w
(Al^{3+})	$= (H^+)^3$	$(Al(OH)_3(s))$			K_1
$(AlOH^{2+})$	$= (H^+)^2$	$(Al(OH)_3(s))$			K_2
$(Al(OH)_2{}^+)$	$= (H^+)^1$	$(Al(OH)_3(s))$			K_3
$(Al(OH)_3{}^0)$	$=$	$(Al(OH)_3(s))$			K_4
$(Al(OH)_4{}^-)$	$= (H^+)^{-1}$	$(Al(OH)_3(s))$			K_5
$Al_2(OH)_2{}^{4+}$	$= (H^+)^4$	$(Al(OH)_3(s))^2$			K_6
$Al_3(OH)_4{}^{5+}$	$= (H^+)^5$	$(Al(OH)_3(s))^3$			K_7
$(\equiv AlOH_2{}^+)$	$= (H^+)$		$(\equiv AlOH^0)$	P	$1/K_{a1}^{int}$
$(\equiv AlO^-)$	$= (H^+)^{-1}$		$(\equiv AlOH^0)$	P^{-1}	K_{a2}^{int}

where $P = \exp(-F\Psi/RT)$

Mole Balance Equations

TOTH $= (H^+) - (OH^-) + 3(Al^{3+}) + 2(AlOH^{2+}) + (Al(OH)_2{}^+) - (Al(OH)_4{}^-) + 4(Al_2(OH)_2{}^{4+}) + 5(Al_3(OH)_4{}^{5+}) + (\equiv AlOH_2{}^+) - (\equiv AlO^-) = C_A - C_B$

TOTAl(OH)$_3$ $= (Al^{3+}) + (AlOH^{2+}) + (Al(OH)_2{}^+) + (Al(OH)_3{}^0) + (Al(OH)_4{}^-) + 2(Al_2(OH)_2{}^{4+}) + 3(Al_3(OH)_4{}^{5+}) + (Al(OH)_3(s))$

TOT($\equiv$AlOH) $= (\equiv AlOH_2{}^+) + (\equiv AlOH^0) + (\equiv AlO^-)$

TOTP $= (\equiv AlOH_2{}^+) - (\equiv AlO^-) = \sigma(AS/F)$

Charge–Potential Relationship

$\sigma = 0.1174c^{1/2}\sinh(Z\Psi \times 19.46)$

TABLE 5.9 Tableau for surface acid–base chemistry of a soluble hydrous aluminum oxide

	H^+	$Al(OH)_3(s)$	$\equiv AlOH^0$	$P = \exp(-F\Psi/RT)$
H^+	1			
OH^-	−1			
Al^{3+}	3	1		
$AlOH^{2+}$	2	1		
$Al(OH)_2{}^+$	1	1		
$Al(OH)_3{}^0$		1		
$Al(OH)_4{}^-$	−1	1		
$Al_2(OH)_2{}^{4+}$	4	2		
$Al_3(OH)_4{}^{5+}$	5	3		
$\equiv AlOH_2{}^+$	1		1	1
$\equiv AlOH^0$			1	
$\equiv AlO^-$	−1		1	−1
$Al(OH)_3(s)$		1		
	TOTH	$TOTAl(OH)_3$	$TOT(\equiv AlOH)$	$TOTP = \sigma(AS/F)$

as this is the usual formulation for consideration of equilibrium dissolution of a solid phase (Morel and Hering, 1993), which is done here.

In order to consider the effect gibbsite solubility has on the system, the total dissolved concentration of aluminum species must be measured or calculated assuming equilibrium for each titration point, and then the concentration of each aluminum species can be calculated. The reason for doing that is that upon addition of acid or base to the suspension the concentration of the dissolved aluminum changes.

In some test modeling efforts, the total dissolved aluminum concentration was calculated for each pH value assuming equilibrium at each pH. The calculated concentrations of dissolved aluminum species were then considered in the TOTH equation. The results showed that the TOTH equation is dominated by the concentrations of charged surface species ($\equiv AlOH_2{}^+$, $\equiv AlO^-$), H^+, and OH^-. The contributions of the concentrations of the dissolved aluminum species to the TOTH equation were determined to be negligible under environmental conditions ($4 < pH < 10$). The total concentration of dissolved aluminum species contributes somewhat significantly to the TOTH equation at pH values below 4.5, depending on the ionic strength of the solution. As the ionic strength decreases the difference between considering or not considering dissolution of gibbsite decreases as well, which can be attributed to lesser maximum dissolved aluminum. Similar results also have been reported by Hiemstra et al. (1987). A comparison of fitting titration data with or without the consideration of gibbsite solubility is given in Figures 5.2 and 5.3. It was thus concluded that the dissolved aluminum species can be ignored in surface-complexation modeling calculations for gibbsite in the pH range of 4–10.

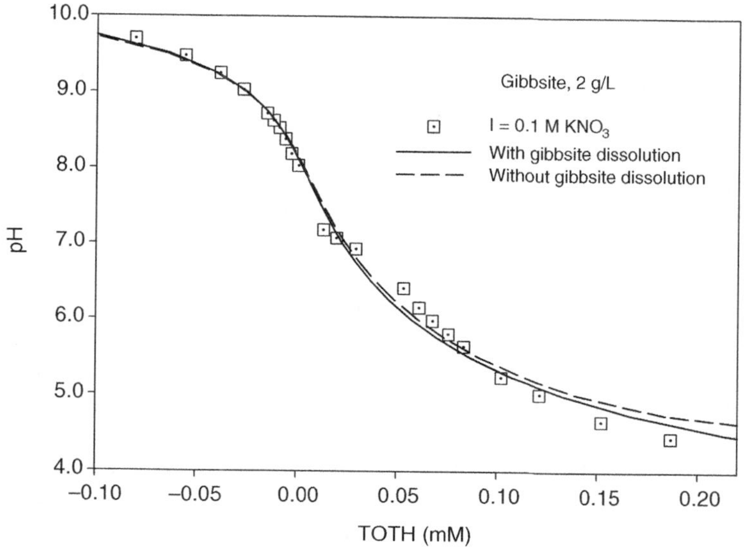

FIGURE 5.2 Comparison of FITEQL fittings considering and not considering gibbsite dissolution, at high ionic strength and low concentration of suspension (calculations based on 4T7 data).

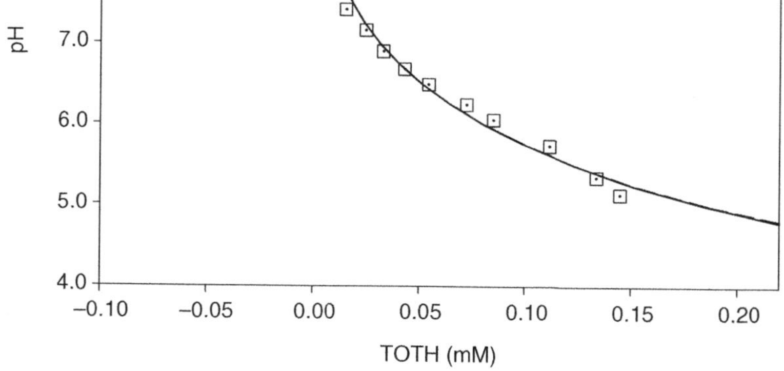

FIGURE 5.3 Comparison of FITEQL fittings considering and not considering gibbsite dissolution, at low ionic strength and low concentration of suspension (calculations based on 4T6 data).

5.6 SUMMARY

Best estimates for the basic surface properties of gibbsite (A, N_s, K_{a1}, K_{a2}) have been determined through compilation and evaluation of available data. A specific surface area, A, of 32 m^2/g was determined as the average of available BET surface-area data for gibbsite. The site density N_s was set at 8 sites/nm^2 based on values reported in the literature and by trial fitting of available acid–base titration data. This value is consistent with crystallographic determinations of N_s for gibbsite. The intrinsic surface acidity constants, pK_{a1}^{int} and pK_{a2}^{int}, were determined to be 7.17 and 11.18, respectively, from fitting of acid–base titration data for gibbsite suspensions less than 30 g/L. The PPZC for gibbsite was estimated as 9.0 from various kinds of data that have been used to determine this parameter for gibbsite. The PPZC value of 9.17 implied by the pK_{a1}^{int} and pK_{a2}^{int} values is close to the value of 9.0 determined by other methods.

Modeling of the gibbsite acid–base titration data was performed without consideration of gibbsite dissolution. Modeling trials with consideration of equilibrium gibbsite dissolution revealed that because the solubility of gibbsite is low, the total concentration of Al species (including complexes and polymers) dissolved from the solid phase was negligible compared to gibbsite surface species concentrations in the pH range 4.5 to 10.

The derived surface parameters for gibbsite were used for the extraction of the cation and anion surface-complexation constants presented in Chapters 6 and 7, respectively.

6

CATION SORPTION ON GIBBSITE

In this chapter, the reactions and sorption equilibrium constants obtained from the fitting of available cation-sorption data for gibbsite with the generalized two-layer model are presented. The chapter has separate sections for each of the cations investigated, and each of these sections contains the following: (1) the sources of the sorption data used for parameter extraction, (2) the data sets not used, along with the reasons for exclusion, (3) the surface-complexation reactions used in fitting the data, (4) the optimal equilibrium surface-complexation constants for each data set, (5) the best estimates (and associated confidence intervals) for the surface-complexation constants, and (6) a plot of each data set along with the individual best fit (solid line) and the fit corresponding to the best estimate surface-complexation constants (dashed line). In the separate sections for the various cations studied, every table and figure has been assigned with a specific identification code as explained previously in Chapter 4 (Table 4.3).

6.1 MODELING METHODOLOGY AND REACTIONS

In all modeling conducted, a consistent set of aqueous-phase reactions and associated equilibrium constants was used. Table 6.1 presents the solution reactions considered in fitting the cation-sorption data sets.

The methodology used for parameter extraction in the application of the generalized two-layer model to cation-sorption data was described in Chapter 4. The nonlinear regression program FITEQL (Herbelin and Westall, 1999) was used to

TABLE 6.1 Solution reactions and mass law constants used in fitting cation sorption data for gibbsite

Reaction	Log K^a
$H^+ + OH^- = H_2O$	14.00
$H^+ + CO_3^{2-} = HCO_3^-$	10.33
$2H^+ + CO_3^{2-} = H_2CO_3*$	16.68
$CO_2(g) + H_2 = H_2CO_3*$	−1.46
$Pb^{2+} + H_2O = PbOH^+ + H^+$	−7.597
$Pb^{2+} + 2H_2O = Pb(OH)_2^0 + 2H^+$	−17.094
$Pb^{2+} + 3H_2O = Pb(OH)_3^- + 3H^+$	−28.091
$Zn^{2+} + H_2O = ZnOH^+ + H^+$	−8.997
$Zn^{2+} + 2H_2O = Zn(OH)_2^0 + 2H^+$	−16.894
$Zn^{2+} + 3H_2O = Zn(OH)_3^- + 3H^+$	−28.391
$Cd^{2+} + H_2O = CdOH^+ + H^+$	−10.097
$Cd^{2+} + 2H_2O = Cd(OH)_2^0 + 2H^+$	−20.294
$Cd^{2+} + 3H_2O = Cd(OH)_3^- + 3H^+$	−33.3
$Cd^{2+} + Cl^- = CdCl^+$	1.98
$Cd^{2+} + 2Cl^- = CdCl_2^0$	2.6
$Cu^{2+} + H_2O = CuOH^+ + H^+$	−7.497
$Cu^{2+} + 2H_2O = Cu(OH)_2^0 + 2H^+$	−16.23
$Cu^{2+} + 4H_2O = Cu(OH)_4^{2-} + 4H^+$	−39.73
$Co^{2+} + H_2O = CoOH^+ + H^+$	−9.697
$Co^{2+} + 2H_2O = Co(OH)_2^0 + 2H^+$	−18.794
$Co^{2+} + 3H_2O = Co(OH)_3^- + 3H^+$	−31.491
$Co^{2+} + Cl^- = CoCl^+$	−0.35
$Co^{2+} + NO_3^- = CoNO_3^+$	0.2
$Co^{2+} + 2NO_3^- = Co(NO_3)_2^0$	0.51
$Ca^{2+} + H_2O = CaOH^+ + H^+$	−12.697
$Mn^{2+} + H_2O = MnOH^+ + H^+$	−10.597
$Mn^{2+} + 4H_2O = Mn(OH)_4^{2-} + 4H^+$	−48.288
$Fe^{2+} + H_2O = FeOH^+ + H^+$	−9.397
$Fe^{2+} + 2H_2O = Fe(OH)_2^0 + 2H^+$	−20.494
$Fe^{2+} + 3H_2O = Fe(OH)_3^- + 3H^+$	−30.991
$Hg(OH)_2 + H^+ = HgOH^+ + H_2O$	2.797
$Hg(OH)_2 + 2H^+ = Hg^{2+} + 2H_2O$	6.164
$Hg(OH)_2 + Cl^- + 2H^+ = HgCl^+ + 2H_2O$	13.494
$Hg(OH)_2 + 2Cl^- + 2H^+ = HgCl_2^0 + 2H_2O$	20.194
$Hg(OH)_2 + 3Cl^- + 2H^+ = HgCl_3^- + 2H_2O$	21.194
$Hg(OH)_2 + 4Cl^- + 2H^+ = HgCl_4^{2-} + 2H_2O$	21.794
$Hg(OH)_2 + Cl^- + H^+ = HgOCl^+ + H_2O$	10.444
$Hg(OH)_2 + NO_3^- + 2H^+ = HgNO_3^+ + 2H_2O$	5.7613

TABLE 6.1 (*Continued*)

Reaction	Log K^a
$Hg(OH)_2 + 2NO_3^- + 2H^+ = Hg(NO_3)_2^0 + 2H_2O$	5.38
$UO_2^{2+} + H_2O = UO_2OH^+ + H^+$	−5.25
$UO_2^{2+} + 2H_2O = UO_2(OH)_2(aq) + 2H^+$	−12.15
$UO_2^{2+} + 3H_2O = UO_2(OH)_3^- + 3H^+$	−20.25
$UO_2^{2+} + 4H_2O = UO_2(OH)_4^{2-} + 4H^+$	−32.4
$2UO_2^{2+} + 2H_2O = (UO_2)_2(OH)_2^{2+} + 2H^+$	−5.62
$2UO_2^{2+} + H_2O = (UO_2)_2(OH)^{3+} + H^+$	−2.7
$3UO_2^{2+} + 4H_2O = (UO_2)_3(OH)_4^{2+} + 4H^+$	−11.9
$3UO_2^{2+} + 5H_2O = (UO_2)_3(OH)_5^+ + 5H^+$	−15.55
$Th^{4+} + H_2O = Th(OH)^{3+} + H^+$	−3.197
$Th^{4+} + 2H_2O = Th(OH)_2^{2+} + 2H^+$	−6.894
$2Th^{4+} + 2H_2O = Th_2(OH)_2^{6+} + 2H^+$	−6.094
$Al^{3+} + H_2O = Al(OH)^{2+} + H^+$	−4.997
$Al^{3+} + 2H_2O = Al(OH)_2^+ + 2H^+$	−10.294
$Al^{3+} + 3H_2O = Al(OH)_3^0 + 3H^+$	−16.691
$Al^{3+} + 4H_2O = Al(OH)_4^- + 4H^+$	−23.0
$2Al^{3+} + 2H_2O = Al_2(OH)_2^{4+} + 2H^+$	−7.694
$3Al^{3+} + 4H_2O = Al_3(OH)_4^{5+} + 4H^+$	−13.888

Source: Smith, Martell and Motekaitis (NIST 46.6, 2001; NIST 46.7, 2003).

[a] Mass law constants correspond to $T = 20$–25°C and $I = 0$, and where necessary have been corrected to $I = 0$ using the activity coefficients in Table 4.1.

calculate the optimal surface-complexation constants from the individual data sets. Various surface reactions and combinations of surface reactions were tried for each cation, with the simplest reactions and the smallest number of reactions able to fit the data selected for use. When insufficient data were available in a particular data set to extract a surface-complexation constant for a particular reaction with FITEQL, or when convergence problems were experienced with FITEQL for other reasons, manual adjustment of surface-complexation constants was performed using Visual MINTEQ (Gustafsson, 2009).

Best-estimate surface-complexation constants were determined from the optimal constants fitted to the various data sets for a particular cation. These were assessed by means of weighted average of log K values in which $1/\sigma_{\log K}$ was used as the weighting factor. The standard deviation $\sigma_{\log K}$ reflects the quantity and variability of data in a particular data set. All surface-complexation constants and $\sigma_{\log K}$ values are presented in the following sections for individual cations. Surface-complexation constants determined by manual fitting were assigned a $\sigma_{\log K}$ value of 0.15.

All surface-complexation constants and $\sigma_{\log K}$ values are presented in the following sections for individual anions. When a particular surface species was not used to fit individual data sets was indicated in tables of intrinsic surface-complexation constants with a blank space.

In this study, a single site type was used to fit all the available cation-sorption data sets for gibbsite. Surface-complexation modeling of cation sorption on hydrous ferric oxide required consideration of two site types with the generalized two-layer model (Dzombak and Morel, 1990). However, the two-layer model was successfully applied to cation-sorption data for goethite using a single site type (Mathur and Dzombak, 2006) and the same approach was adequate to fit available cation-sorption data for gibbsite.

For modeling cation sorption on gibbsite, reasonably good fits could be achieved in nearly all cases considering one surface species, $\equiv AlOM^+$. Although for most divalent cations this surface species was sufficient to describe surface complexation, for Hg^{2+}, $UO_2{}^{2+}$, and Th^{4+} some other species were required.

Surface complexation of cations by gibbsite involves the release of protons from the surface and the formation of bonds with surface oxygen atoms which can be represented, for example, by:

$$\equiv AlOH^0 + M^{2+} = \equiv AlOM^+ + H^+ \quad K_M^{app} \tag{6.1}$$

where M^{2+} represents a divalent cation. Sorption of cations onto gibbsite surface sites is strongly pH-dependent. Increasing the pH promotes sorption of cations on hydrous oxides, mainly because the reactive surface sites are hydroxyl groups that can coordinate and dissociate protons. For surface complexation of divalent cations on gibbsite, reaction 6.1 was adequate to describe the available data and was used throughout this study. For thorium (Th^{4+}), mercury (Hg^{2+}), and uranyl ($UO_2{}^{+}$), it was necessary to consider some additional surface species and reactions. When considering surface precipitation of Hg^{2+} yet another reaction had to be used:

$$\equiv AlOH^0 + M^{2+} + H_2O = \equiv AlOMOH_2^+ + H^+ \quad K_M^{app} \tag{6.2}$$

As in the case of protons, cations sorb onto oxide surfaces in response to chemical and coulombic forces. Therefore, the inclusion of a coulombic term in the mass law expressions for sorption reactions, theoretically, should be taken into account in order to describe the long-range electrostatic effects because of the surface charge.

$$K_M^{int} = K_M^{app} \exp(\Delta Z F \Psi / RT) \tag{6.3}$$

where K_M^{int} is the intrinsic equilibrium constant, K_M^{app} the apparent equilibrium constant, ΔZ the net change in the charge number of the surface species, and Ψ the surface potential.

The net charge on an oxide surface is changed if a cation is sorbed. For example, if the charged surface complex $\equiv AlOM^+$ is formed by the sorption of a cation M^{2+} to an oxide, then the net surface charge density is expressed by the following equation:

$$\sigma = (F/AS)[(\equiv AlOH_2^+) - (\equiv AlO^-) + (\equiv AlOM^+)] \tag{6.4}$$

A complete set of equations for the surface complexation of M^{2+} on gibbsite is given in Table 6.2 and the corresponding tableau in Table 6.3. Note that all proton-

TABLE 6.2 Cation surface complexation on gibbsite

Species

H^+, OH^-, M^{2+}, MOH^+, $M(OH)_2{}^0$, $\equiv AlOH^0$, $\equiv AlOH_2{}^+$, $\equiv AlO^-$, $\equiv AlOM^+$

Mass Law Equations

(OH^-)	$= (H^+)^{-1}$				$\gamma_1^{-2} \times 10^{-14}$
(MOH^+)	$= (H^+)^{-1}$	(M^{2+})			$\gamma_2\gamma_1^{-2}K_1$
$(M(OH)_2{}^0)$	$= (H^+)^{-1}$	(M^{2+})			$\gamma_2\gamma_1^{-2}K_2$
$(\equiv AlOH_2{}^+)$	$= (H^+)$		$(\equiv AlOH^0)$	$\exp(-F\Psi/RT)$	$\gamma_1(K_{a1}^{int})^{-1}$
$(\equiv AlO^-)$	$= (H^+)^{-1}$		$(\equiv AlOH^0)$	$\exp(F\Psi/RT)$	$\gamma_1{}^{-1}K_{a2}^{int}$
$(\equiv AlOM^+)$	$= (H^+)^{-1}$	(M^{2+})	$(\equiv AlOH^0)$	$\exp(-F\Psi/RT)$	$\gamma_2\gamma_1^{-1}K_{1M}^{int}$

Mole Balance Equations

TOTH	$= (H^+)-(OH^-)-(MOH^+)-2(M(OH)_2{}^0) + (\equiv AlOH_2{}^+)-(\equiv AlO^-)-(\equiv AlOM^+)$
TOTM	$= (M^{2+}) + (MOH^+) + (M(OH)_2{}^0) + (\equiv AlOM^+)$
TOT(≡AlOH)	$= (\equiv AlOH_2{}^+) + (\equiv AlOH) + (\equiv AlO^-) + (\equiv AlOM^+)$

Surface Charge

$$\sigma = (F/AS)\text{TOTP} = (F/AS)\ [(\equiv AlOH_2{}^+)-(\equiv AlO^-) + (\equiv AlOM^+)]$$

Charge–Potential Relationship

$$\sigma = 0.1174c^{1/2}\sinh(Z\Psi \times 19.46)$$

binding sites are considered available for cation binding and that all sites are considered to be uniform with respect to cation binding as one site type. For tri- and tetravalent cations, different surface sites were considered. The surface complexation reactions considered for such cations (e.g., Th^{4+}) can be found at their respective sections in this chapter.

TABLE 6.3 Tableau for cation surface complexation on gibbsite

	H^+	M^{2+}	$\equiv AlOH^0$	$P = \exp(-F\Psi/RT)$
H^+	1			
OH^-	−1			
M^{2+}		1		
MOH^+	−1	1		
$M(OH)_2{}^0$	−2	1		
$\equiv AlOH_2{}^+$	1		1	1
$\equiv AlOH^0$			1	
$\equiv AlO^-$	−1		1	−1
$\equiv AlOM^+$	−1	1	1	1
	TOTH	TOTM	TOT(≡AlOH)	TOTP = $\sigma(AS/F)$

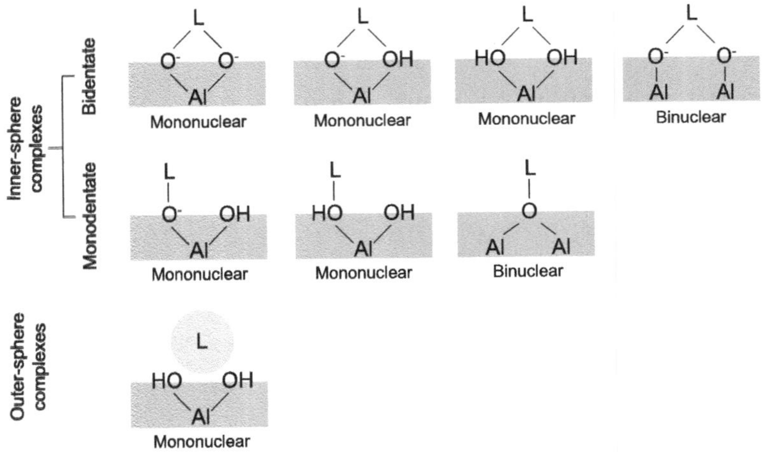

FIGURE 6.1 Schematic representation of possible surface complexes on gibbsite. The two-layer model does not distinguish between inner and outer-sphere complexes, but such classification is presented here for completeness. L, ligand (cation or anion); O, oxygen; Al, aluminum; H, hydrogen (after Ladeira et al., 2001; Kremleva et al., 2008; Palmer et al., 2009).

6.2 AVAILABLE SPECTROSCOPIC DATA AND USE IN MODELING

The potential to constrain modeling of cation sorption on gibbsite using available spectroscopic data was investigated. A search of the literature revealed that some work has been done in the identification of cation surface complexes on the gibbsite surface, but spectroscopic data are limited.

In this study, all divalent cations were treated as monodentate mononuclear surface complexes with gibbsite, expressed mostly as $\equiv AlOM^{+}$. Different surface species were used in the case of polynuclear cations of uranium and thorium. Possible modes of ligand coordination to the gibbsite surface are shown in Figure 6.1. Monodentate, bidentate, and polydentate complexes comprise a ligand, either a cation or anion, which form one, two, or more bonds with a surface Al atom, respectively. Mononuclear, binuclear, and polynuclear complexes are those involving one, two, or more surface Al atoms, respectively.

The surface-complexation modeling performed in this study was guided only partially with available spectroscopic evidence. Lack of sufficient spectroscopic data for specific cations and the uncertainty associated with the interpretation of the existing data limited the usefulness of spectroscopic data in constraining choice of surface species. This section presents an overview of available spectroscopic data for the various cations studied, and discusses how these data were considered in the surface-complexation modeling for the particular cations.

6.2.1 Copper

Pokrovsky et al. (2008) successfully simulated copper sorption on gibbsite using the constant capacitance model considering copper sorption in the form of mononuclear complex ($\equiv AlOCu^+$). Based on extended X-ray absorption fine structure (EXAFS) spectra, they concluded that copper sorption onto kaolinite occurs via inner-sphere mononuclear complexes with two or three edge-sharing $Al(O,OH)_6$ polyhedra (Peacock and Sherman, 2005). Further, with the assumption that kaolinite sites can be considered analogous to gibbsite surface sites, they concluded that the structure of adsorbed Cu complexes is more compact with a smaller number of first shell neighbors. Thus, they proposed a formation of tetrahedral Cu–O–Al complexes with shorter Cu–O bonds than in aqueous solution and consequently the Cu coordination number reduced from 6 to 4 or 5 (Peacock and Sherman, 2005; Pokrovsky et al., 2008).

Weerasooriya et al. (2000) tried to use a modification of the Pauling's bond valence theory as an indirect method, due to lack of evidence for the molecular properties of gibbsite–copper complexes, to elucidate the most stable surface structures of copper with gibbsite. They concluded that the charge of the surface oxygen is oversaturated in the monodentate surface complex and thus this complex probably is not stable assuming that the copper surface complex should involve more than a $\equiv AlOH^{0.5-}$ sites.

EXAFS data for copper sorbed on gibbsite are available (Chang et al., 2005); however, from the interpretation of these data it is not clear whether Cu(II) forms monodentate or bidentate surface complexes. Electron spin resonance studies (Ottaviani and Martini, 1980; McBride, 1982) suggested that the bonding of copper to a single surface oxygen may occur with hydrogen bonding holding the Cu ion in a rigid arrangement, along with a bidentate complex. Transmission electron microscopy revealed chemisorption of Cu^{2+} on the (001) faces of gibbsite via one Al–OH or Al–OH_2 group to form a planar Cu–hydroxy species (McBride et al., 1984).

In the absence of sufficient evidence about the copper surface complex on gibbsite, in this study we treated the copper surface complex as a monodentate, mononuclear complex, which was the simplest complex to give good fits and also coincided with some of the reported spectroscopy findings.

6.2.2 Lead

Spectroscopic evidence of lead sorption on gibbsite is limited. Gräfe et al. (2007) studied the sorption of lead on kaolinite and synthesized gibbsite, which was proven to contain noncrystalline material. Based on EXAFS analysis, Gräfe et al. (2007) concluded Pb ions sorbed as polynuclear complexes to the gibbsite surface. They also concluded that inner-sphere complexation to the gibbsite sheet was evident from Al backscattering atoms at 3.13–3.42 Å on gibbsite and lead backscattering atoms at 4.56–4.65 Å.

Weerasooriya et al. (2001a) examined the stability of lead surface complexes based on the calculation of charge saturation of the surface oxygen according to the modified Pauling bond valence theory. Their calculations suggest that the most energetically stable surface complex of lead on gibbsite is the bidentate complex (Weerasooriya et al., 2001a). The studies of Gräfe et al. (2007) and Weerasooriya et al. (2001a) seem to converge, but in the first case sorption was applied on an "amorphous like" gibbsite, and amorphous and crystalline forms of aluminum oxide are known to have different types of surface reactions (Kim et al., 2004b). In the work of Weerasooriya et al. (2001a), conclusions were developed from calculations.

The available spectroscopic data for Pb sorption on gibbsite are not sufficient to guide surface-complexation modeling to choice of particular species. Thus, in this study a simple monodentate Pb surface species, $\equiv AlOPb^{+}$, was used.

6.2.3 Cobalt

EXAFS data for Co sorption on aluminum oxides with crystallinity similar to gibbsite were indicative of inner-sphere complexes (Wiesner et al., 2006). In the work of Wiesner et al. (2006), surface-complexation modeling was used to assist with data interpretation, and cobalt sorption was described with a monodentate complex, $\equiv AlOCo^{+}$.

6.2.4 Cadmium

Gräfe et al. (2007) attempted to study cadmium surface speciation and bonding to gibbsite using EXAFS, but the sorbent was not entirely crystalline gibbsite. The EXAFS data indicated that Cd sorbed on "amorphous" gibbsite as a double-corner-sharing complex, but it was unclear whether Cd was sorbed as a hydrated dimer or if small polynuclear complexes contributed to the EXAFS data (Gräfe et al., 2007). The indication of bidentate cadmium surface complex on the amorphous aluminum hydroxide is the only available spectroscopic data.

Weerasooriya et al. (2002) examined the stability of cadmium complexes based on the calculation of charge saturation of the surface oxygen according to the modified Pauling bond valence theory. The calculations suggested that the most energetically stable surface complex of cadmium on gibbsite is the bidentate complex (Weerasooriya et al., 2002).

The available spectroscopic data for cadmium sorption on gibbsite are not sufficient to guide surface-complexation modeling to choice of particular species. Thus, in this study a simple monodentate surface species, $\equiv AlOCd^{+}$ was used.

6.2.5 Manganese

No spectroscopic data for manganese sorption on gibbsite are available.

6.2.6 Iron(II)

No spectroscopic data for iron(II) sorption on gibbsite are available.

6.2.7 Calcium

No spectroscopic data for calcium sorption on gibbsite are available.

6.2.8 Zinc

Results of stable isotopic fractionation of sorbed zinc on gibbsite have revealed that the gibbsite surface was enriched by the heavy isotope ^{66}Zn (Pokrovsky et al., 2005) and accompanied by change of its coordination and shortening the Zn–O distances on the surface compared to aqueous solution, as has been observed with X-ray Absorption Spectroscopy (XAS) measurements (Roberts et al., 2003). Although it could be assumed that the change of Zn coordination from 6 ($[Zn(H_2O)_6]^{2+}$(aq)) in aqueous solution to a coordination number of 5 on the mineral face means that Zn is bound to the ≡AlOH groups of the gibbsite surface; these isotopic variations cannot be directly related to thermodynamic stability, coordination, and first-neighbor distances of surface complexes formed (Pokrovsky et al., 2005). To describe the sorption of Zn on gibbsite, Pokrovsky et al. (2005) used a monodentate mononuclear complex, $\equiv AlOZn^+$, based on the modeling approach of Dzombak and Morel (1990) for hydrous ferric oxide.

The available spectroscopic data for Zn sorption on gibbsite are not sufficient to guide complexation modeling to choice of particular surface species. Thus, in this study a simple monodentate Zn surface species, $\equiv AlOZn^+$, was used.

6.2.9 Mercury

Spectroscopic data and sorption data to elucidate the chemical behavior of mercury–gibbsite system are limited (Sarkar et al., 1999; Kim et al., 2004a, 2004b; Weerasooriya et al., 2006, 2007b). It is known that gibbsite is an efficient sink for Hg(II) and its sorption is endothermic and spontaneous (Weerasooriya et al., 2007b), but little is known about mercury surface species on gibbsite. Infrared (IR) spectra collected for Hg(II) sorbed on gibbsite implied direct surface bonding of Hg(II) with the interlayer OH groups of gibbsite surface (Weerasooriya et al., 2007a).

EXAFS studies on γ-alumina and bayerite showed an intermediate Hg–O distance of 2.20 Å which can be correlated with sorption of the $Hg(OH)_2$ species in monodentate or bidentate corner-sharing modes to Al octahedral (Kim et al., 2004a, 2004b). Three possible inner-sphere complexes were proposed for bayerite based on EXAFS analysis: Hg(II) sorbed as a bidentate edge-sharing (mononuclear) complex to two singly coordinated oxygens, as a bidentate corner-sharing (binuclear) complex to two singly coordinated oxygens, and as a monodentate (mononuclear) complex to a singly coordinated (i.e., terminal or nonbridging) oxygen site (Kim et al., 2004a). Although the structure of bayerite has some differences with that of gibbsite (see Chapter 2), it similarly comprises AlOH octahedra arranged such that the oxygens are hexagonal close-packed, with two thirds of the interstitial octahedral sites occupied by Al atoms (Bargar et al., 1997). Thus, surfaces of hydrated Al_2O_3, and Al(hydr)oxides should have similar functional groups and therefore Hg surface complexes found for bayerite may be indicative of possible Hg surfaces complexes on gibbsite.

Taking into consideration the available spectroscopic data for mercury–gibbsite interactions, a monodentate surface-complexation species of mercury, $\equiv AlOHg^+$, was used to fit all extracted data sets. Other surface species, such as $\equiv AlOHHg(OH)_2{}^0$, $\equiv AlOHgOH^0$, $\equiv AlOHg(OH)_2{}^-$, were also tested, even though the existence of the latter surface species has been questioned (Lützenkirchen, 2001). Their contribution to the improvement of fits was marginal. Thus, for the sake of simplicity only one surface species was maintained throughout modeling. In most cases, using the $\equiv AlOHg^+$ surface species was adequate to produce good fits for all experimental data sets, except for those reported by Sarkar et al. (1999). Even when surface species reported by Sarkar et al. (1999) were implemented in the two-layer model they were unable to produce good fits at pH > 5.

Among experimental data sets used in this study, some were produced in the presence of chloride and new surface species had to be included to describe the aqueous-phase mercury–gibbsite–chloride system. In this system, Hg(II) complexes involving two ligands (Cl^- or OH^-) are the most stable ones over wide pH ranges (Lützenkirchen, 2001). Spectroscopic evidence of mercury surface complexes on gibbsite in the presence of chloride would possibly provide valuable information for constraining modeling of cation sorption on gibbsite, but such data are not available. Thus, based on available information for the mercury–bayerite–chloride system (Kim et al., 2004b) the surface species $\equiv$AlOH–HgOHCl was used along with $\equiv AlOHg^+$ to describe mercury sorption on gibbsite.

6.2.10 Uranium

Spectroscopic data for the sorption of uranyl on gibbsite are limited and most surface species used in the literature are chosen based on modeling attempts rather than spectroscopic evidence (McKinley et al., 1995; Zhang et al., 2005).

Time-resolved laser-induced fluorescence spectroscopy (TRLIFS) under cryogenic conditions in the absence of CO_2 provided evidence that the predominant, pH-dependent, forms of gibbsite-adsorbed uranyl are associated with aluminol sites and can be denoted as $\equiv AlO\text{–}(UO_2)^+$ and $\equiv AlO\text{–}(UO_2)OH^0$, respectively (Chang et al., 2006). Others (Baumann et al., 2005) using the same technique, in the presence of CO_2, attributed their findings to a bidentate mononuclear inner-sphere surface complex in which the uranyl(VI) is bound to two reactive OH– groups at the gibbsite edge linked to one Al at lower pH values. Small sorbed clusters consisting of polynuclear uranyl(VI) surface species were identified at higher pH values.

Veilly et al. (2008), using IR spectroscopy measurements, tried to determine uranyl interaction with the (001) face of gibbsite. To study the interaction of water molecules and uranyl cations with this face, and due to weak interactions existing between the gibbsite layers, they used the two-layer model to assist data interpretation. Their findings and density functional theory (DFT) calculations showed that uranyl formed a unique bidentate surface complex, which was confirmed by Raman spectroscopy and TRLFS measurements.

Froideval et al. (2006) concluded via EXAFS studies that uranium surface speciation at pH 3 is dominated by a mononuclear, inner-sphere uranyl surface complex. For sorption at pH 4–5, EXAFS data indicated a dinuclear, bidentate, inner-sphere uranyl surface complex as an average of surface structures. The same study reported that TRLFS data complement such structural information and provide evidence that the uranyl surface speciation is complex (Froideval et al., 2006). Due to the abundance of strong sites and the complex surface structure of Al-phases, minor uranyl surface species may result from polymerization and surface precipitation. Also it cannot be ruled out that U colloids are formed, despite inner-sphere surface complexation, and are sorbed on the Al-hydroxide (Froideval et al., 2006). A study by Hattori et al. (2009) of the structures of uranyl sorption complexes on gibbsite using DFT calculations and EXAFS spectroscopy at pH 5.6 and 9.7 showed contradictory results. DFT calculations suggested sorption of mononuclear uranyl to gibbsite edge faces via formation of corner-sharing inner-sphere complexes, while EXAFS results were suggestive of polymer formation and/or surface precipitates. Combination of the two techniques concluded that at pH 5.6 both corner and edge-sharing uranyl surface complexes coexist at the edge faces of gibbsite and at pH 9.7, $(UO_2)_2(OH)_2^{2+}$, uranyl dimmers, form corner-sharing complexes at the gibbsite surface (Hattori et al., 2009).

Previous attempts at modeling sorption of UO_2^{2+} on α-$Al(OH)_3(s)$ and α-$SiO_2(s)$ also considered the aforementioned minor surface species. Based on goodness-of-fit to the macroscopic sorption data rather than spectroscopic evidence, Turner et al. (1996) proposed $\equiv SO{-}UO_2^+$ and $\equiv SO{-}(UO_2)_3(OH)_5$ (where $\equiv S$ represents $\equiv Al$ or $\equiv Si$) as the most likely surface-sorption scheme for both solids. This surface-speciation scheme is similar to that used by Hsi and Langmuir (1985) on ferrihydrite ($\equiv FeO{-}UO_2OH$ and $\equiv FeO{-}(UO_2)_3(OH)_5$). Waite et al. (1994) argued against the formation of $\equiv FeO{-}(UO_2)_3(OH)_5$ complex, but their data interpretation was not conclusive. Spectroscopic data collected in a study of uranyl sorption on smectites and on gibbsite (Chisholm-Brause et al., 2004) indicated two surface-complexation species: $\equiv AlOUO_2^+$ at pH 4, and $\equiv AlO(UO_2)_3(OH)_5$ at pH 8. The first is a monomeric species and the second is a trimer, a more hydrolyzed species that formed on the edge sites, and its importance increases with increasing pH (Chisholm-Brause et al., 2004).

Due to limited spectroscopic evidence for uranyl sorption on gibbsite and also to the importance of uranyl solution species, this study considered, along with surface species found in the literature, the most predominant aqueous uranyl species in the pH range of the extracted data sets and incorporated these species into the modeling efforts. Although the dominant species in solution are not necessarily the species with the greatest impact on surface-complexation formation, a viewpoint also embraced by others (Bargar et al., 2000; Lützenkirchen, 2001), it was a hypothesis that was explored. Thus, three major uranyl surface species were included: $\equiv AlOUO_2^+$, $\equiv AlO(UO_2)(OH)_3$, $\equiv AlO(UO_2)_3(OH)_5$. The surface species $\equiv AlOUO_2OH^0$ was tested and it was found that it did not contribute significantly to the overall fitting of the data sets, and thus was not used. Other mononuclear and polynuclear species in solution were also considered, but they demonstrated negligible improvement of fits, and therefore were not used.

6.2.11 Thorium

Very few data for sorption of thorium on gibbsite exist and there is no spectroscopic data. In general, several monodentate surface-complexation reactions have been proposed for thorium (Zhang et al., 2006) some of which were also considered in this study (e.g., $\equiv AlOTh^{3+}$ and $\equiv AlOTh(OH)^{2+}$). Available spectroscopic data for thorium sorption on $SiO_2(s)$ have been modeled with bidentate surface species (Östhols, 1995), but such species were not useful in fitting of data for sorption of thorium on gibbsite in this study.

6.3 COPPER

TABLE 6.4 Cu/gibbsite sorption data

ID	Source	Data Set
4ECU1	McBride (1982, Fig. 6)	TOTCu = 5.00×10^{-4} M
4ECU2	Weerasooriya et al. (2000, Fig. 3A)	TOTCu = 7.80×10^{-7} M
4ECU3	Weerasooriya et al. (2000, Fig. 3A)	TOTCu = 7.80×10^{-7} M
4ECU4	Weerasooriya et al. (2000, Fig. 3A)	TOTCu = 7.80×10^{-7} M
4ECU5	Weerasooriya et al. (2000, Fig. 3B)	TOTCu = 7.87×10^{-6} M
4ECU6	Weerasooriya et al. (2000, Fig. 3B)	TOTCu = 7.87×10^{-6} M
4ECU7	Weerasooriya et al. (2000, Fig. 3B)	TOTCu = 7.87×10^{-6} M
4ICU8	Weerasooriya et al. (2000, Fig. 5)	TOTAl = 8.5×10^{-4} M; pH = 4.01
4ICU9	Weerasooriya et al. (2000, Fig. 5)	TOTAl = 8.5×10^{-4} M; pH = 5.41
4ICU10	Weerasooriya et al. (2000, Fig. 5)	TOTAl = 8.5×10^{-4} M; pH = 7.37
4ECU11	Saeki (2004, Fig. 6)	TOTCu = 3.0×10^{-6} M
4ECU12	Pokrovsky et al. (2008, Fig. 2)	TOTCu = 8.2×10^{-6} M

TABLE 6.5 Cu/gibbsite sorption data not used

Source	Data Set	Reason
Okazaki et al. (1986, Fig. 3)	TOTCu = 1.11×10^{-4} M	High ionic strength ($I = 1$ M)
McBride et al. (1984, Fig. 2A)	TOTCu = 5×10^{-5} M	Consideration only of (001) face of gibbsite as reactive
McBride et al. (1984, Fig. 2B)	TOTCu = 5×10^{-4} M	Consideration only of (001) face of gibbsite as reactive

TABLE 6.6 Cu/gibbsite surface complexation reaction

$\equiv AlOH^0 + Cu^{2+} = \equiv AlOCu^+ + H^+$	K_1

TABLE 6.7 Cu/gibbsite intrinsic surface complexation constants

ID	$\log K_1^{int}$	$\sigma_{\log K}$	WSOS/DF
4ECU1	1.33	0.026	181
4ECU2	0.42	0.027	28.4
4ECU3	0.57	0.029	200
4ECU4	1.84	0.039	8.4
4ECU5	−0.25	0.043	2.5
4ECU6	0.01	0.031	1.6
4ECU7	0.39	0.053	2.6
4ICU8	1.88	0.104	0.6
4ICU9	0.94	0.026	15.6
4ICU10	−0.44	0.014	51.8
4ECU11	−0.61	0.025	34.0
4ECU12	−1.94	0.052	0.2

TABLE 6.8 Best estimates for Cu/gibbsite surface complexation constants

Best-Estimate K	Confidence Level	$\log K^{int}(-)$	$\log K^{int}(+)$
$\log \overline{K_1^{int}} = 0.25$	0.99	0.20	0.30
	0.95	0.21	0.29
	0.90	0.21	0.28
	0.80	0.22	0.27

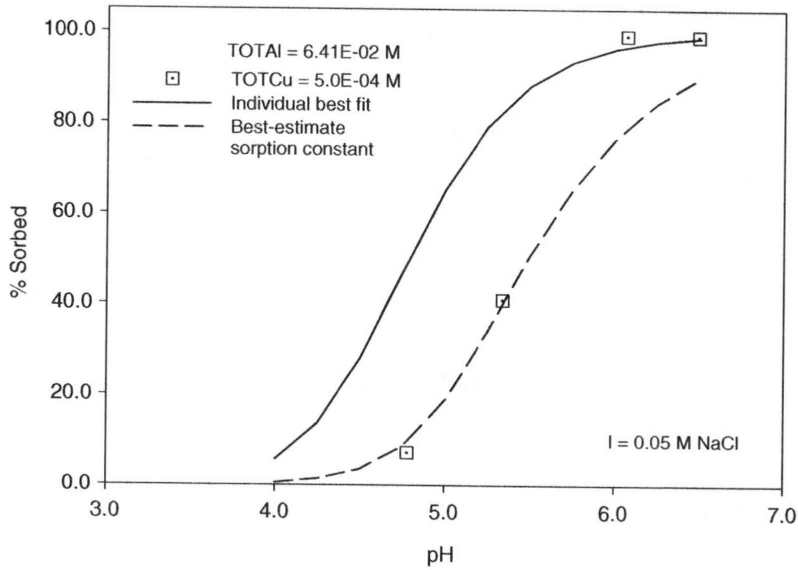

FIGURE 4ECU1

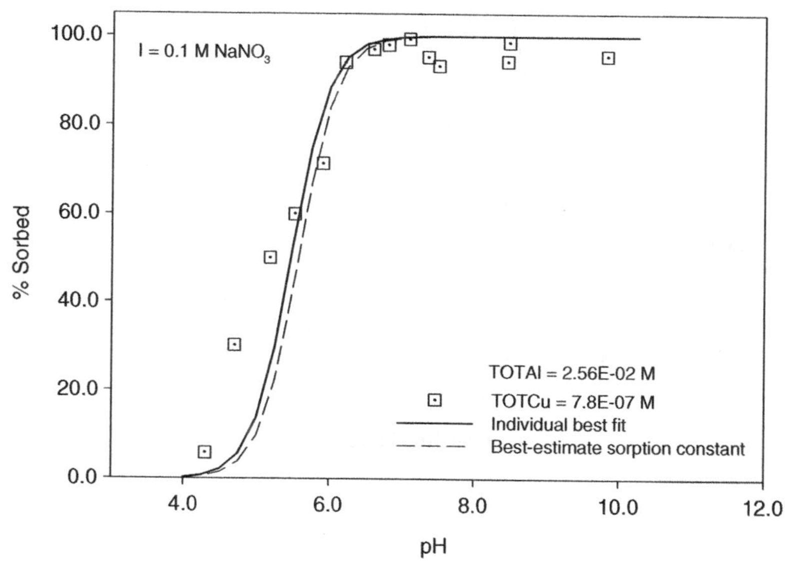

FIGURE 4ECU2

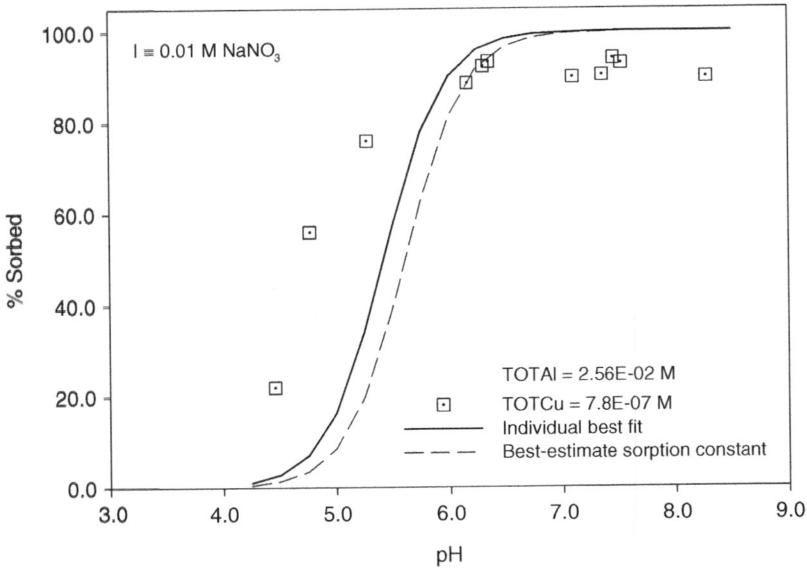

FIGURE 4ECU3

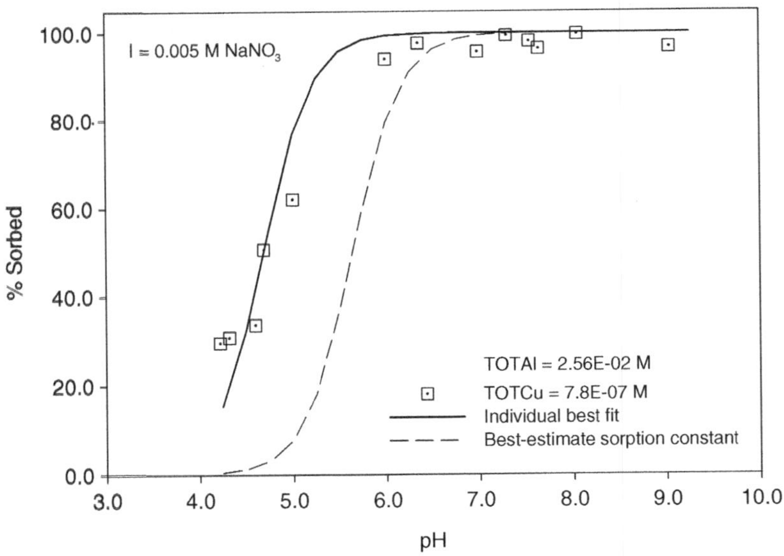

FIGURE 4ECU4

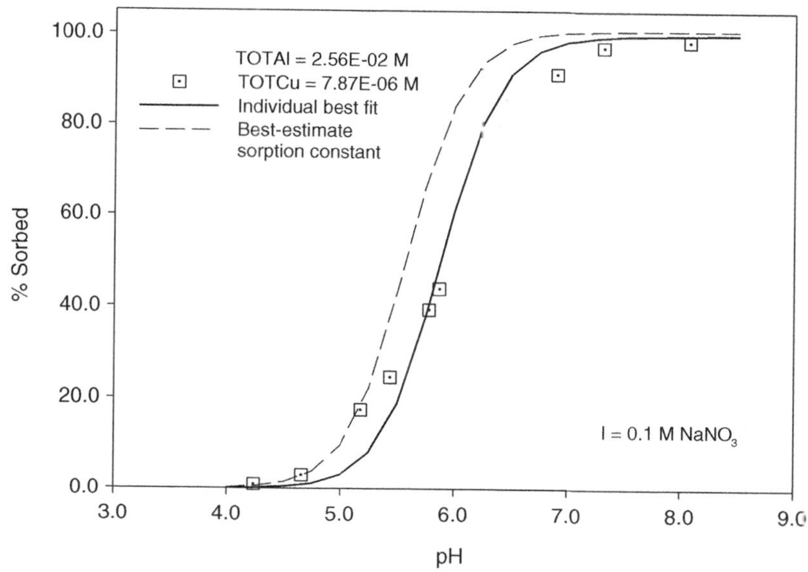

FIGURE 4ECU5

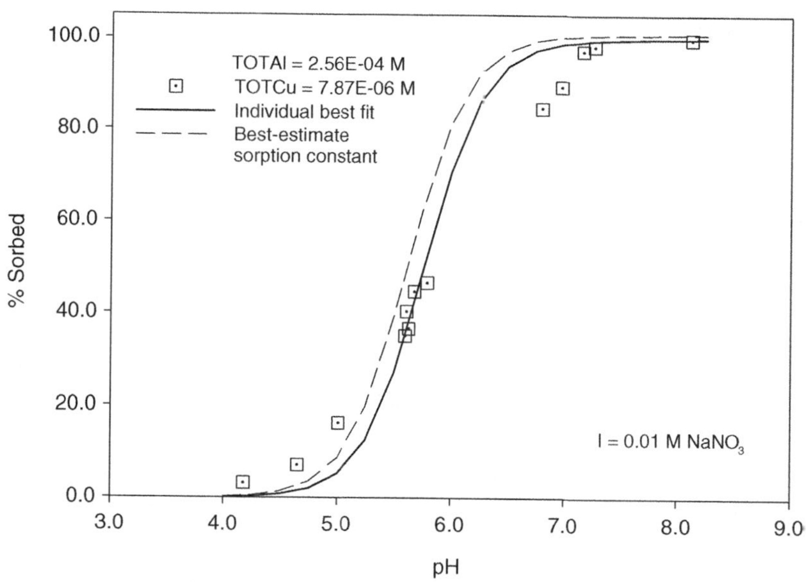

FIGURE 4ECU6

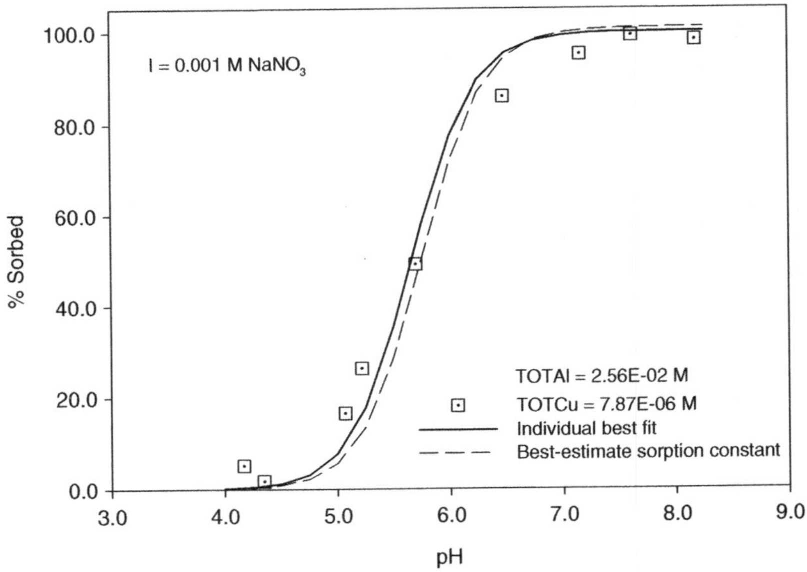

FIGURE 4ECU7

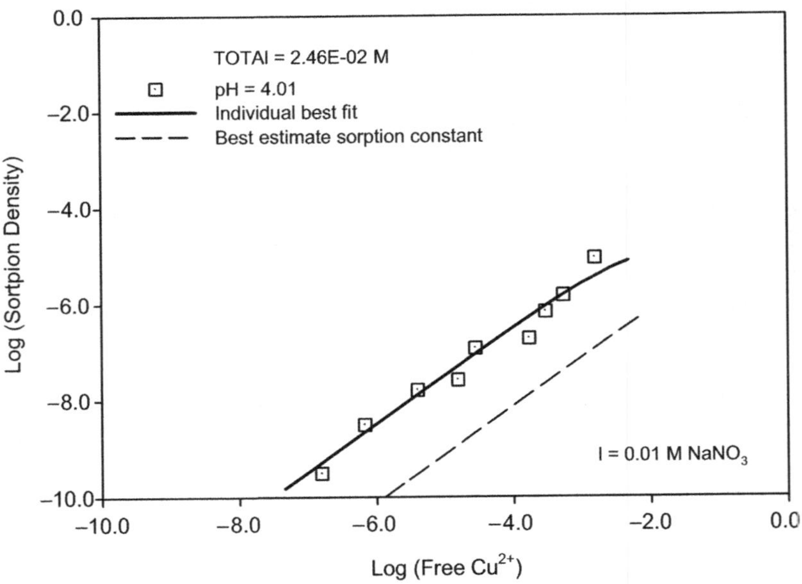

FIGURE 4ICU8

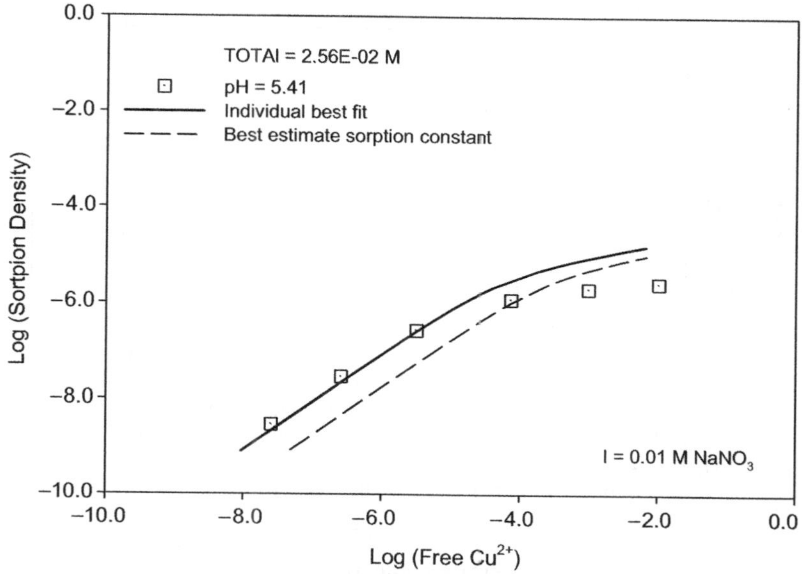

FIGURE 4ICU9

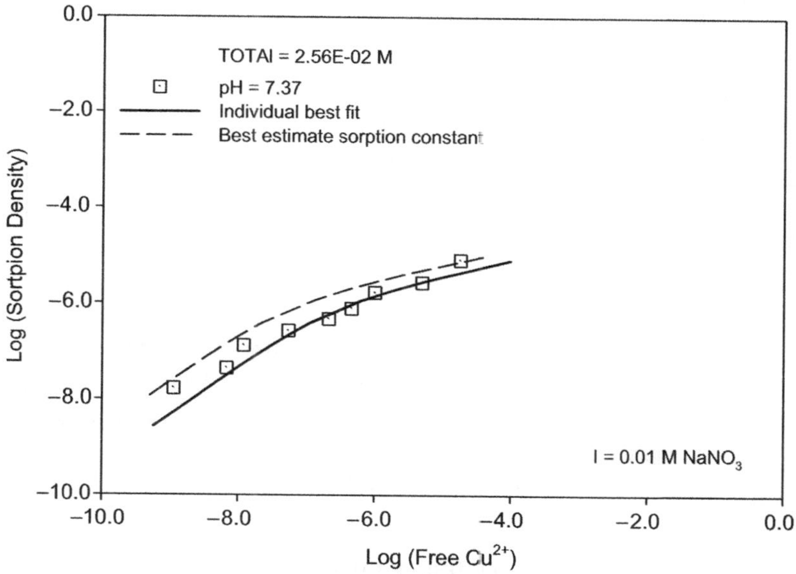

FIGURE 4ICU10

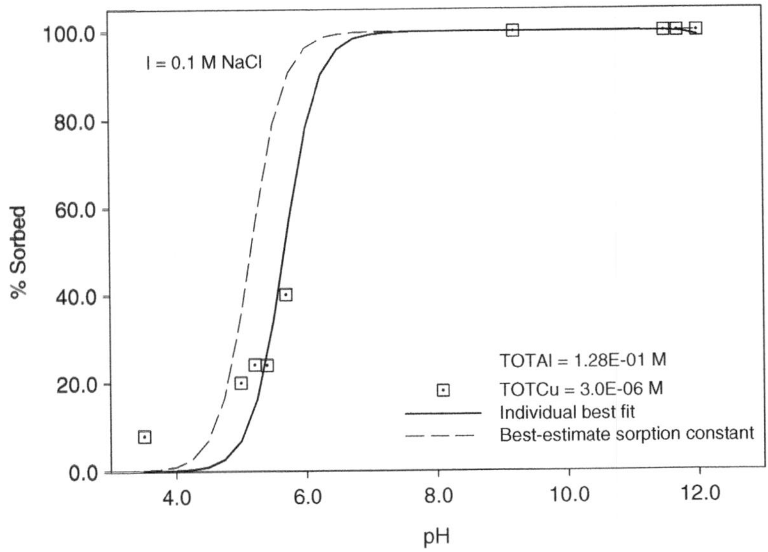

FIGURE 4ECU11

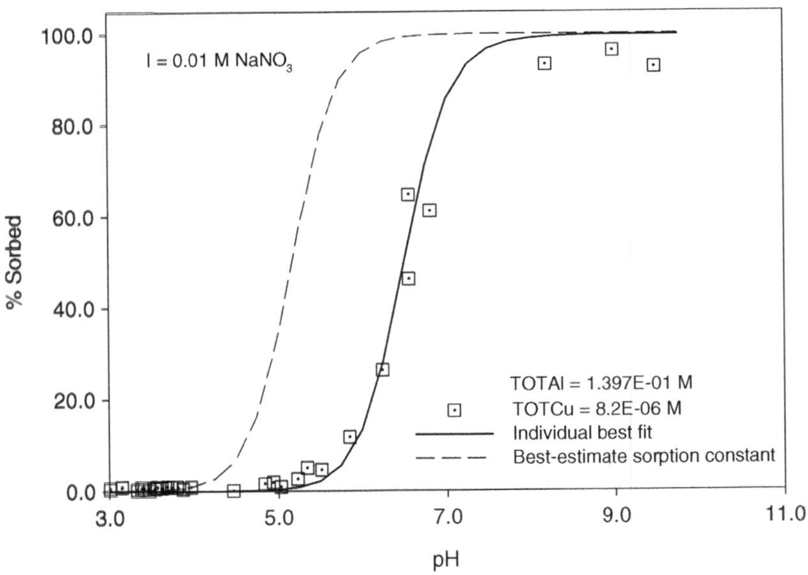

FIGURE 4ECU12

6.4 LEAD

TABLE 6.9 Pb/gibbsite sorption data

ID	Source	Data Set
4EPB1	Weerasooriya et al. (2001a, Fig. 3A)	$TOTPb = 7.8 \times 10^{-7}$ M
4EPB2	Weerasooriya et al. (2001a, Fig. 3A)	$TOTPb = 7.8 \times 10^{-7}$ M
4EPB3	Weerasooriya et al. (2001a, Fig. 3A)	$TOTPb = 7.8 \times 10^{-7}$ M
4EPB4	Weerasooriya et al. (2001a, Fig. 3B)	$TOTPb = 7.78 \times 10^{-6}$ M
4EPB5	Weerasooriya et al. (2001a, Fig. 3B)	$TOTPb = 7.78 \times 10^{-6}$ M
4EPB6	Weerasooriya et al. (2001a, Fig. 3B)	$TOTPb = 7.78 \times 10^{-6}$ M
4EPB7	Weerasooriya et al. (2001a, Fig. 3C)	$TOTPb = 7.78 \times 10^{-5}$ M
4EPB8	Weerasooriya et al. (2001a, Fig. 3C)	$TOTPb = 7.78 \times 10^{-5}$ M
4EPB9	Weerasooriya et al. (2001a, Fig. 3C)	$TOTPb = 7.78 \times 10^{-5}$ M
4IPB10	Weerasooriya et al. (2001a, Fig. 2)	$I = 0.01$ M NaCl; pH = 5.4
4IPB11	Weerasooriya et al. (2001a, Fig. 2)	$I = 0.01$ M NaCl; pH = 6.4
4IPB12	Weerasooriya et al. (2001a, Fig. 2)	$I = 0.01$ M NaCl; pH = 7 2
4IPB13	Weerasooriya et al. (2001a, Fig. 2)	$I = 0.01$ M $NaNO_3$; pH = 5.4
4IPB14	Weerasooriya et al. (2001a, Fig. 2)	$I = 0.01$ M $NaNO_3$; pH = 6.4

TABLE 6.10 Pb/gibbsite surface complexation reaction

$\equiv AlOH^0 + Pb^{2+} = \equiv AlOPb^+ + H^+$	K_1

TABLE 6.11 Pb/gibbsite intrinsic surface complexation constants

ID	$\log K_1^{int}$	$\sigma_{\log K}$	WSOS/DF
4EPB1	0.25	0.032	4.9
4EPB2	0.35	0.029	8.5
4EPB3	0.47	0.033	3.3
4EPB4	−0.20	0.027	3.7
4EPB5	0.49	0.059	2.2
4EPB6	0.71	0.020	17.2
4EPB7	−0.41	0.048	2.5
4EPB8	−0.02	0.018	20.8
4EPB9	0.39	0.033	2.6
4IPB10	0.44	0.021	6.0
4IPB11	0.32	0.011	30.1
4IPB12	0.88	0.016	48.2
4IPB13	0.48	0.024	40.1
4IPB14	0.45	0.017	47.8

TABLE 6.12 Best estimates for Pb/gibbsite surface complexation constants

Best-Estimate K	Confidence Level	$\log K_1^{int}(-)$	$\log K_1^{int}(+)$
$\log \overline{K_1^{int}} = 0.37$	0.99	0.35	0.39
	0.95	0.35	0.38
	0.90	0.36	0.38
	0.80	0.36	0.38

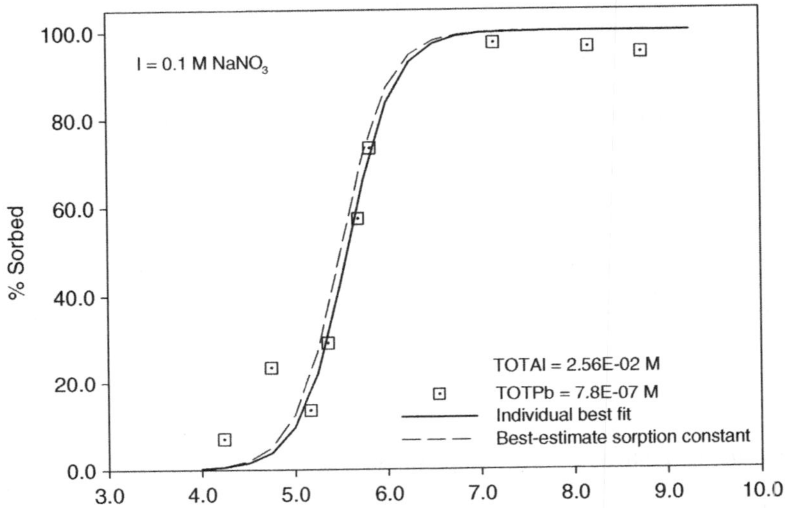

FIGURE 4EPB1

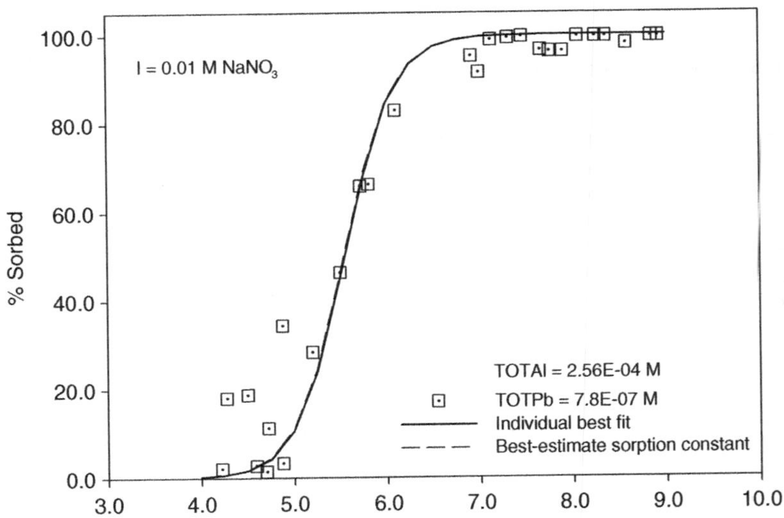

FIGURE 4EPB2

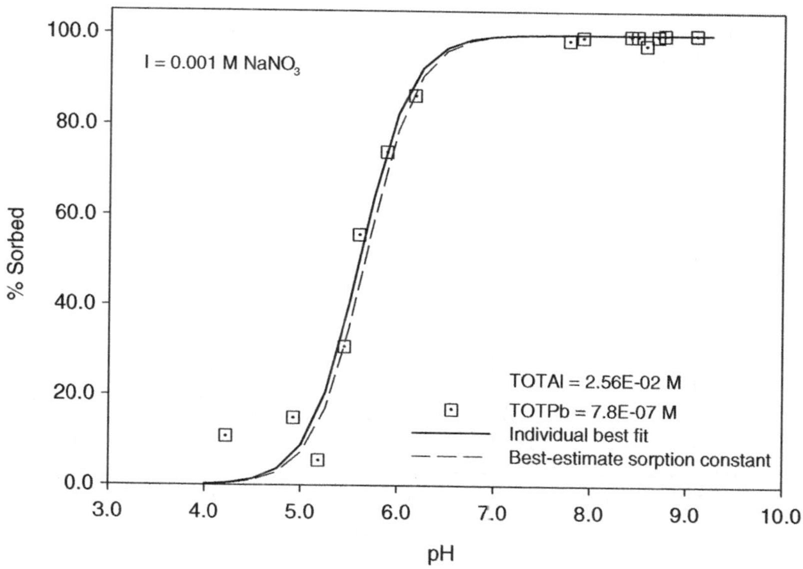

FIGURE 4EPB3

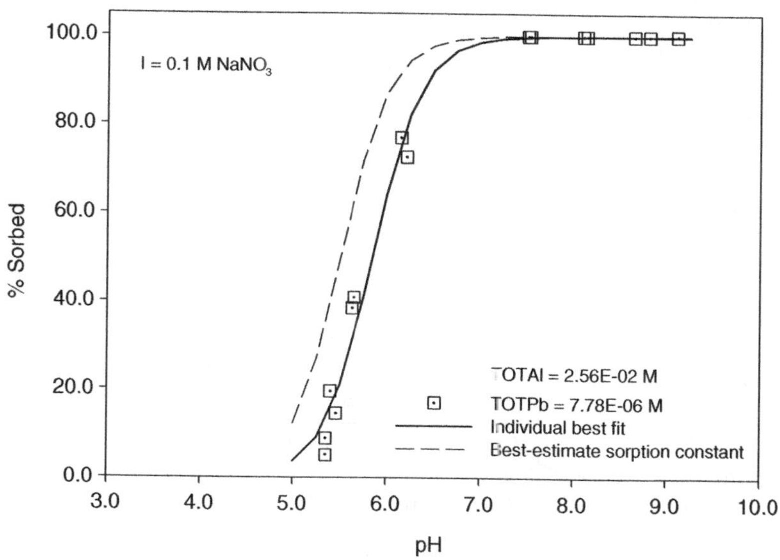

FIGURE 4EPB4

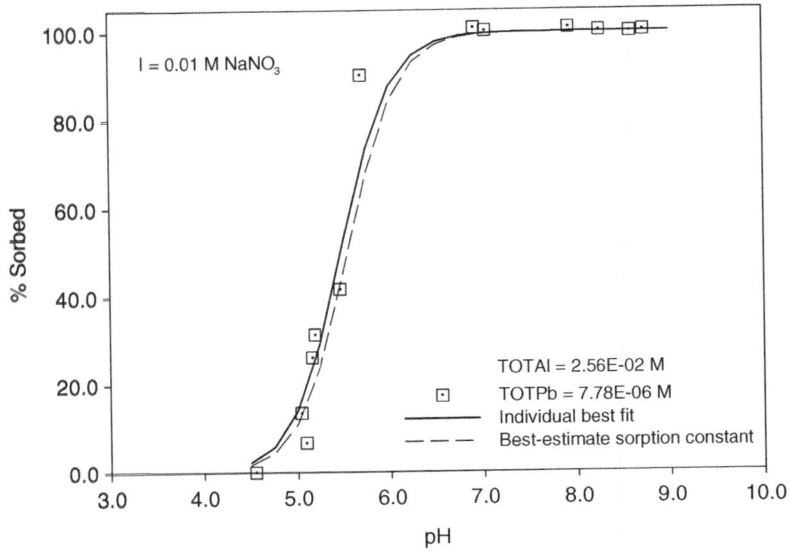

FIGURE 4EPB5

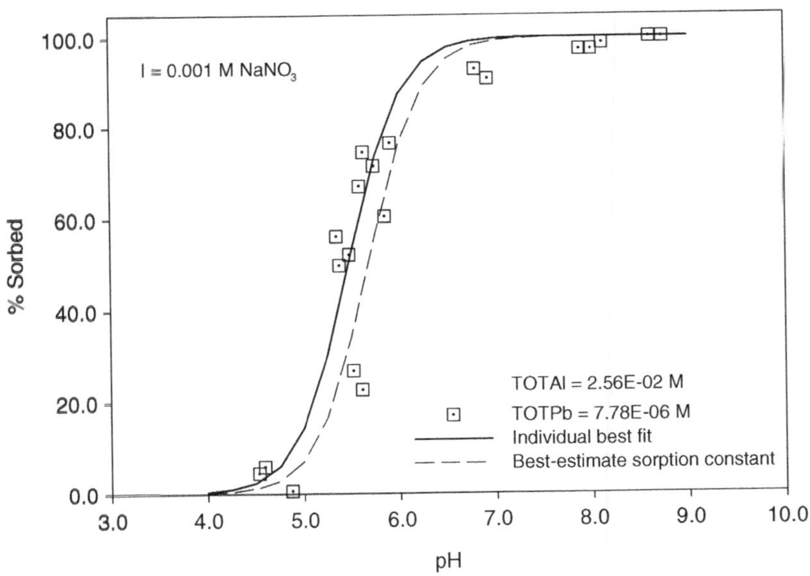

FIGURE 4EPB6

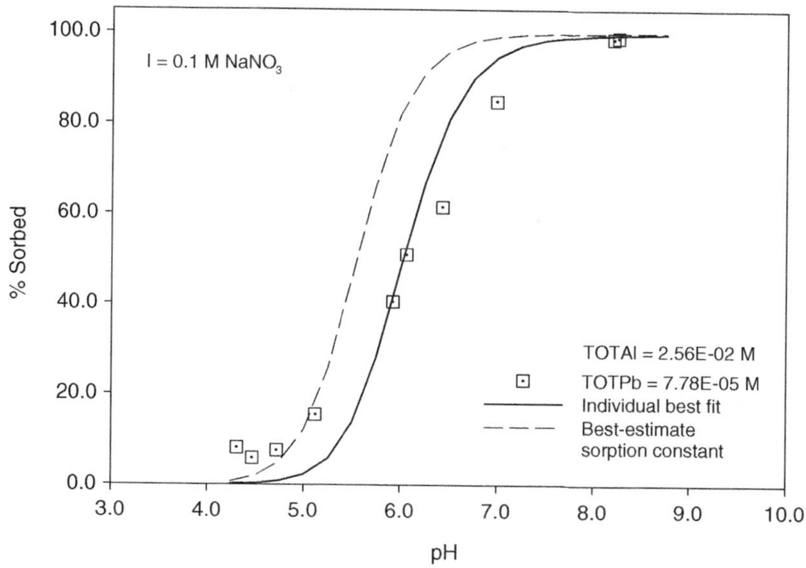

FIGURE 4EPB7

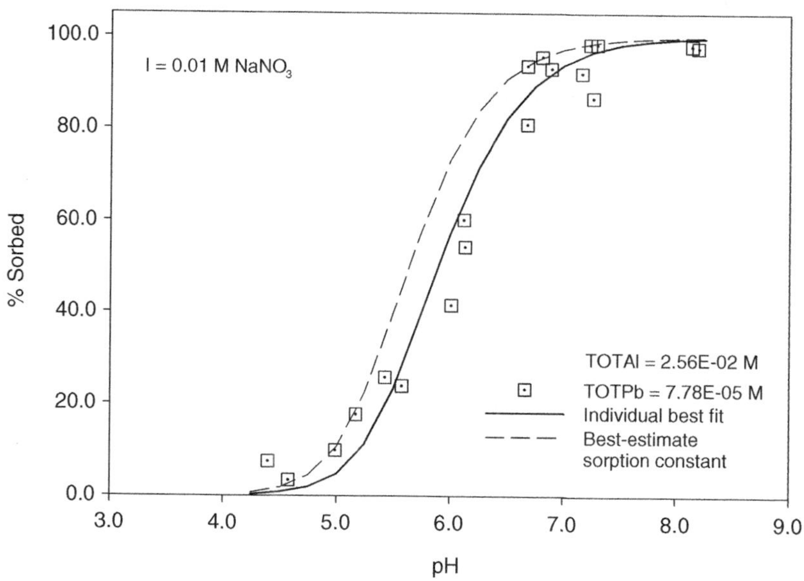

FIGURE 4EPB8

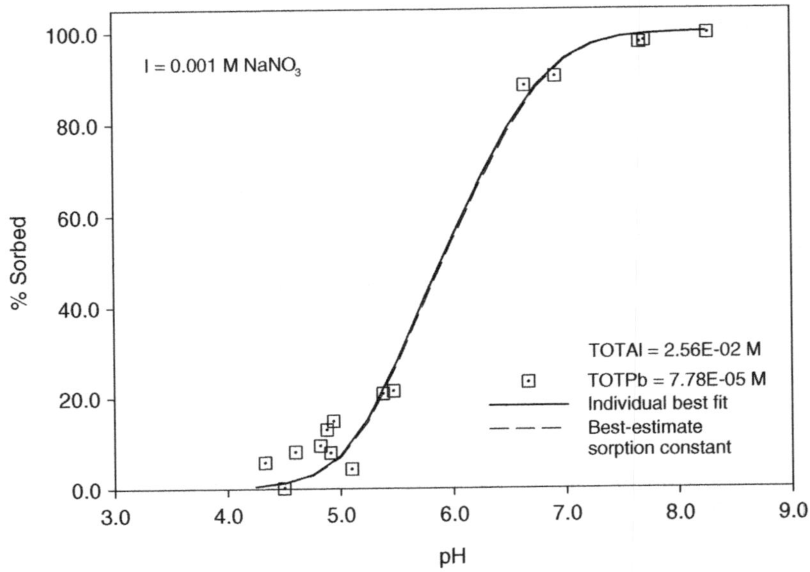

FIGURE 4EPB9

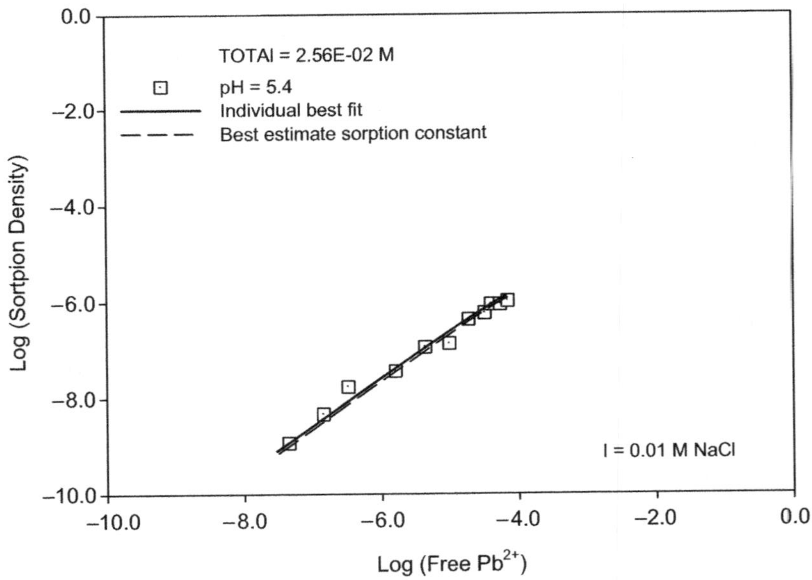

FIGURE 4IPB10

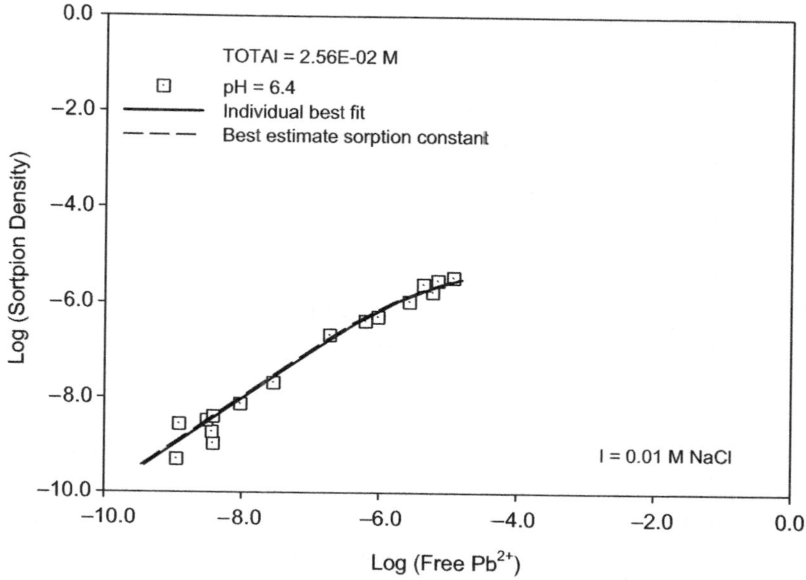

FIGURE 4IPB11

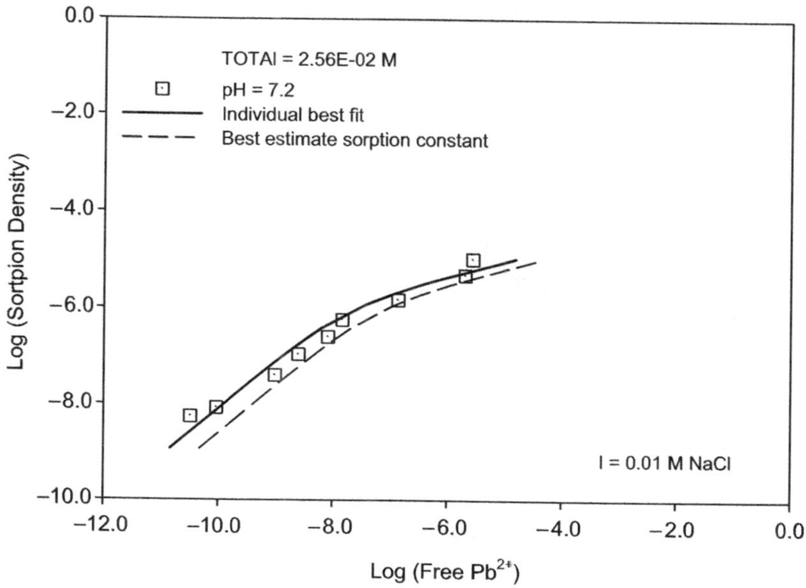

FIGURE 4IPB12

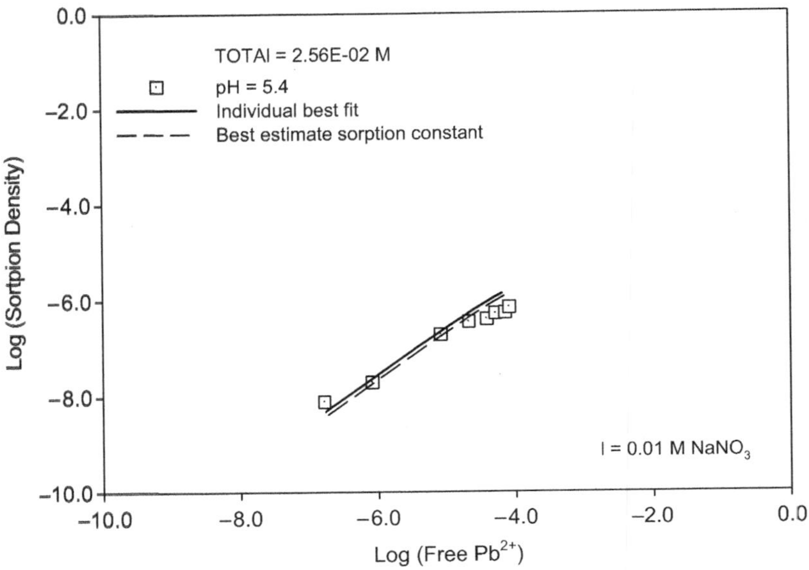

FIGURE 4IPB13

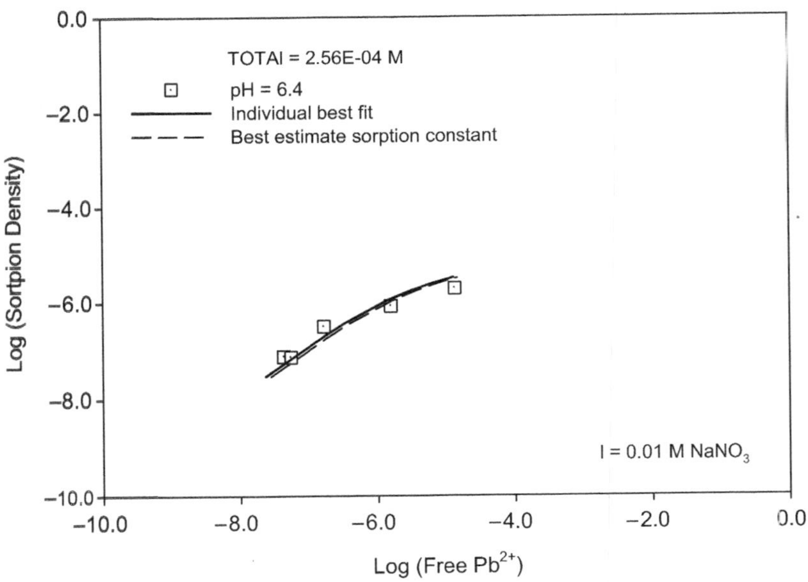

FIGURE 4IPB14

6.5 COBALT

TABLE 6.13 Co/gibbsite sorption data

ID	Source	Data Set
4ECO1	Wiesner et al. (2006, Fig. 2A)	$TOTCo = 1 \times 10^{-5}$ M
4ECO2	Wiesner et al. (2006, Fig. 2A)	$TOTCo = 1 \times 10^{-5}$ M
4ECO3	Wiesner et al. (2006, Fig. 2A)	$TOTCo = 1 \times 10^{-5}$ M
4ECO4	Wiesner et al. (2006, Fig. 2A)	$TOTCo = 1 \times 10^{-5}$ M
4ECO5	Wiesner et al. (2006, Fig. 2B)	$TOTCo = 1 \times 10^{-5}$ M
4ECO6	Wiesner et al. (2006, Fig. 2B)	$TOTCo = 1 \times 10^{-5}$ M
4ECO7	Wiesner et al. (2006, Fig. 2B)	$TOTCo = 1 \times 10^{-5}$ M
4ECO8	Wiesner et al. (2006, Fig. 2B)	$TOTCo = 1 \times 10^{-5}$ M
4ECO9	Wiesner et al. (2006, Fig. 2C)	$TOTCo = 1 \times 10^{-5}$ M
4ECO10	Wiesner et al. (2006, Fig. 2C)	$TOTCo = 1 \times 10^{-5}$ M
4ECO11	Wiesner et al. (2006, Fig. 2C)	$TOTCo = 1 \times 10^{-5}$ M
4ECO12	Wiesner et al. (2006, Fig. 2C)	$TOTCo = 1 \times 10^{-5}$ M
4ECO13	Girvin et al. (1996, Fig. 2)	$TOTCo = 1 \times 10^{-6}$ M
4ECO14	Girvin et al. (1996, Fig. 6)	$TOTCo = 1 \times 10^{-6}$ M
4ECO15	Girvin et al. (1996, Fig. 6)	$TOTCo = 1 \times 10^{-6}$ M
4ECO16	Zachara et al. (1994, Fig. 8D)	$TOTCo = 1 \times 10^{-6}$ M
4ECO17	Zachara et al. (1994, Fig. 8D)	$TOTCo = 1 \times 10^{-6}$ M

TABLE 6.14 Co/gibbsite sorption data not used

Source	Data Set	Reason
Wiesner et al. (2006, Fig. 2A)	$TOTCo = 1 \times 10^{-5}$ M; 1 M $NaNO_3$ electrolyte; Standard buffer $TOTCo = 1 \times 10^{-5}$ M; 1 M $NaNO_3$ electrolyte; Corrected buffer	High ionic strength. Davies equation cannot be applied
Wiesner et al. (2006, Fig. 2B)	$TOTCo = 1 \times 10^{-5}$ M; 1 M NaCl electrolyte; Standard buffer $TOTCo = 1 \times 10^{-5}$ M; 1 M NaCl electrolyte; Corrected buffer	High ionic strength. Davies equation cannot be applied
Wiesner et al. (2006, Fig. 2C)	$TOTCo = 1 \times 10^{-5}$ M; 1 M $NaClO_4$ electrolyte; Standard buffer $TOTCo = 1 \times 10^{-5}$ M; 1 M $NaClO_4$ electrolyte; Corrected buffer	High ionic strength. Davies equation cannot be applied
Girvin et al. (1996, Fig. 5A)	$TOTCo = 1 \times 10^{-6}$ M; 1 M $NaClO_4$	High ionic strength. Davies equation cannot be applied
Girvin et al. (1996, Fig. 6)	$TOTCo = 1 \times 10^{-6}$ M; 1 M $NaClO_4$, low surface area $TOTCo = 1 \times 10^{-6}$ M; 1 M $NaClO_4$, high surface area	High ionic strength. Davies equation cannot be applied

(Continued)

TABLE 6.14 (*Continued*)

Source	Data Set	Reason
Tochiyama et al. (1996, Fig. 2d)	TOTCo = 1×10^{-8} M; 1×10^{-6} M; 1×10^{-5} M; 1×10^{-4} M; 1×10^{-3} M	No exclusion of CO_2 was attempted. Ionic strength was not specified
Zachara et al. (1994, Fig. 8D)	TOTCo = 1×10^{-6} M; I = 0.1 M TOTCo = 1×10^{-6} M; I = 0.01 M	Experiments conducted in the presence of leonardite humic acid

TABLE 6.15 Co/gibbsite surface complexation reaction

$\equiv AlOH^0 + Co^{2+} = \equiv AlOCo^+ + H^+$	K_1

TABLE 6.16 Co/gibbsite intrinsic surface complexation constants

ID	$\log K_1^{int}$	$\sigma_{\log K}$	WSOS/DF
4ECO1	−3.08	0.024	1.6
4ECO2	−2.77	0.031	3.3
4ECO3	−2.77	0.036	2.6
4ECO4	−2.81	0.025	1.4
4ECO5	−2.74	0.042	0.7
4ECO6	−2.76	0.026	4.7
4ECO7	−2.85	0.031	3.7
4ECO8	−2.35	0.042	0.5
4ECO9	−2.71	0.026	9.7
4ECO10	−2.32	0.063	0.8
4ECO11	−2.46	0.029	1.0
4ECO12	−2.00	0.026	10.4
4ECO13	−2.72	0.021	1.6
4ECO14	−2.89	0.026	2.9
4ECO15	−1.50	0.022	0.7
4ECO16	−2.10	0.021	3.3
4ECO17	−2.13	0.045	1.0

TABLE 6.17 Best estimates for Co/gibbsite surface complexation constants

Best-Estimate K	Confidence Level	$\log K_1^{int}(-)$	$\log K_1^{int}(+)$
$\log \overline{K_1^{int}} = -2.52$	0.99	−2.55	−2.50
	0.95	−2.55	−2.50
	0.90	−2.54	−2.50
	0.80	−2.54	−2.51

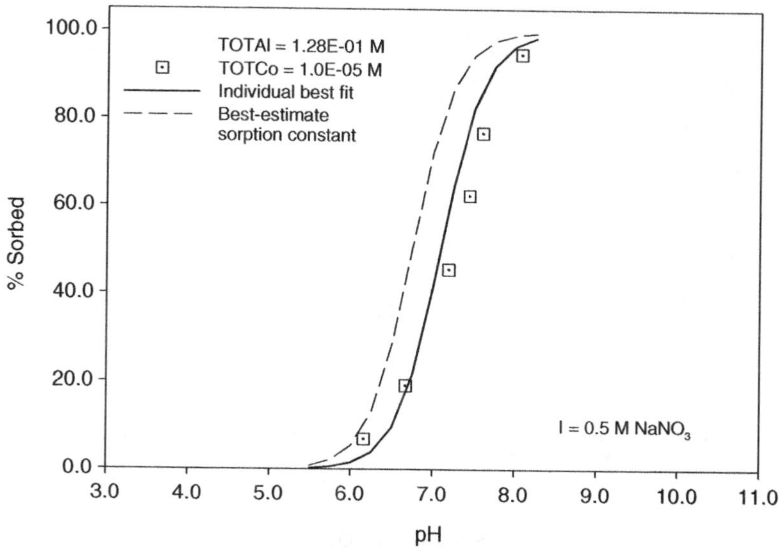

FIGURE 4ECO1

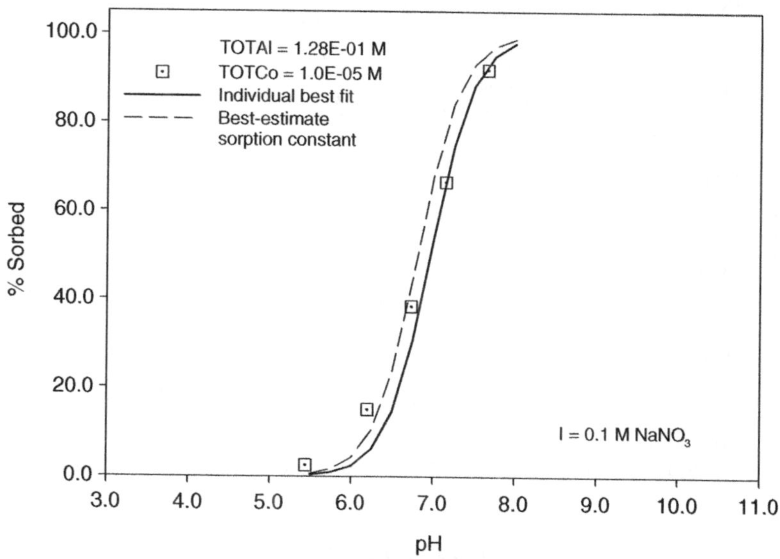

FIGURE 4ECO2

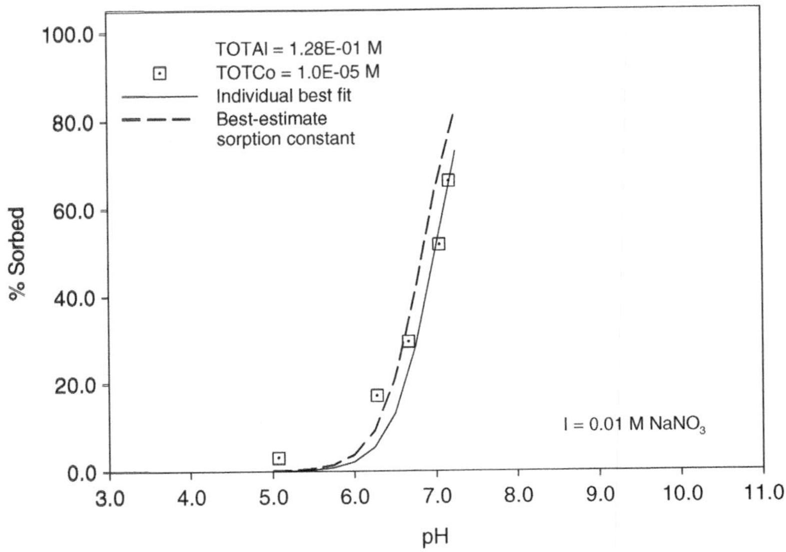

FIGURE 4ECO3

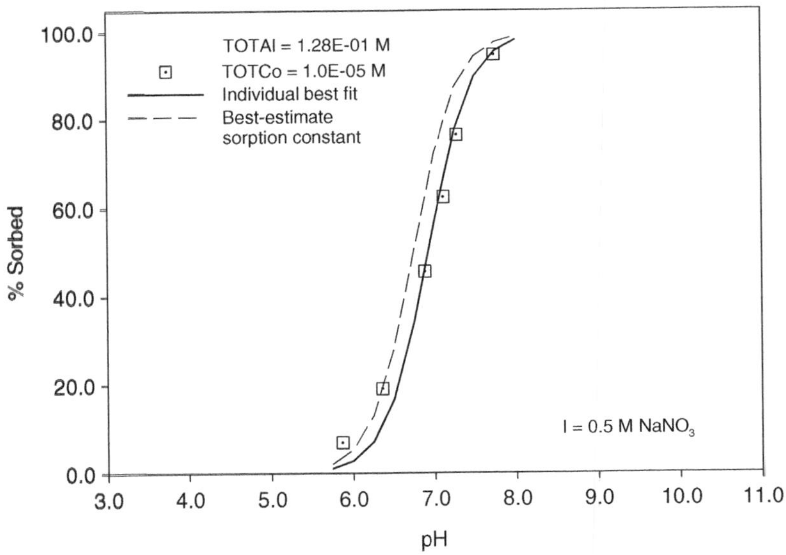

FIGURE 4ECO4

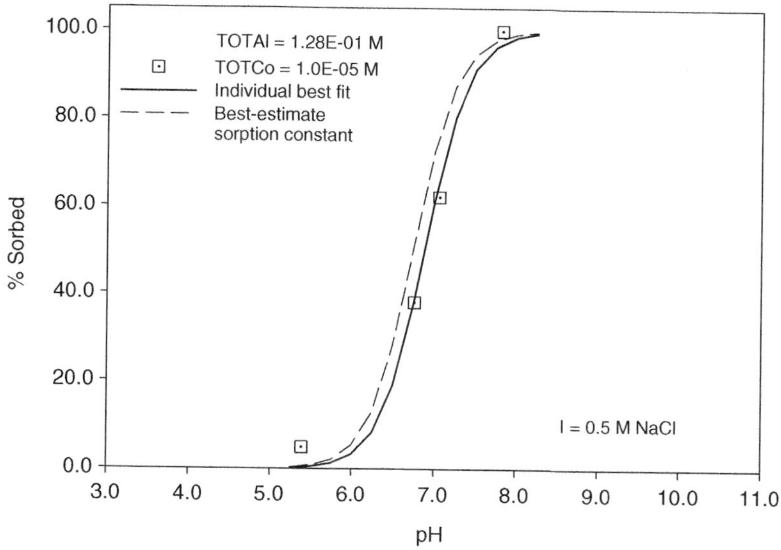

FIGURE 4ECO5

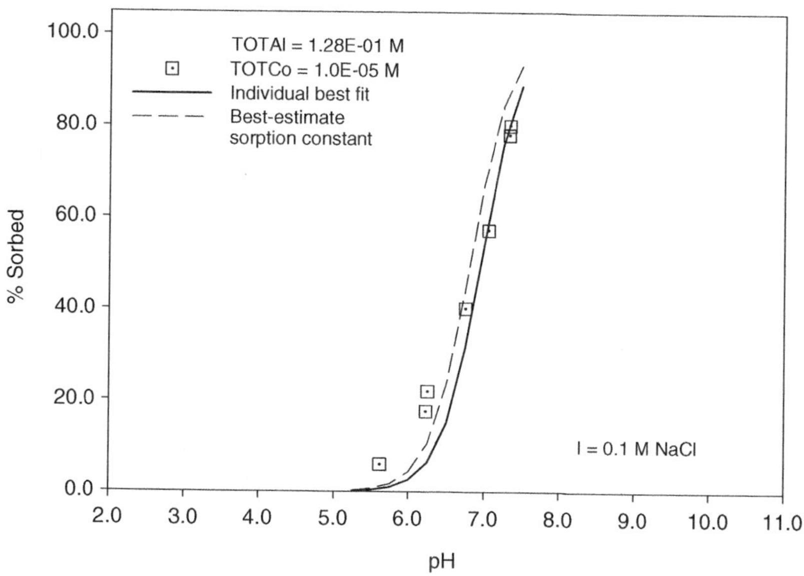

FIGURE 4ECO6

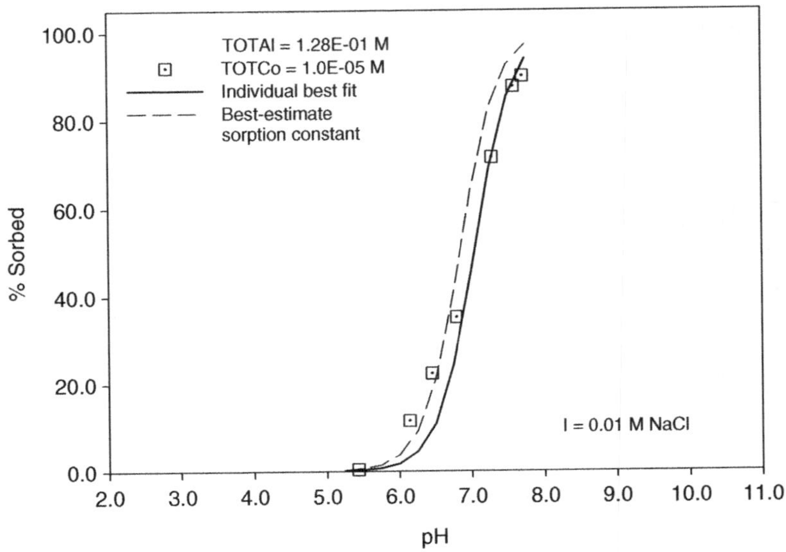

FIGURE 4ECO7

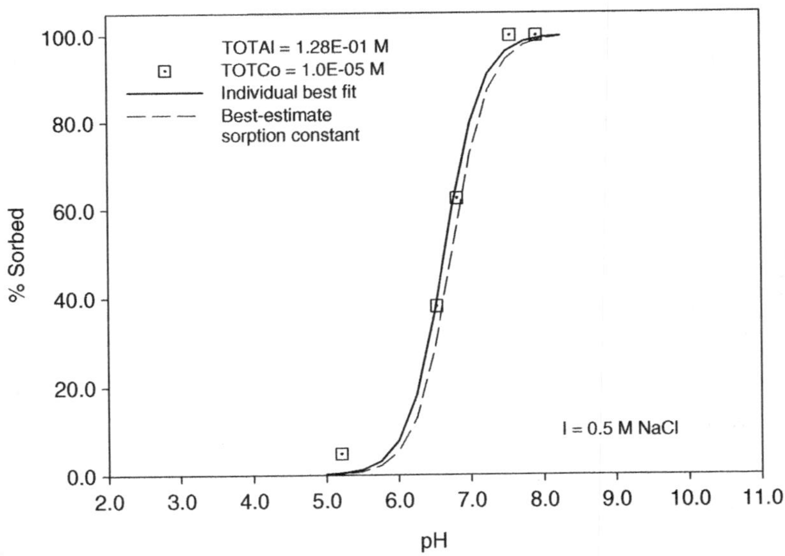

FIGURE 4ECO8

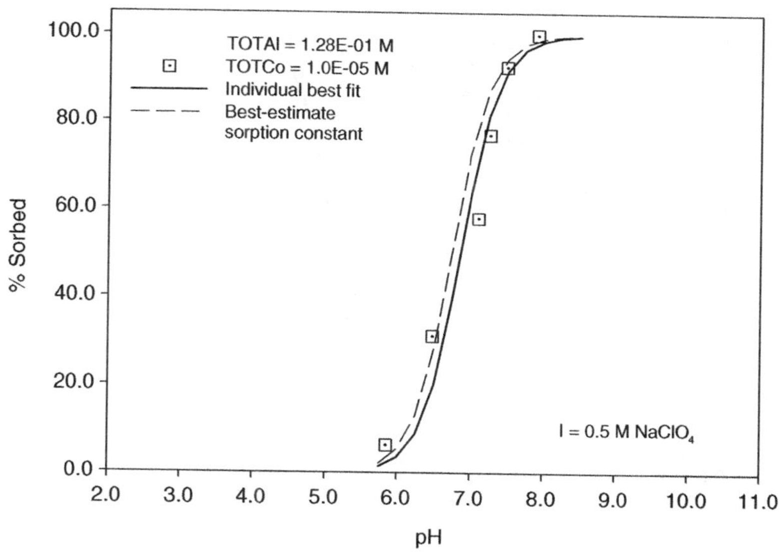

FIGURE 4ECO9

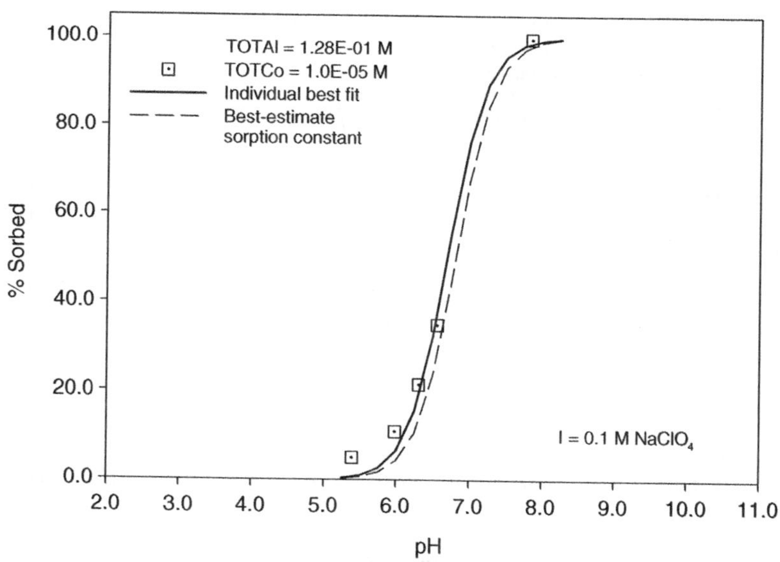

FIGURE 4ECO10

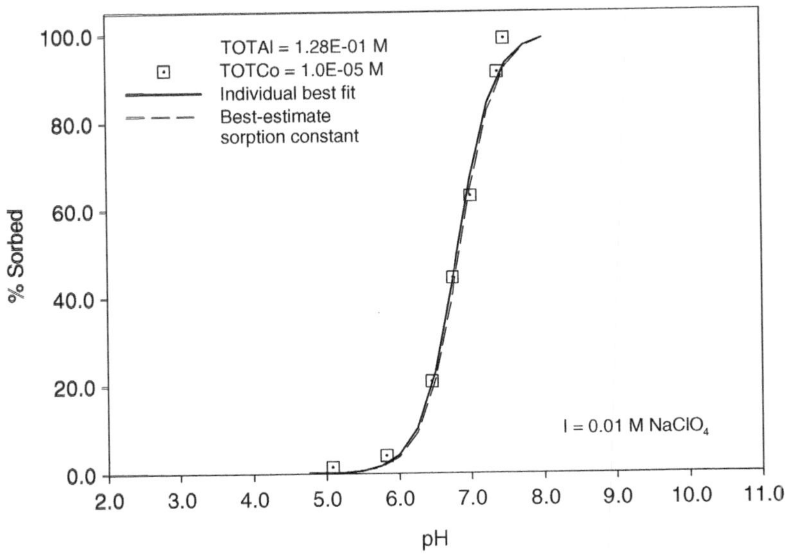

FIGURE 4ECO11

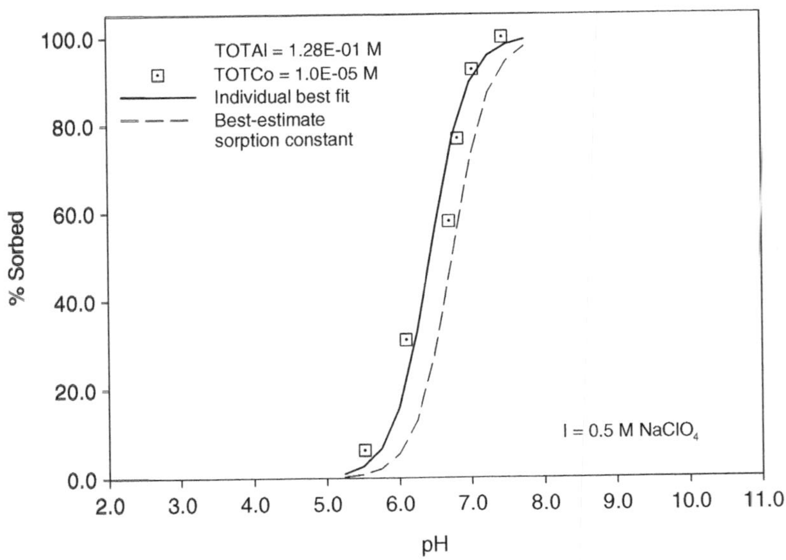

FIGURE 4ECO12

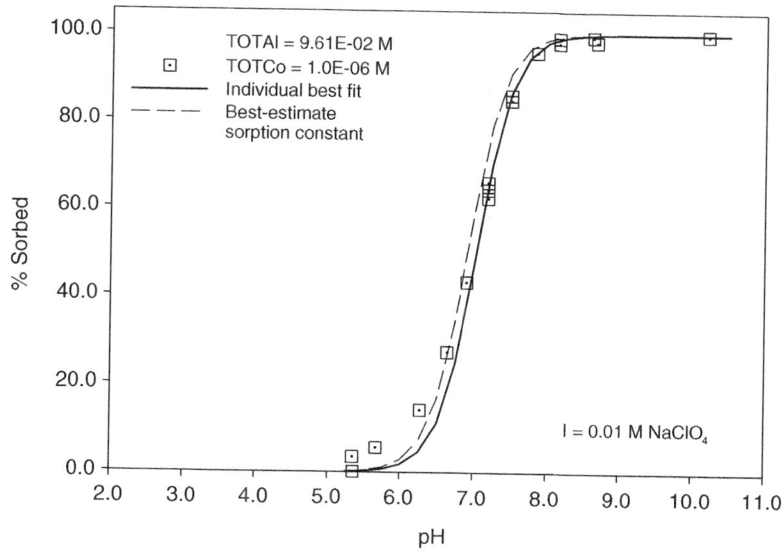

FIGURE 4ECO13

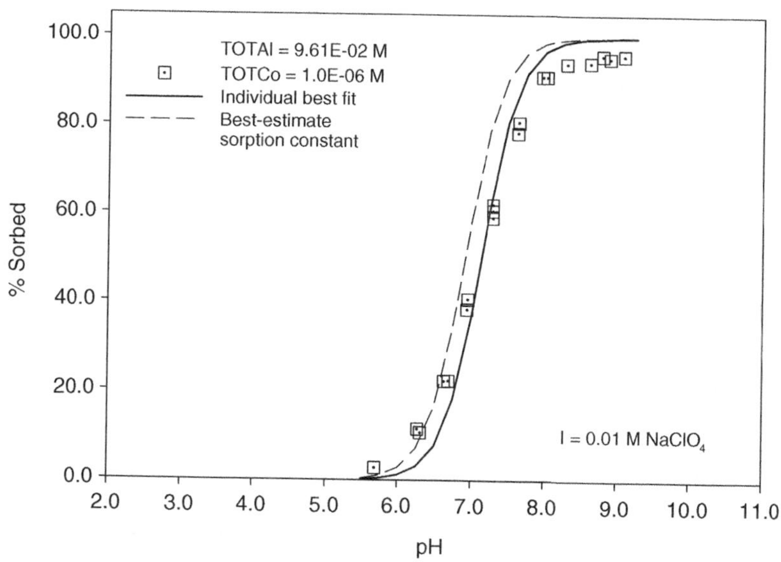

FIGURE 4ECO14

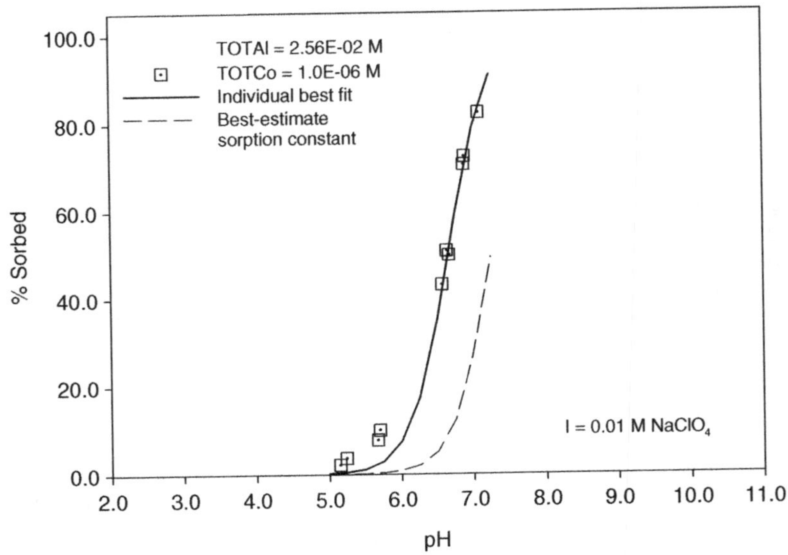

FIGURE 4ECO15

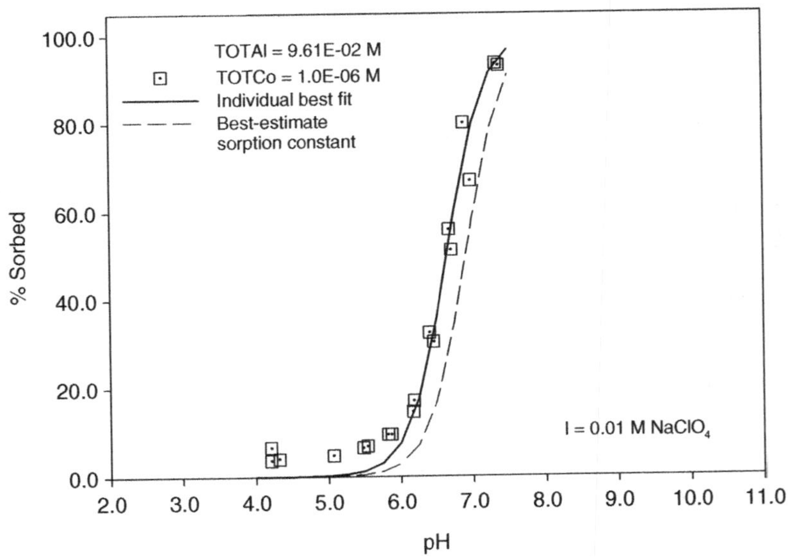

FIGURE 4ECO16

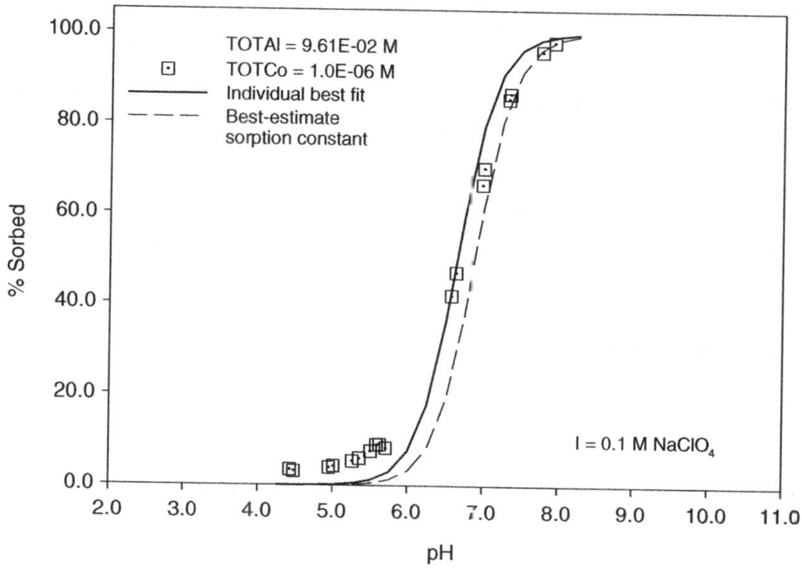

FIGURE 4ECO17

6.6 CADMIUM

TABLE 6.18 Cd/gibbsite sorption data

ID	Source	Data Set
4ECD1	Saeki (2004, Fig. 6)	TOTCd = 3×10^{-6} M
4ECD2	Weerasooriya et al. (2002, Fig. 3A)	TOTCd = 7.7×10^{-7} M
4ECD3	Weerasooriya et al. (2002, Fig. 3A)	TOTCd = 7.7×10^{-7} M
4ECD4	Weerasooriya et al. (2002, Fig. 3A)	TOTCd = 7.7×10^{-7} M
4ECD5	Weerasooriya et al. (2002, Fig. 3B)	TOTCd = 7.74×10^{-6} M
4ECD6	Weerasooriya et al. (2002, Fig. 3B)	TOTCd = 7.74×10^{-6} M
4ECD7	Weerasooriya et al. (2002, Fig. 3B)	TOTCd = 7.74×10^{-6} M
4ECD8	Weerasooriya et al. (2002, Fig. 3B)	TOTCd = 7.74×10^{-6} M
4ECD9	Weerasooriya et al. (2002, Fig. 3B)	TOTCd = 7.74×10^{-6} M
4ECD10	Weerasooriya et al. (2002, Fig. 3B)	TOTCd = 7.74×10^{-6} M
4ECD11	Weerasooriya et al. (2002, Fig. 3C)	TOTCd = 7.74×10^{-5} M
4ECD12	Weerasooriya et al. (2002, Fig. 3C)	TOTCd = 7.74×10^{-5} M
4ECD13	Weerasooriya et al. (2002, Fig. 3C)	TOTCd = 7.74×10^{-5} M
4ICD14	Weerasooriya et al. (2002, Fig. 2)	$I = 0.01$ M $NaNO_3$; pH = 6.0
4ICD15	Weerasooriya et al. (2002, Fig. 2)	$I = 0.01$ M $NaNO_3$; pH = 7.1
4ICD16	Weerasooriya et al. (2002, Fig. 2)	$I = 0.01$ M $NaNO_3$; pH = 7.4

TABLE 6.19 Cd/gibbsite sorption data not used

Source	Data Set	Reason
Meng and Letterman (1993b, Fig. 7)	TOTCd = 1.78×10^{-5} M, $I = 0.04$ M KNO_3	Adsorption experiment in a suspension of 5 mmol of Al/of SiO_2
Meng and Letterman (1993a, Fig. 6; 1993b, Fig. 4)	TOTCd = 1.78×10^{-5} M, $I = 0.04$ M KNO_3	TEM micrographs verified that the $Al(OH)_3$ precipitate is in amorphous form

TABLE 6.20 Cd/gibbsite surface complexation reaction

$\equiv AlOH^0 + Cd^{2+} = \equiv AlOCd^+ + H^+$	K_1

TABLE 6.21 Cd/gibbsite intrinsic surface complexation constants

ID	$\log K_1^{int}$	$\sigma_{\log K}$	WSOS/DF
4ECD1	−2.67	0.032	5.5
4ECD2	−2.67	0.051	3.4
4ECD3	−2.70	0.030	1.9
4ECD4	−2.22	0.052	1.6
4ECD5	−3.23	0.055	1.1
4ECD6	−3.14	0.049	4.0
4ECD7	−2.33	0.091	2.6
4ECD8	−2.83	0.046	1.0
4ECD9	−2.24	0.034	12.0
4ECD10	−2.20	0.037	3.5
4ECD11	−3.28	0.067	1.0
4ECD12	−3.18	0.030	30.6
4ECD13	−3.10	0.061	4.1
4ICD14	−2.79[a]	0.15[b]	—
4ICD15	−2.64	0.032	9.3
4ICD16	−2.77	0.021	6.1

[a] Constant was fixed manually.
[b] Fixed at this value by convention.

TABLE 6.22 Best estimates for Cd/gibbsite surface complexation constants

Best-Estimate K	Confidence Level	$\log K_1^{int}(-)$	$\log K_1^{int}(+)$
$\log \overline{K_1^{int}} = -2.73$	0.99	−2.78	−2.68
	0.95	−2.77	−2.69
	0.90	−2.77	−2.70
	0.80	−2.76	−2.71

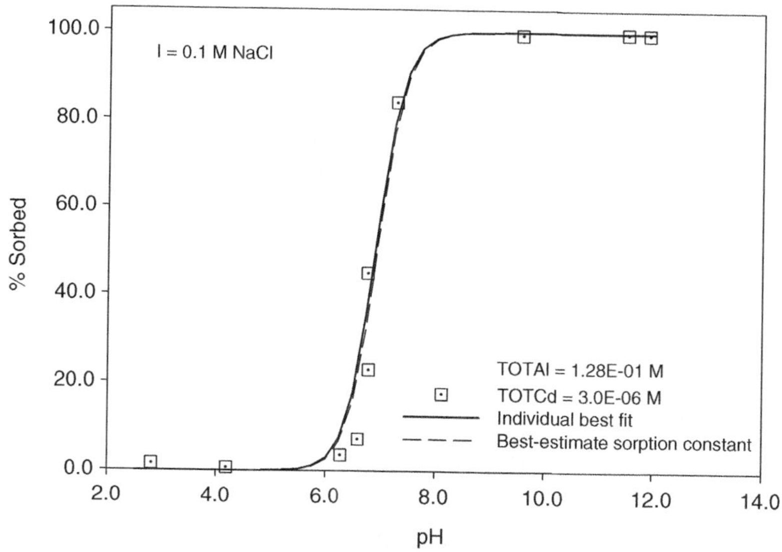

FIGURE 4ECD1

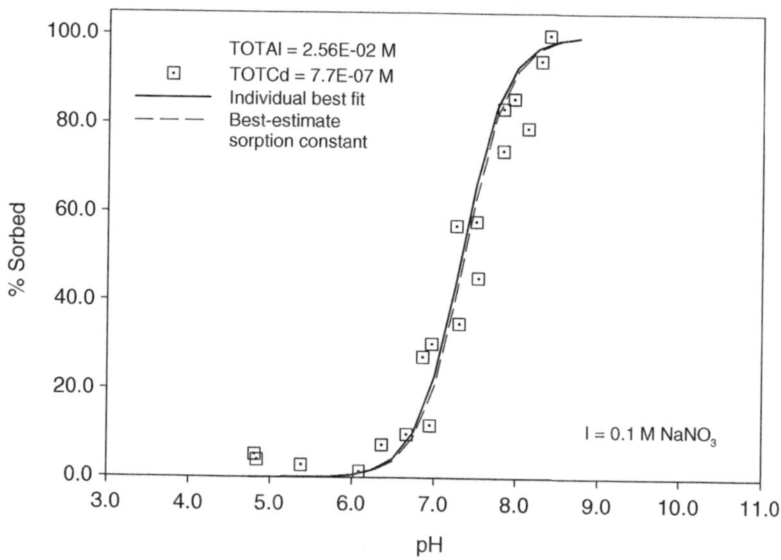

FIGURE 4ECD2

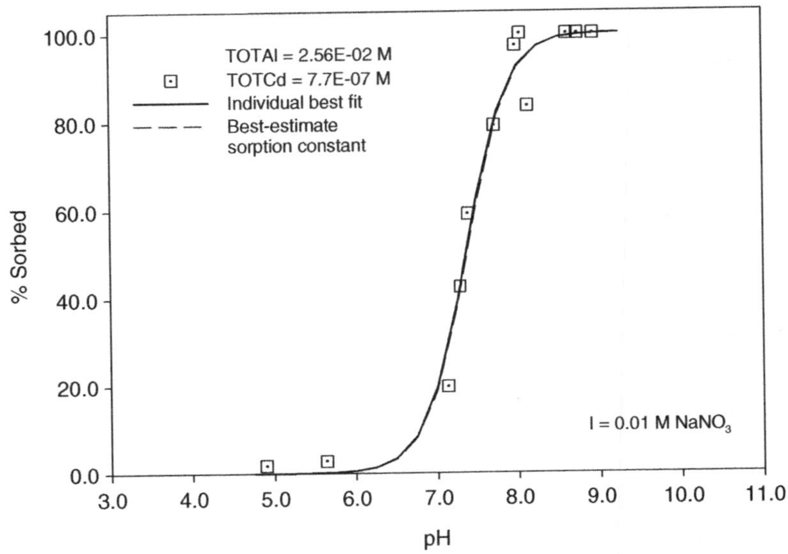

FIGURE 4ECD3

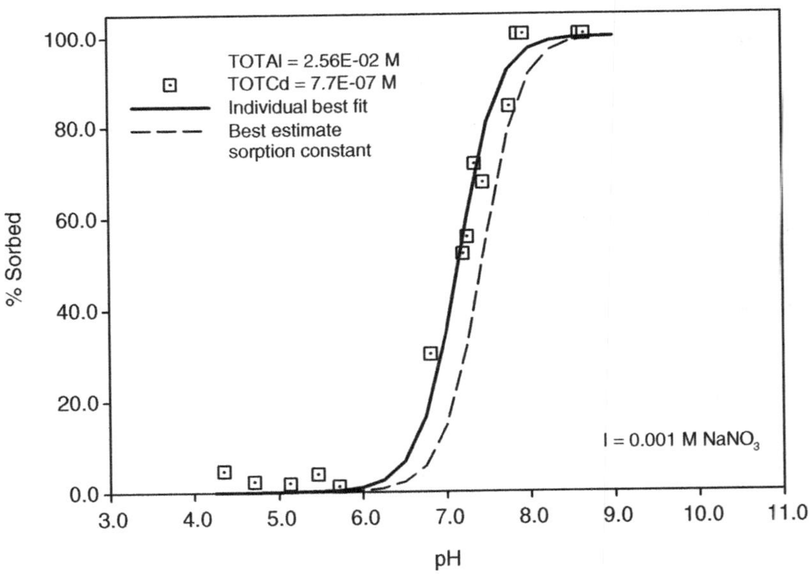

FIGURE 4ECD4

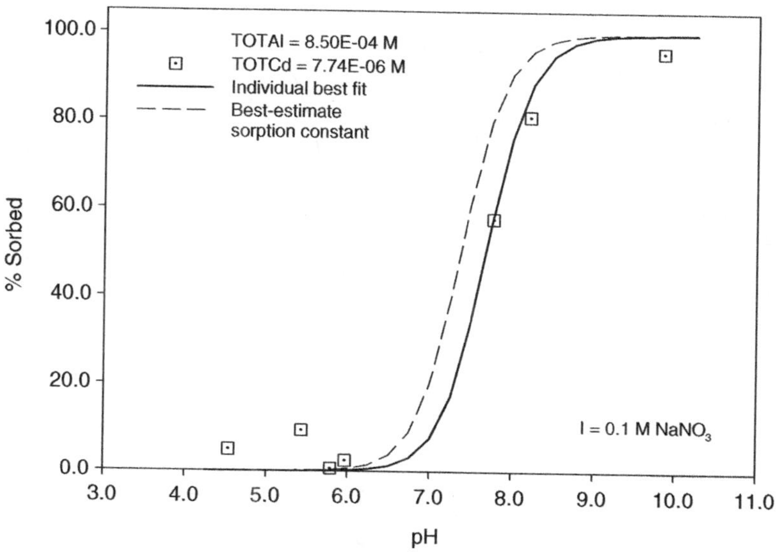

FIGURE 4ECD5

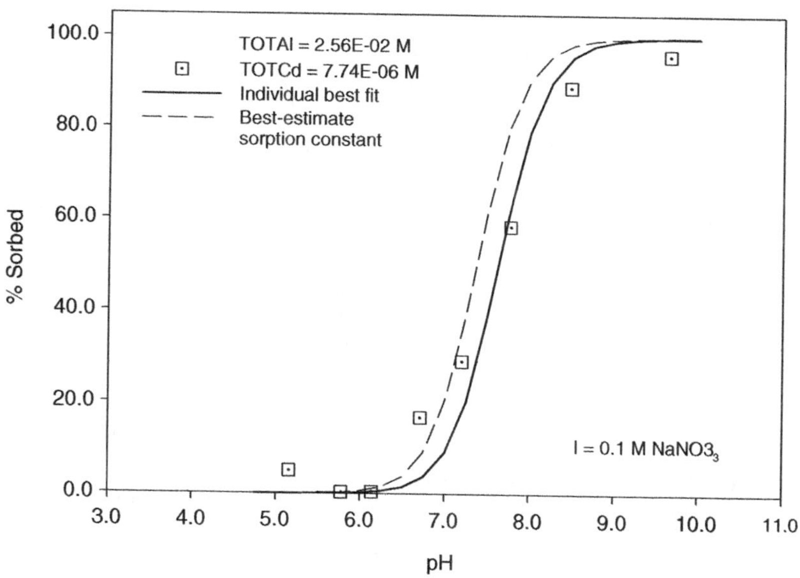

FIGURE 4ECD6

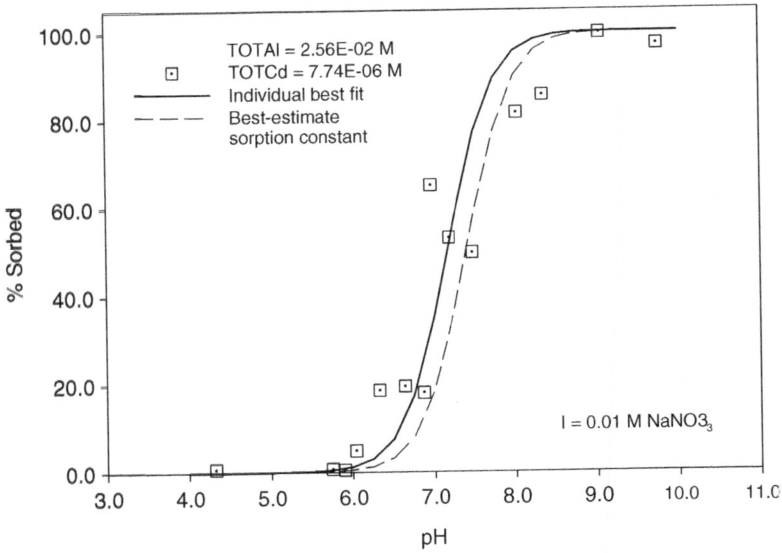

FIGURE 4ECD7

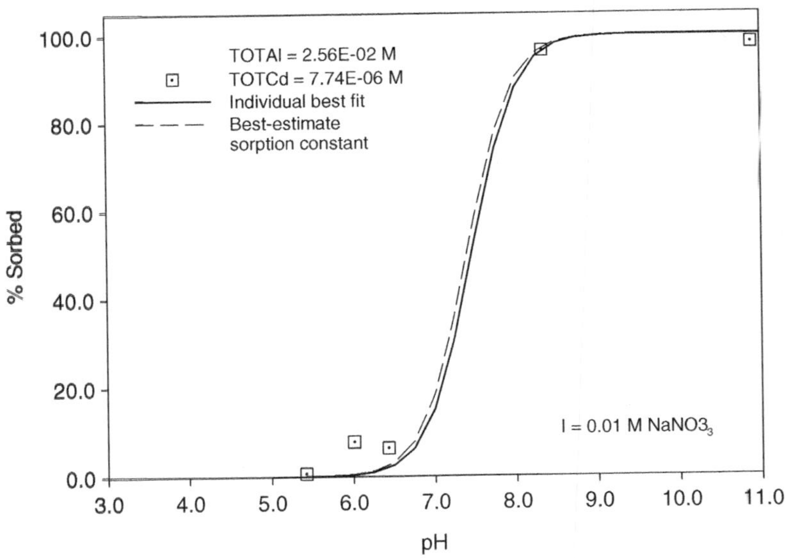

FIGURE 4ECD8

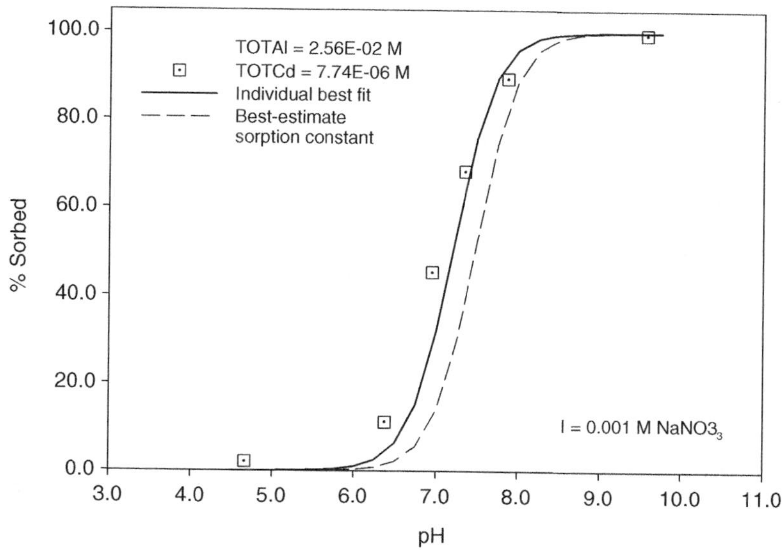

FIGURE 4ECD9

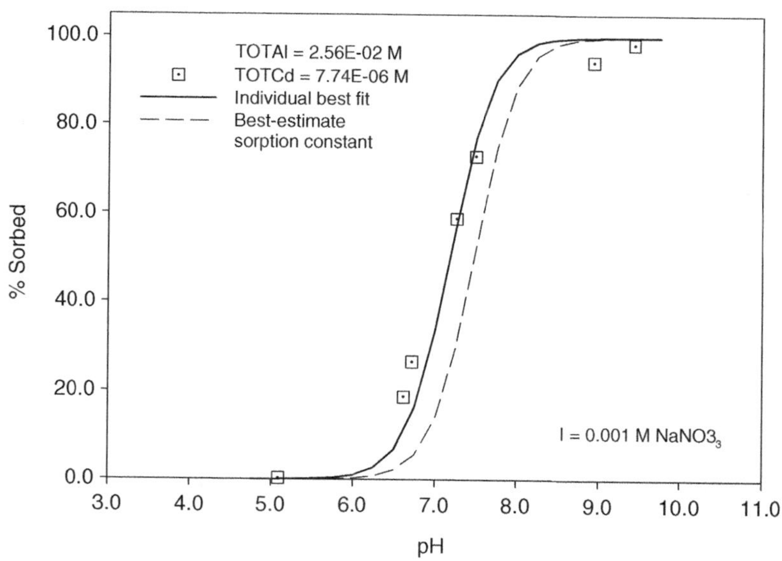

FIGURE 4ECD10

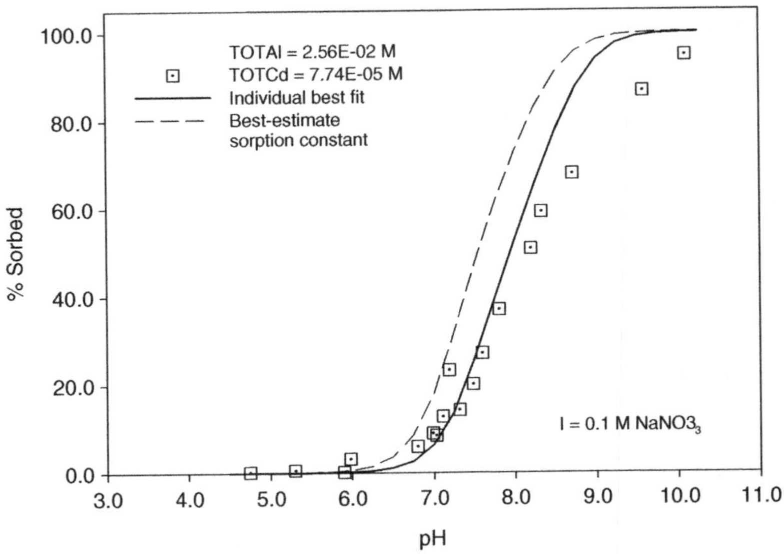

FIGURE 4ECD11

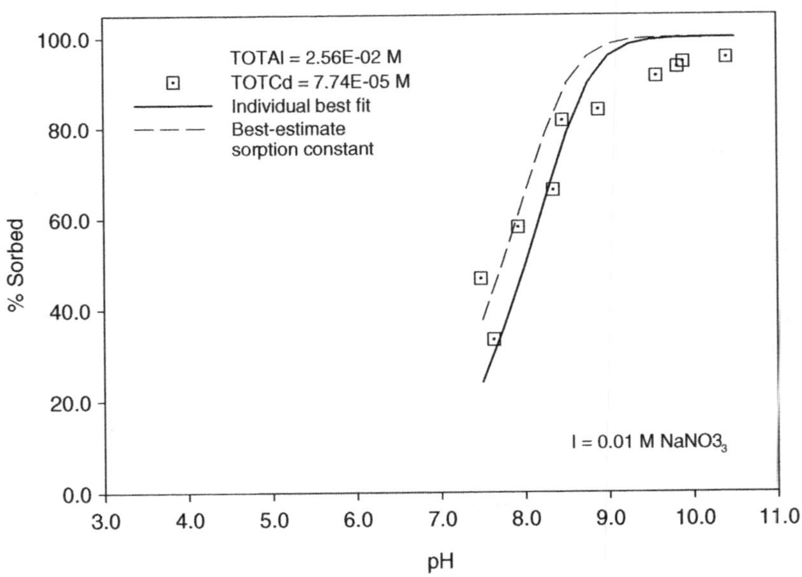

FIGURE 4ECD12

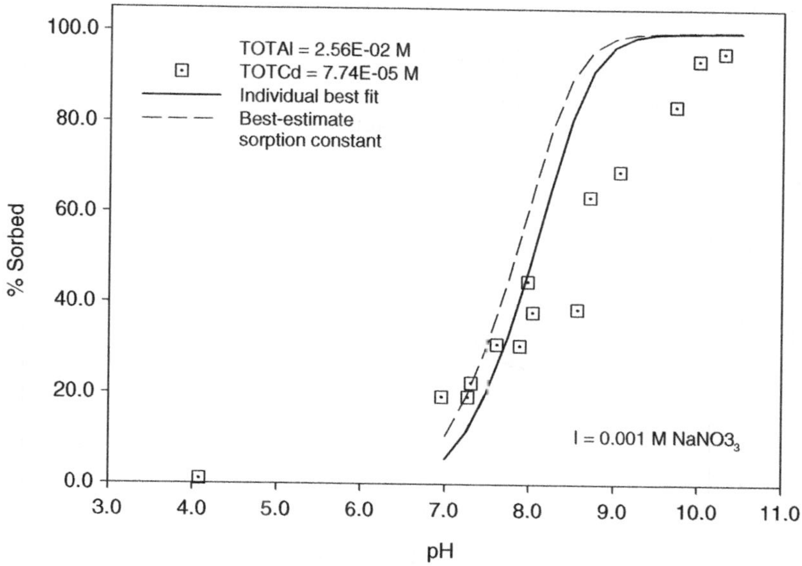

FIGURE 4ECD13

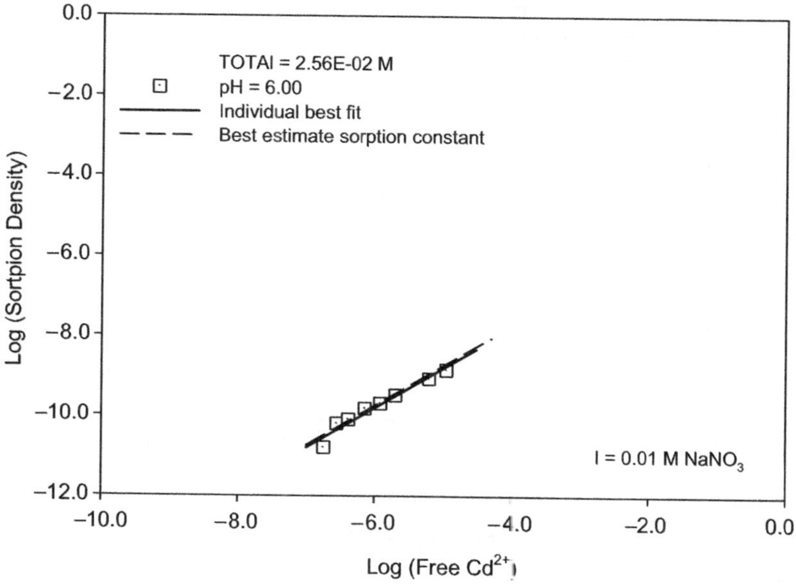

FIGURE 4ICD14

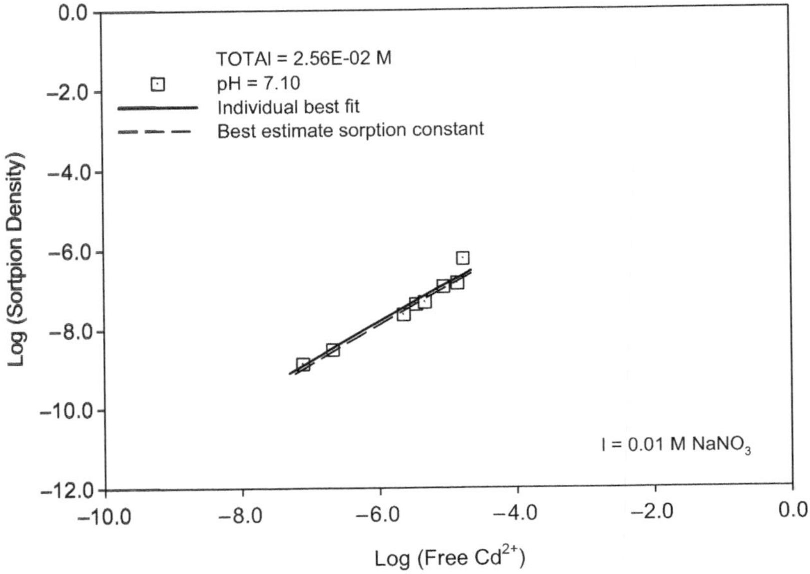

FIGURE 4ICD15

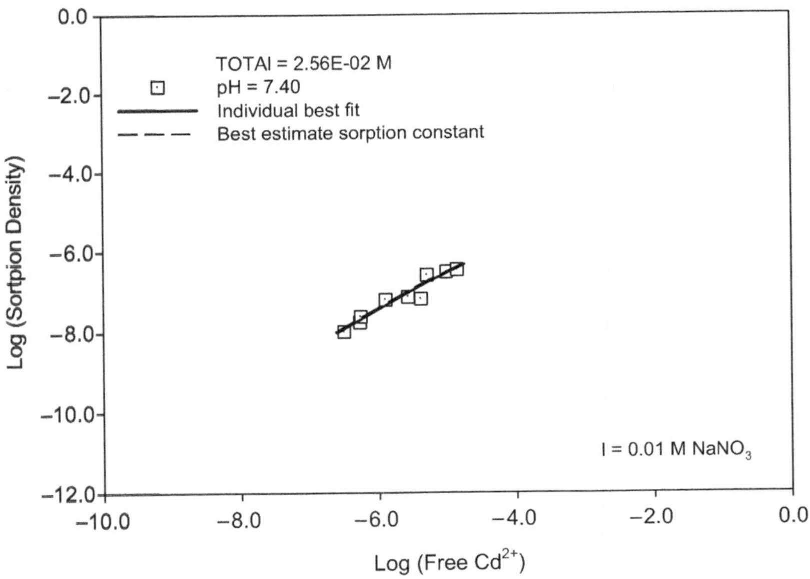

FIGURE 4ICD16

6.7 MANGANESE

TABLE 6.23 Mn/gibbsite sorption data

ID	Source	Data Set
4EMN1	Saeki (2004, Fig. 6)	TOTMn = 3×10^{-6} M

TABLE 6.24 Mn/gibbsite surface complexation reaction

$\equiv AlOH^{\circ} + Mn^{2+} = \equiv AlOMn^{+} + H^{+}$	K_1

TABLE 6.25 Mn/gibbsite intrinsic surface complexation constants

ID	$\log K_1^{int}$	$\sigma_{\log}K$	WSOS/DF
4EMN1	−5.49	0.025	10.7

TABLE 6.26 Best estimates for Mn/gibbsite surface complexation constants

Best-Estimate K	Confidence Level	$\log K_1^{int}(-)$	$\log K_1^{int}(+)$
$\log \overline{K_1^{int}} = -5.49$	0.99		
	0.95		
	0.90		
	0.80		

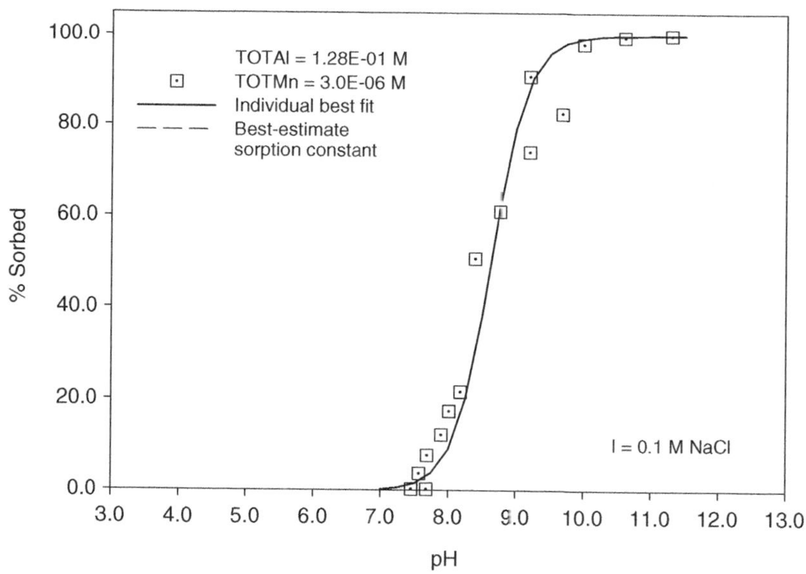

FIGURE 4EMN1

6.8 IRON (II)

TABLE 6.27 Fe(II)/gibbsite sorption data

ID	Source	Data Set
4EFE1	Saeki (2004, Fig. 6)	$TOTFe = 3 \times 10^{-6}$ M

TABLE 6.28 Fe(II)/gibbsite surface complexation reaction

$\equiv AlOH^0 + Fe^{2+} = \equiv AlOFe^+ + H^+$	K_1

TABLE 6.29 Fe(II)/gibbsite intrinsic surface complexation constants

ID	$\log K_1^{int}$	$\sigma_{\log K}$	WSOS/DF
4EFE1	−3.77	0.025	65.1

TABLE 6.30 Best estimates for Fe(II)/gibbsite surface complexation constants

Best-Estimate K	Confidence Level	$\log K_1^{int}(-)$	$\log K_1^{int}(+)$
$\log \overline{K_1^{int}} = -3.77$	0.99		
	0.95		
	0.90		
	0.80		

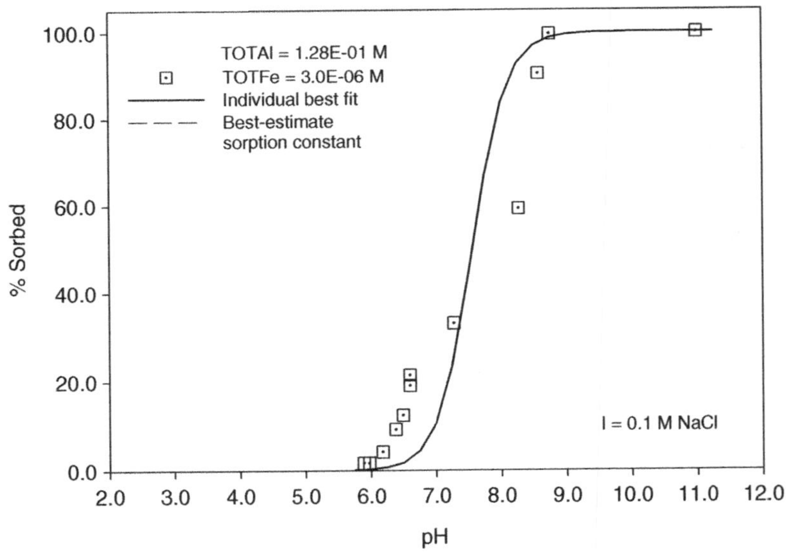

FIGURE 4EFE1

6.9 CALCIUM

TABLE 6.31 Ca/gibbsite sorption data

ID	Source	Data Set
4ECA1	Saeki (2004, Fig. 6)	$TOTCa = 3 \times 10^{-6}$ M

TABLE 6.32 Ca/gibbsite sorption data not used

Source	Data Set	Reason
Meng and Letterman (1993a, Fig. 8)	TOTCa = 1×10^{-4} M	TEM micrographs verified that the $Al(OH)_3$ precipitate is in amorphous form
Kinniburgh et al. (1975, Fig. 2)	TOTCa = 2×10^{-6} M, $I = 1$ M $NaNO_3$	The aluminum (hydr)oxide (called "Al gel") is an amorphous form. The "heated gibbsite" is probably χ-Al_2O_3. High ionic strength. Davies equation cannot be applied
Kinniburgh et al. (1975, Fig. 3)	pH = 7, $I = 1$ M $NaNO_3$	The aluminum (hydr)oxide (called "Al gel") is an amorphous form. The "heated gibbsite" is probably χ-Al_2O_3. High ionic strength, Davies equation cannot be applied

TABLE 6.33 Ca/gibbsite surface complexation reaction

$\equiv AlOH^0 + Ca^{2+} = \equiv AlOCa^+ + H^+$	K_1

TABLE 6.34 Ca/gibbsite intrinsic surface complexationconstants

ID	$\log K_1^{int}$	$\sigma_{\log K}$	WSOS/DF
4ECA1	−10.49	0.043	4.0

TABLE 6.35 Best estimates for Ca/gibbsite surface complexation constants

Best-Estimate K	Confidence Level	$\log K_1^{int}(-)$	$\log K_1^{int}(+)$
$\overline{\log K_1^{int}} = -10.49$	0.99		
	0.95		
	0.90		
	0.80		

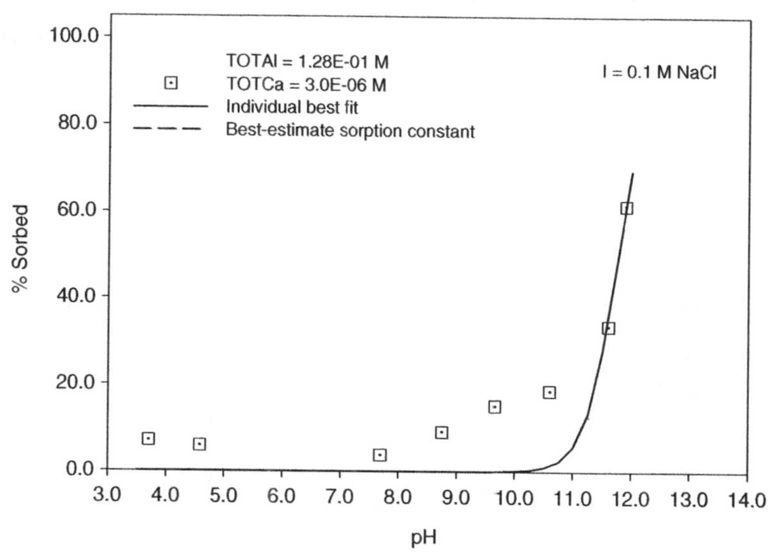

FIGURE 4ECA1

6.10 ZINC

TABLE 6.36 Zn/gibbsite sorption data

ID	Source	Data Set
4EZN1	Pokrovsky et al. (2005, Fig. 1F)	$TOTZn = 7.65 \times 10^{-6}$ M
4EZN2	Roberts et al. (2003, Fig. 1)	$TOTZn = 1.0 \times 10^{-3}$ M
4EZN3	Roberts et al. (2003, Fig. 1)	$TOTZn = 1.0 \times 10^{-3}$ M

TABLE 6.37 Zn/gibbsite sorption data not used

Source	Data Set	Reason
Micera et al. (1986)	—	The aluminum hydroxide that was used for the sorption experiments was precipitated at high pH and characterized by XRD as bayerite

TABLE 6.38 Zn/gibbsite surface complexation reaction

$\equiv AlOH^\circ + Zn^{2+} = \equiv AlOZn^+ + H^+$	K_1

TABLE 6.39 Zn/gibbsite intrinsic surface complexation constants

ID	$\log K_1^{int}$	$\sigma_{\log K}$	WSOS/DF
4EZN1	−1.65	0.039	2.3
4EZN2	−0.80	0.035	38.7
4EZN3	−0.46	0.038	54.1

TABLE 6.40 Best estimates for Zn/gibbsite surface complexation constants

Best-Estimate K	Confidence Level	$\log K_1^{int}(-)$	$\log K_1^{int}(+)$
$\log \overline{K_1^{int}} = -0.96$	0.99	−1.09	−0.83
	0.95	−1.06	−0.86
	0.90	−1.04	−0.87
	0.80	−1.02	−0.89

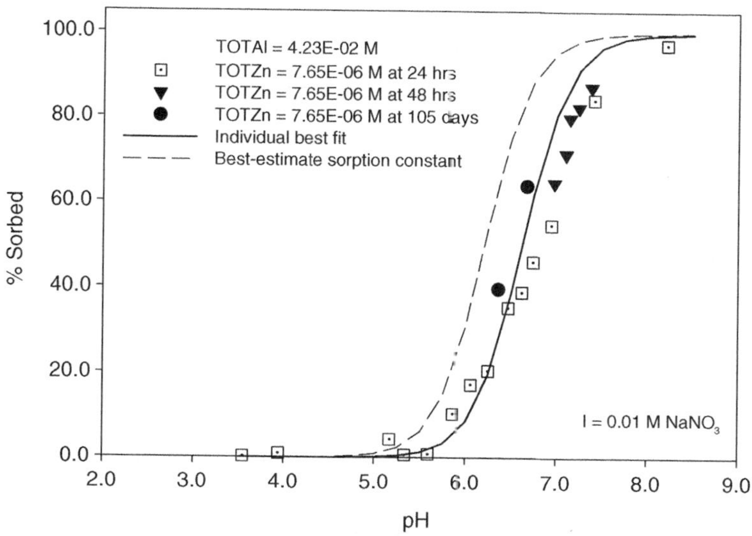

FIGURE 4EZN1

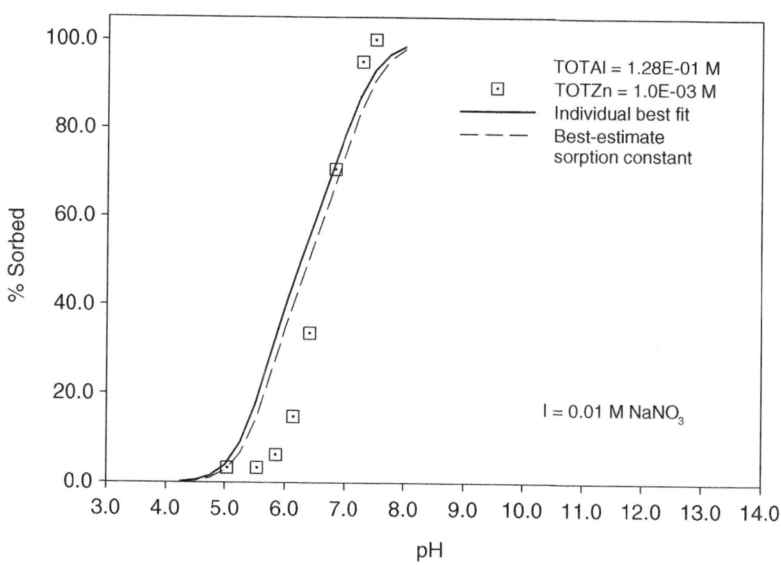

FIGURE 4EZN2

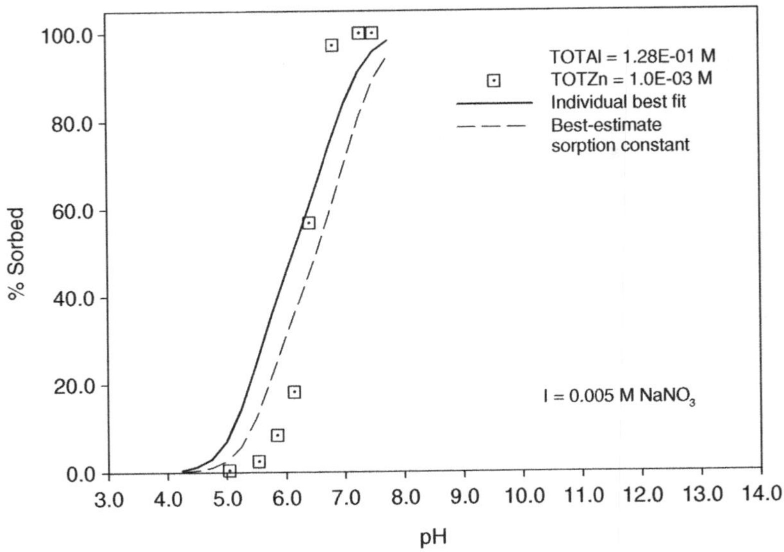

FIGURE 4EZN3

6.11 MERCURY

TABLE 6.41 Hg/gibbsite sorption data

ID	Source	Data Set
4IHG1	Weerasooriya et al. (2006, Fig. 2A)	$I = 0.01$ M $NaNO_3$; pH = 4.0
4IHG2	Weerasooriya et al. (2006, Fig. 2A)	$I = 0.01$ M $NaNO_3$; pH = 8.0
4IHG3	Weerasooriya et al. (2006, Fig. 2B)	$I = 0.01$ M NaCl; pH = 4.0
4IHG4	Weerasooriya et al. (2006, Fig. 2B)	$I = 0.01$ M NaCl; pH = 8.0
4IHG5	Weerasooriya et al. (2006, Fig. 2C)	$I = 0.01$ M $NaClO_4$; pH = 4.0
4IHG6	Weerasooriya et al. (2006, Fig. 2C)	$I = 0.01$ M $NaClO_4$; pH = 8.0
4EHG7	Weerasooriya et al. (2007a, Fig. 1A)	TOTHg = 1.00×10^{-7} M
4EHG8	Weerasooriya et al. (2007a, Fig. 1A)	TOTHg = 1.00×10^{-7} M
4EHG9	Weerasooriya et al. (2007a, Fig. 1A)	TOTHg = 1.00×10^{-7} M
4EHG10	Weerasooriya et al. (2007a, Fig. 1B)	TOTHg = 1.00×10^{-7} M
4EHG11	Weerasooriya et al. (2007a, Fig. 1B)	TOTHg = 1.00×10^{-7} M
4EHG12	Weerasooriya et al. (2007a, Fig. 1B)	TOTHg = 1.00×10^{-7} M
4EHG13	Sarkar et al. (1999, Fig. 1B)	TOTHg = 6.00×10^{-7} M
4EHG14	Sarkar et al. (1999, Fig. 1B)	TOTHg = 6.00×10^{-7} M
4EHG15	Sarkar et al. (1999, Fig. 1B)	TOTHg = 6.00×10^{-7} M

TABLE 6.42 Hg/gibbsite sorption data not used

Source	Data Set	Reason
Weerasooriya et al. (2006, Fig. 3A)	$I = 0.01$ M $NaClO_4$, pH = 4	Experiments conducted at high temperatures (40 and 50°C)
Weerasooriya et al. (2006, Fig. 3B)	$I = 0.01$ M $NaClO_4$, pH = 8	Experiments conducted at high temperatures (40 and 50°C)
Weerasooriya et al. (2006, Fig. 4A)	$I = 0.01$ M NaCl, pH = 4	Experiments conducted at high temperatures (40 and 50°C)
Weerasooriya et al. (2006, Fig. 4B)	$I = 0.01$ M NaCl, pH = 8	Experiments conducted at high temperatures (40 and 50°C)
Mishra et al. (2007b, Fig. 6)	TOTHg = 1.00×10^{-5} M	The $Al(OH)_3$ is in amorphous form. There is no direct evidence of crystalline gibbsite

TABLE 6.43 Hg/gibbsite surface complexation reactions

$\equiv AlOH^0 + Hg(OH)_2 + H^+ = \equiv AlOHg^+ + H_2O$	K_1
$\equiv AlOH^0 + Hg(OH)_2 + H^+ + Cl^- = \equiv AlOH\text{–}HgOHCl + H_2O$	K_2

Note: See Chapter 8, Table 8.1 for log K_1int values expressed in terms of reactions with Hg^{2+}.

TABLE 6.44 Hg/gibbsite intrinsic surface complexation constants

ID	$\log K_1^{int}$	$\sigma_{\log K}$	$\log K_2^{int}$	$\sigma_{\log K}$	WSOS/DF
4IHG1	12.1[a]	0.15[b]			—
4IHG2	7.6[a]	0.15[b]			—
4IHG3			19.7[a]	0.15[b]	—
4IHG4			12.5[a]	0.15[b]	—
4IHG5	10.6[a]	0.15[b]			—
4IHG6	7.3[a]	0.15[b]			—
4EHG7	9.06[a]	0.15[b]			—
4EHG8	9.38[a]	0.15[b]			—
4EHG9	9.75[a]	0.15[b]			—
4EHG10			14.06[c]	0.15[b]	—
4EHG11			14.06	0.028	78.5
4EHG12			14.14	0.032	4.8
4EHG13	9.12[a]	0.15[b]			—
4EHG14	9.05[a]	0.15[b]			—
4EHG15	9.42[a]	0.15[b]			—

[a] Constant was fixed manually.
[b] Fixed at this value by convention.
[c] Fixed based on the findings of 3EHG11.

TABLE 6.45 Best estimates for Hg/gibbsite surface complexation constants

Best-Estimate K	Confidence Level	$\log K^{int}(-)$	$\log K^{int}(+)$
$\log \overline{K_1^{int}} = 9.34$	0.99	9.24	9.43
	0.95	9.27	9.41
	0.90	9.28	9.39
	0.80	9.29	9.38
$\log \overline{K_2^{int}} = 14.40$	0.99	12.78	16.02
	0.95	13.17	15.63
	0.90	13.37	15.43
	0.80	13.60	15.21

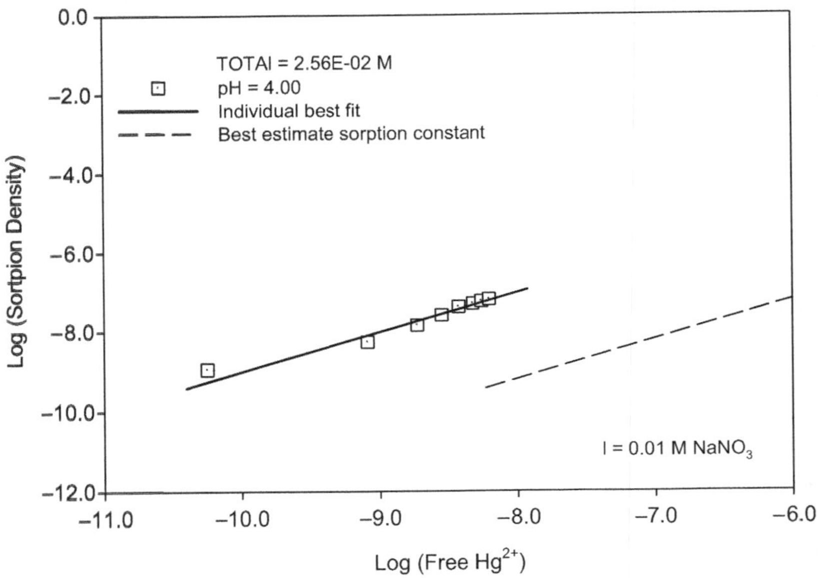

FIGURE 4IHG1

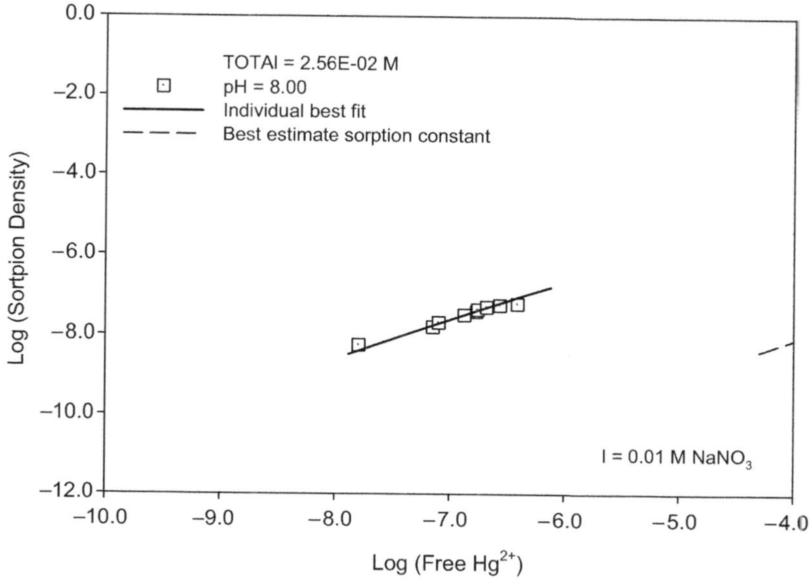

FIGURE 4IHG2

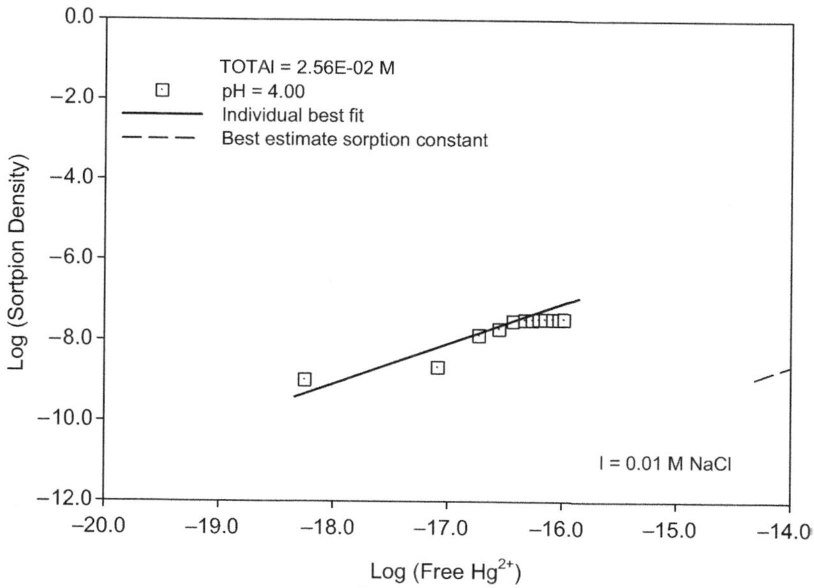

FIGURE 4IHG3

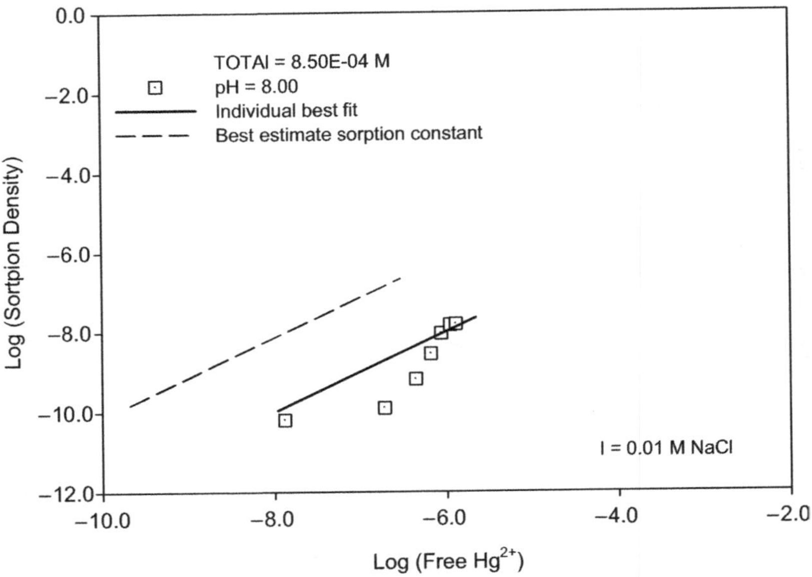

FIGURE 4IHG4

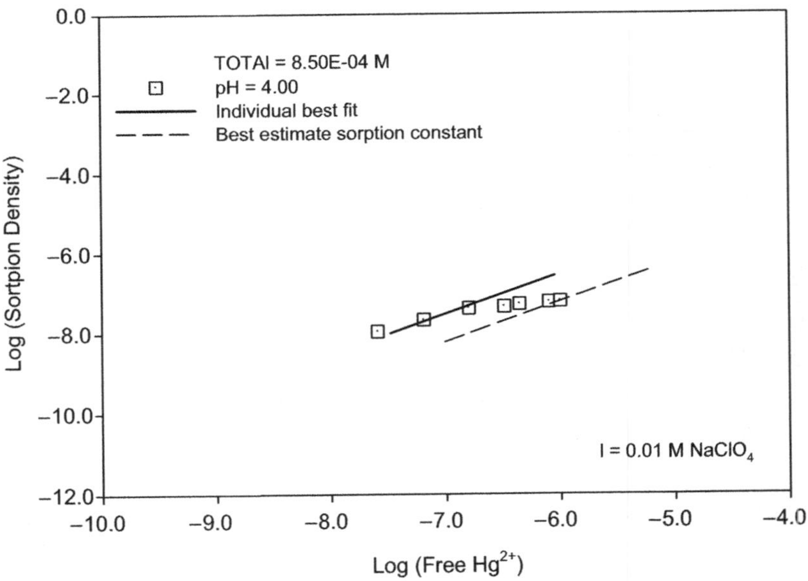

FIGURE 4IHG5

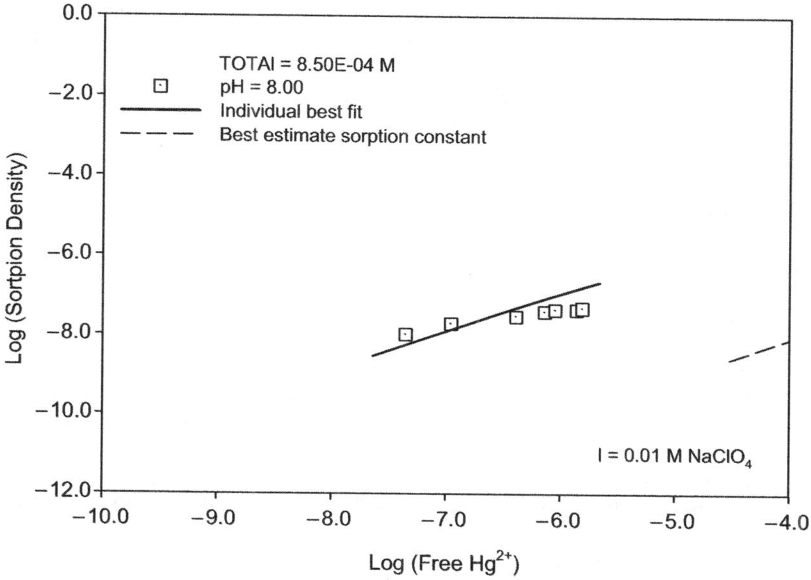

FIGURE 4IHG6

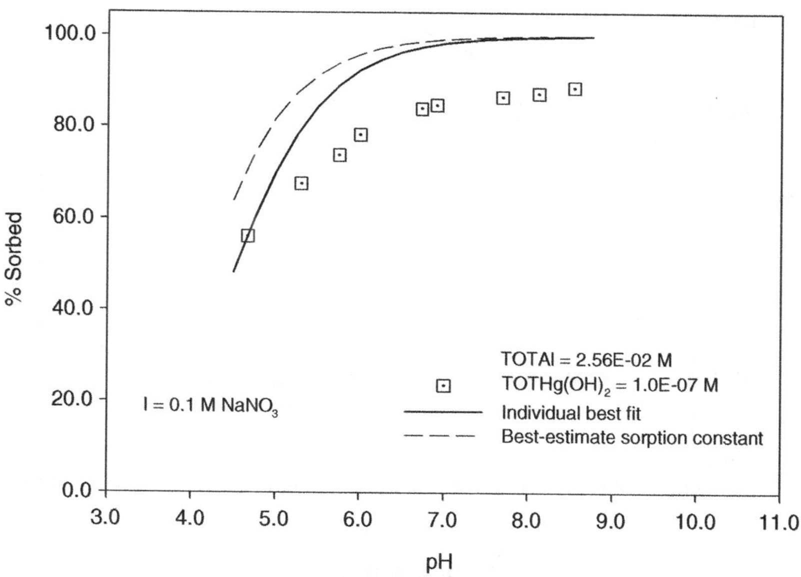

FIGURE 4EHG7

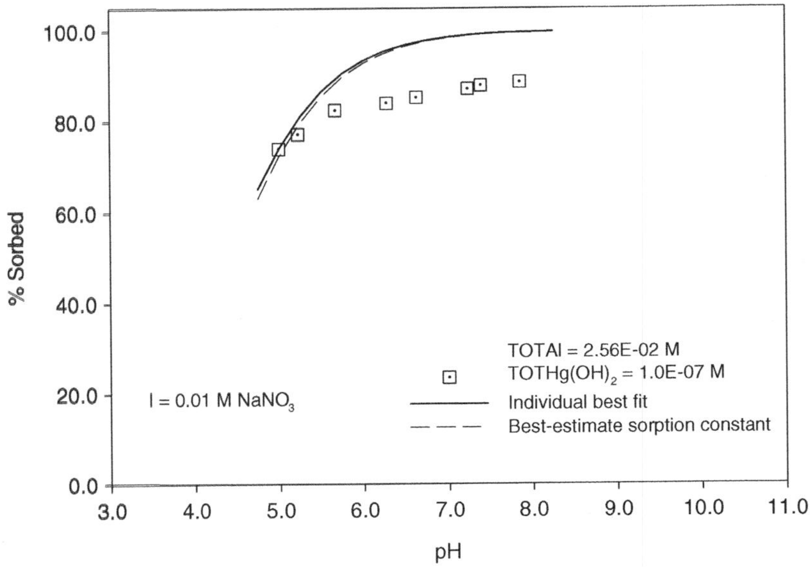

FIGURE 4EHG8

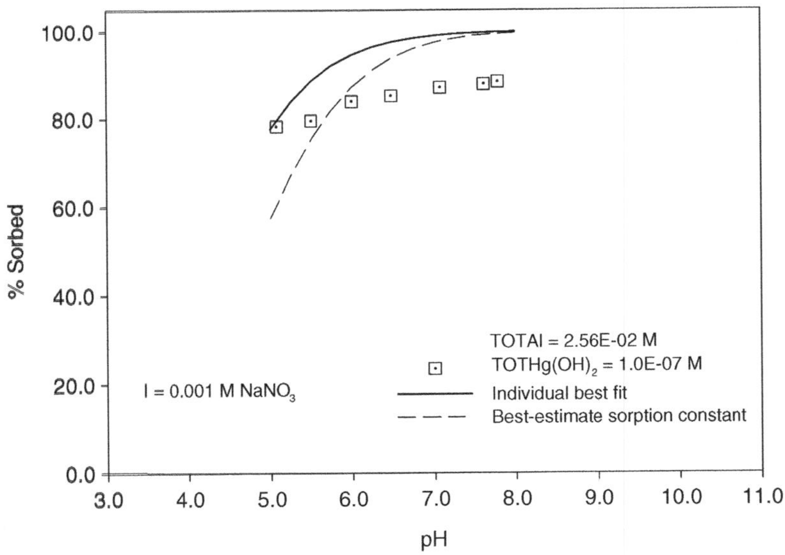

FIGURE 4EHG9

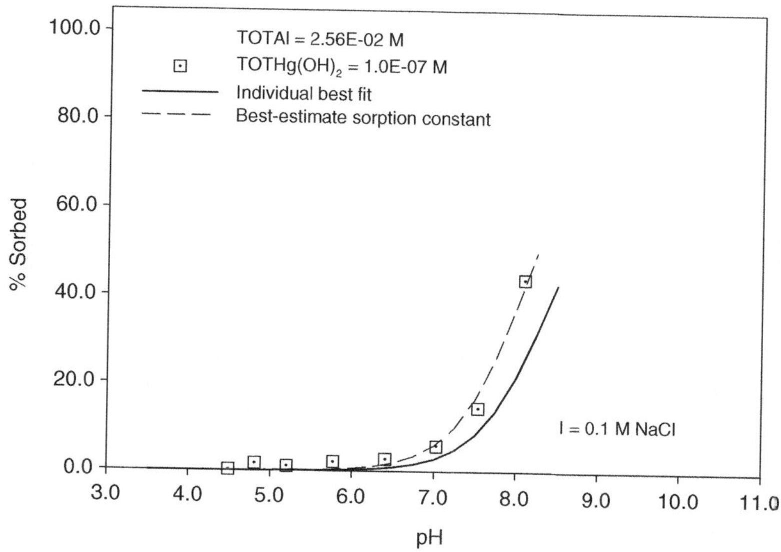

FIGURE 4EHG10

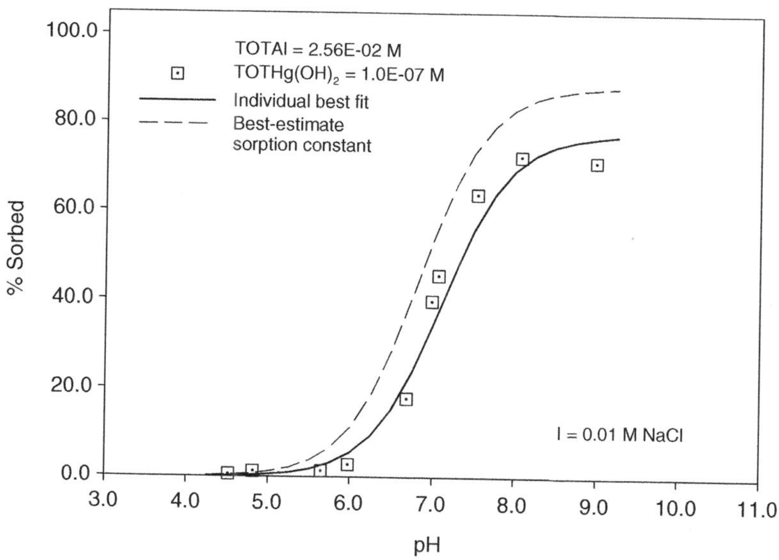

FIGURE 4EHG11

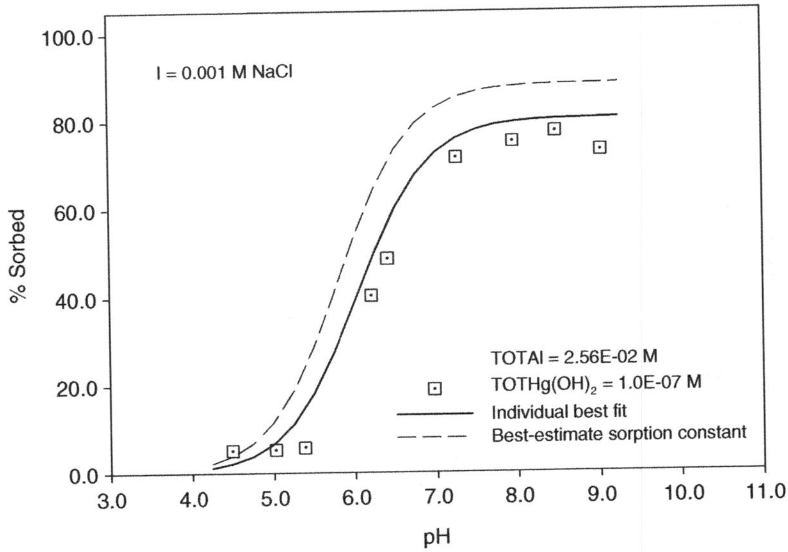

FIGURE 4EHG12

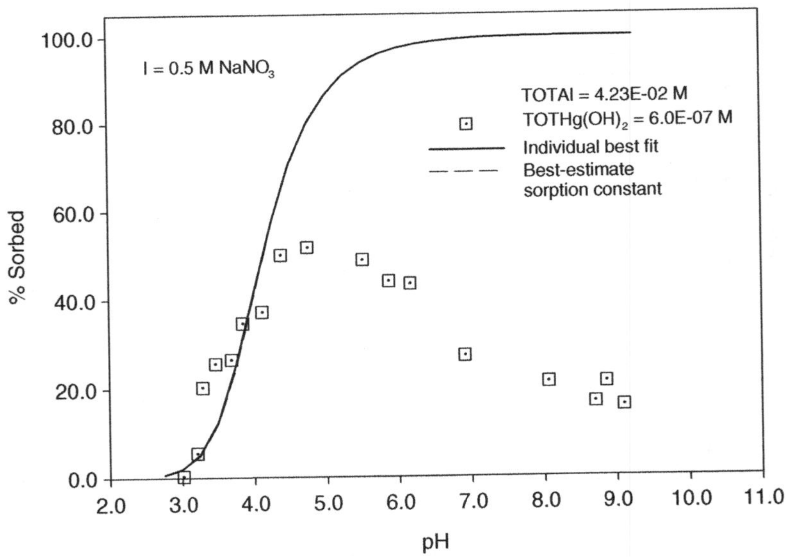

FIGURE 4EHG13

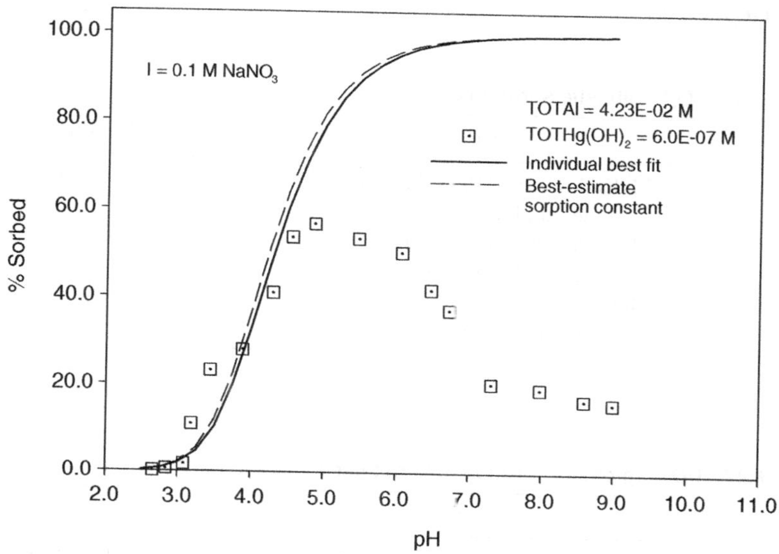

FIGURE 4EHG14

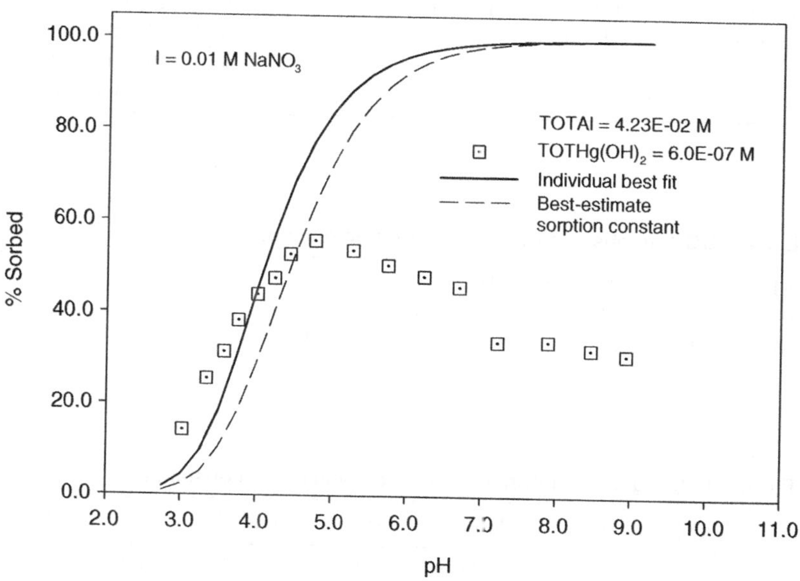

FIGURE 4EHG15

6.12 URANIUM

TABLE 6.46 UO_2^{2+}/gibbsite sorption data

ID	Source	Data Set
4EUO1	McKinley et al. (1995, Fig. 3); Turner et al. (1996, Fig. 5a)	$TOTUO_2^{2+} = 9.30 \times 10^{-6}$ M
4EUO2	McKinley et al. (1995, Fig. 3); Turner et al. (1996, Fig. 5a)	$TOTUO_2^{2+} = 9.10 \times 10^{-6}$ M
4EUO3	McKinley et al. (1995, Fig. 7); Turner et al. (1996, Fig. 5a)	$TOTUO_2^{2+} = 9.10 \times 10^{-6}$ M

TABLE 6.47 UO_2^{2+}/gibbsite sorption data not used

Source	Data Set	Reason
Zhang et al. (2005, Fig. 2)	$TOTUO_2^{2+} = 4.12 \times 10^{-5}$ M, $I = 0.1$ M $TOTUO_2^{2+} = 4.12 \times 10^{-5}$ M, $I = 0.01$ M $TOTUO_2^{2+} = 4.12 \times 10^{-5}$ M, $I = 0.001$ M	No attempts made to exclude CO_2 from the adsorption experiments. Uranyl forms complexes with carbonates
Baumann et al. (2005, Fig. 1)	$TOTUO_2^{2+} = 1.0 \times 10^{-5}$ M, $I = 0.1$ M	No attempts made to exclude CO_2 from the adsorption experiments. Uranyl forms complexes with carbonates

TABLE 6.48 UO_2^{2+}/gibbsite surface complexation reactions

$\equiv AlOH^0 + UO_2^{2+} = \equiv AlOUO_2^+ + H^+$	K_1
$\equiv AlOH^0 + UO_2^{2+} + 3H_2O = \equiv AlO(UO_2)(OH)_3 + 4H^+$	K_3
$\equiv AlOH^0 + 3UO_2^{2+} + 5H_2O = \equiv AlO(UO_2)_3(OH)_5 + 6H^+$	K_4

TABLE 6.49 UO_2^{2+}/gibbsite intrinsic surface complexation constants

ID	$\log K_1^{int}$	$\sigma_{\log K}$	$\log K_3^{int}$	$\sigma_{\log K}$	$\log K_4^{int}$	$\sigma_{\log K}$	WSOS/DF
4EUO1	1.07	0.046	−22.00[a]	0.15[b]	−15.75	0.072	1.1
4EUO2	1.19	0.087	−22.00[a]	0.15[b]	−17.43	0.089	0.8
4EUO3	1.76	0.154	−22.02[a]	0.15[b]	−13.34	0.103	1.0

[a] Constant was fixed manually.
[b] Fixed at this value by convention.

TABLE 6.50 Best estimates for UO_2^{2+}/gibbsite surface complexation constants

Best-Estimate K	Confidence Level	$\log K^{int}(-)$	$\log K^{int}(+)$
$\log \overline{K_1^{int}} = 1.22$	0.99	1.11	1.33
	0.95	1.13	1.30
	0.90	1.15	1.29
	0.80	1.16	1.27
$\log \overline{K_3^{int}} = -22.0$	0.99	−22.01	−22.00
	0.95	−22.01	−22.00
	0.90	−22.01	−22.00
	0.80	−22.01	−22.00
$\log \overline{K_4^{int}} = -15.62$	0.99	−16.54	−14.70
	0.95	−16.32	−14.92
	0.90	−16.20	−15.03
	0.80	−16.08	−15.16

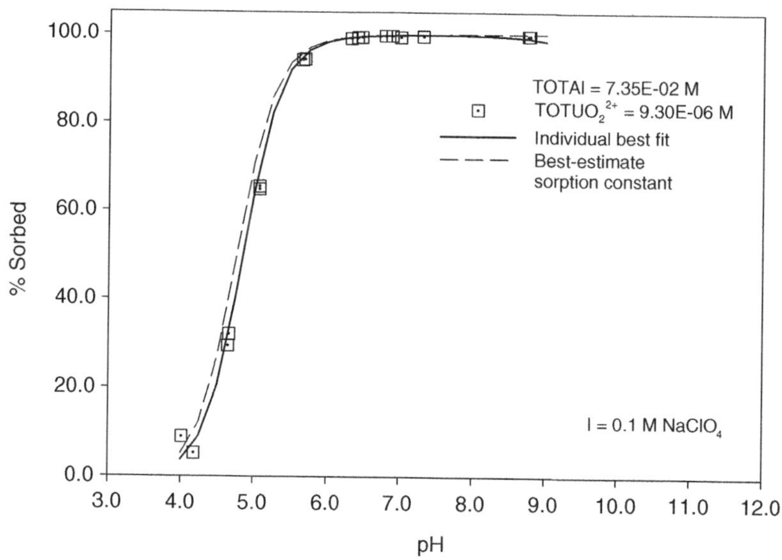

FIGURE 4EUO1

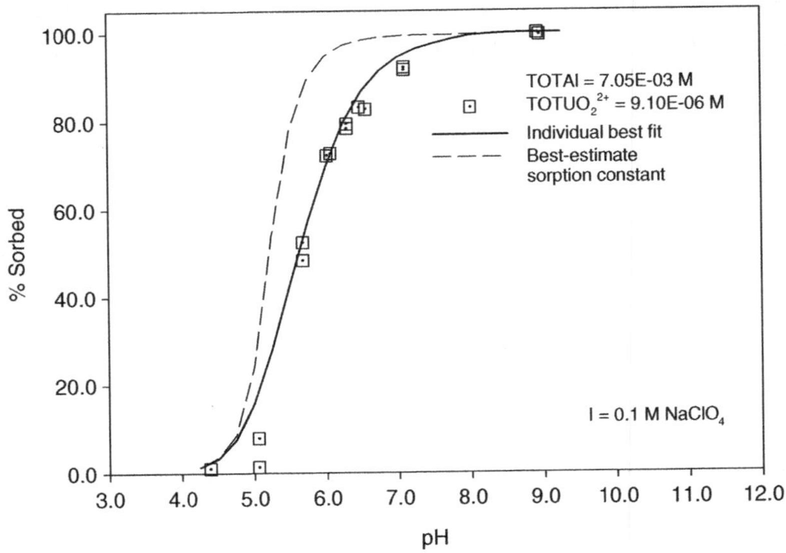

FIGURE 4EUO2

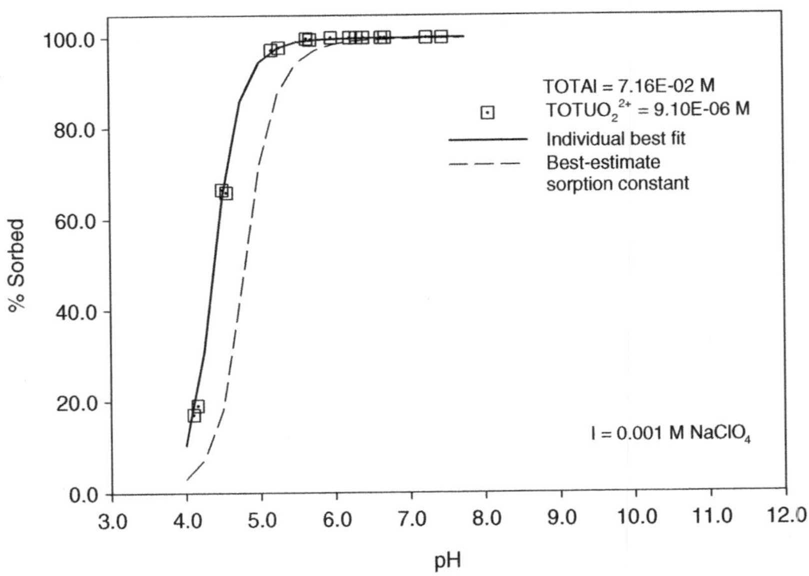

FIGURE 4EUO3

6.13 THORIUM

TABLE 6.51 Th/gibbsite sorption data

ID	Source	Data Set
4ETH1	Zhang et al. (2006, Fig. 2)	$TOTTh = 5.84 \times 10^{-5}$ M
4ETH2	Zhang et al. (2006, Fig. 2)	$TOTTh = 5.84 \times 10^{-5}$ M
4ETH3	Zhang et al. (2006, Fig. 2)	$TOTTh = 5.84 \times 10^{-5}$ M

TABLE 6.52 Th/gibbsite sorption data not used

Source	Data Set	Reason
Zhang et al. (2006, Fig. 4)	$I = 0.1$ M KNO_3, $3.9 < pH < 4.88$, Gibbsite conc. 10 g/L	pH changes almost by 1 during the isotherm experiment
	$I = 0.1$ M KNO_3, $2.91 < pH < 3.29$, Gibbsite conc. 10 g/L	The sorption isotherm was conducted in presence of fulvic acid
Zhang et al. (2006, Fig. 8)	$TOTTh = 5.84 \times 10^{-5}$ M	Initial concentration of phosphate 5.84×10^{-5} M

TABLE 6.53 Th/gibbsite surface complexation reactions

$\equiv AlOH^0 + Th^{4+} = \equiv AlOTh^{3+} + H^+$	K_1
$\equiv AlOH^0 + Th^{4+} + H_2O = \equiv AlOTh(OH)^{2+} + 2H^+$	K_2

TABLE 6.54 Th/gibbsite intrinsic surface complexation constants

ID	$\log K_1^{int}$	$\sigma_{\log K}$	$\log K_2^{int}$	$\sigma_{\log K}$	WSOS/DF
4ETH1	10.25[a]	0.15[b]	3.00[a]	0.15[b]	—
4ETH2	10.65[a]	0.15[b]	3.20[a]	0.15[b]	—
4ETH3	11.60[a]	0.15[b]	4.00[a]	0.15[b]	—

[a] Constant was fixed manually.
[b] Fixed in this value by convention.

TABLE 6.55 Best estimates for Th/gibbsite surface complexation constants

Best-Estimate K	Confidence Level	$\log K^{int}(-)$	$\log K^{int}(+)$
$\log \overline{K_1^{int}} = 10.83$	0.99	10.67	10.99
	0.95	10.71	10.95
	0.90	10.73	10.93
	0.80	10.75	10.91
$\log \overline{K_2^{int}} = 3.40$	0.99	3.28	3.52
	0.95	3.31	3.49
	0.90	3.32	3.48
	0.80	3.34	3.46

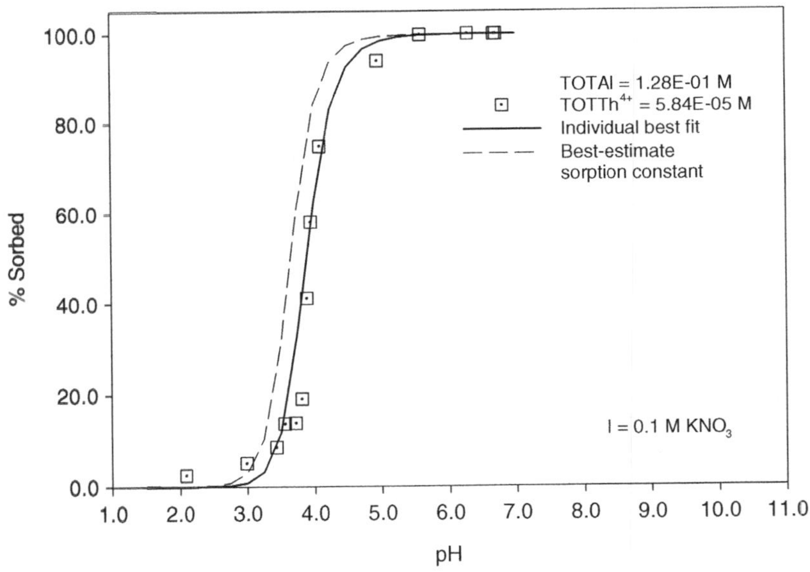

FIGURE 4ETH1

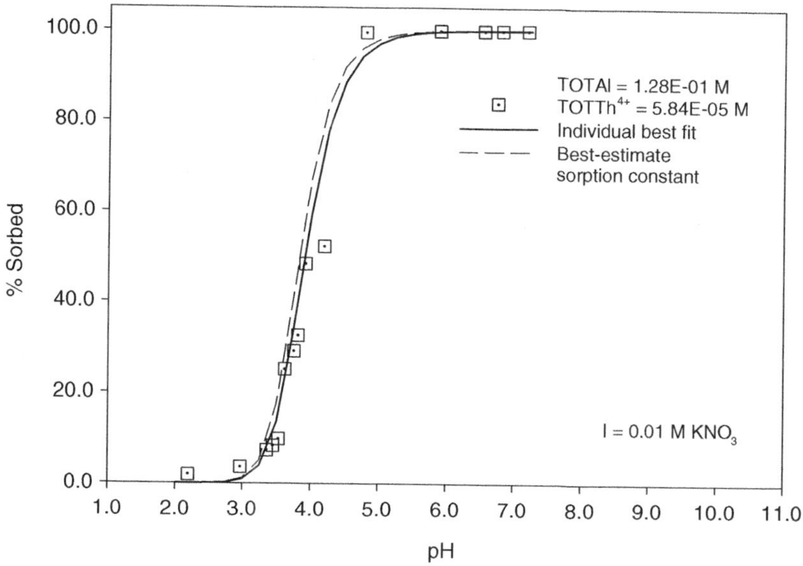

FIGURE 4ETH2

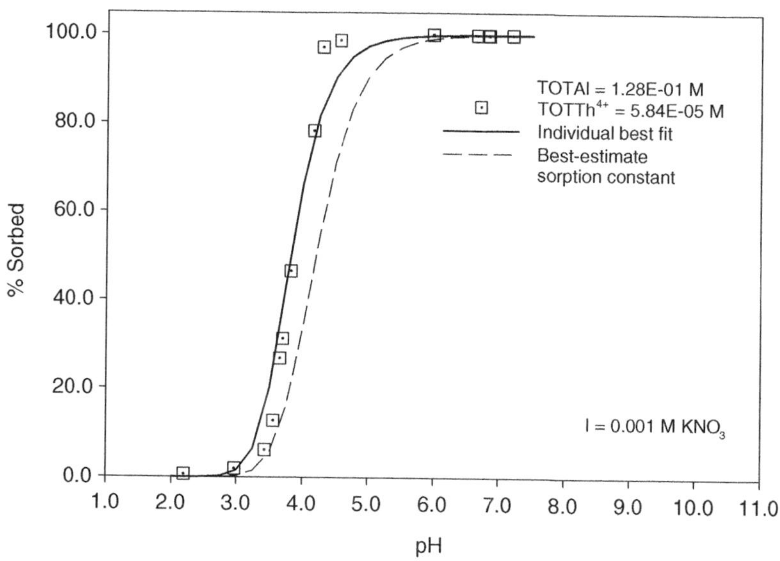

FIGURE 4ETH3

7

ANION SORPTION ON GIBBSITE

In this chapter, the sorption reactions and equilibrium constants obtained from fitting of available anion-sorption data for gibbsite with the generalized two-layer model are presented. The chapter has separate sections for each of the anions investigated, and each of these sections contains the following: (1) the sources of the sorption data used for parameter extraction, (2) the data sets not used, along with reason(s) for exclusion, (3) the surface-complexation reactions used in fitting the data, (4) the optimal surface-complexation constants for each data set, (5) the best estimates (and associated confidence intervals) for the surface-complexation constants, and (6) a plot of each data set along with the optimal fit (solid line) and the fit corresponding to the best estimate of surface-complexation constants (dashed line). In the separate sections for the various anions studies, every table and figure has been assigned a specific identification code as explained previously in Chapter 4 (Table 4.3).

7.1 MODELING METHODOLOGY AND REACTIONS

In all modeling conducted, a consistent set of aqueous-phase reactions and associated equilibrium constants was used. Table 7.1 presents the solution reactions considered in fitting the anion-sorption data sets.

The methodology used for parameter extraction was described in Chapter 4. The nonlinear regression program FITEQL (Herbelin and Westall, 1999) was used to calculate the optimal surface-complexation constants from the individual data sets. Various surface reactions and combinations of surface reactions were tried for each

Surface Complexation Modeling: Gibbsite By Athanasios K. Karamalidis and David A. Dzombak

TABLE 7.1 Solution reactions and mass law constants used in fitting anion sorption data for gibbsite

Reaction	Log K^a
$H^+ + OH^- = H_2O$	14.00
$H^+ + PO_4^{3-} = HPO_4^{2-}$	12.37
$2H^+ + PO_4^{3-} = H_2PO_4^-$	19.57
$3H^+ + PO_4^{3-} = H_3PO_4^0$	21.72
$H^+ + AsO_4^{3-} = HAsO_4^{2-}$	11.80
$2H^+ + AsO_4^{3-} = H_2AsO_4^-$	18.79
$3H^+ + AsO_4^{3-} = H_3AsO_4^0$	21.09
$H_3AsO_3^0 = H_2AsO_3^- + H^+$	−9.17
$H_3AsO_3^0 = HAsO_3^{2-} + 2H^+$	−23.27
$H_3BO_3^0 = H_2BO_3^- + H^+$	−9.236
$H^+ + SO_4^{2-} = HSO_4^-$	1.99
$H^+ + SeO_4^{2-} = HSeO_4^-$	1.70
$H^+ + MoO_4^{2-} = HMoO_4^-$	4.24
$2H^+ + MoO_4^{2-} + 2H_2O = MoO_3(H_2O)_3^0$	8.24
$H^+ + CrO_4^{2-} = HCrO_4^-$	6.51
$2H^+ + CrO_4^{2-} = H_2CrO_4^0$	6.31
$H^+ + F^- = HF^0$	3.18
$H_2SiO_4^{2-} + 2H^+ = H_4SiO_4^0$	23.04
$H_2SiO_4^{2-} + H^+ = H_3SiO_4^-$	13.2
$Al^{3+} + F^- = AlF^{2+}$	7.01
$Al^{3+} + 2F^- = AlF_2^+$	12.63
$Al^{3+} + 3F^- = AlF_{3\,aq}^0$	16.7
$Al^{3+} + 4F^- = AlF_4^-$	19.4

Source: Smith, Martell, and Motekaitis (NIST 46.6, 2001; NIST 46.7, 2003). Nordstrom and Archer (2002), for arsenate and arsenite. Cruywagen (2000), for molybdate.

[a] Mass law constants correspond to $T = 20$–$25°C$ and $I = 0$, and where necessary have been corrected to $I = 0$ using the activity coefficients in Table 4.1.

anion, with the simplest reactions and the smallest number of reactions able to fit the data selected for use. When insufficient data were available in a particular data set to extract a surface-complexation constant for a particular reaction with FITEQL, or when convergence problems were experienced with FITEQL for other reasons, manual adjustment of surface-complexation constants was performed using Visual MINTEQ (Gustafsson, 2009).

Best-estimate surface-complexation constants were determined from the optimal constants fitted to the various data sets for a particular anion. These were assessed by means of weighted average of log K values in which $1/\sigma_{\log K}$ was used as the weighting factor. The standard deviation $\sigma_{\log K}$ reflects the quantity and variability of data in a particular data set. Surface-complexation constants determined by manual fitting were assigned a $\sigma_{\log K}$ value of 0.15.

All surface-complexation constants and $\sigma_{\log K}$ values are presented in the following sections for individual anions. When a particular surface species was not used to fit individual data sets is indicated in tables of intrinsic surface-complexation constants with a blank space.

In this study, a single site type was used to fit all the available anion-sorption data sets for gibbsite. Surface-complexation modeling of anion sorption on hydrous ferric oxide and goethite required consideration of one site type with the generalized two-layer model as well (Dzombak and Morel, 1990; Mathur and Dzombak, 2006). For modeling A^{3-} anions (PO_4^{3-}, AsO_4^{3-}, AsO_3^{3-}) sorption on gibbsite, the surface species $\equiv AlH_2A^0$ and $\equiv AlHA^-$ were used as standard and one more surface species was added to fit AsO_4^{3-} ($\equiv AlOHA^{3-}$) and PO_4^{3-} ($\equiv AlA^{2-}$) sorption data sets, respectively. In the case of A^{2-} anions (MoO_4^{2-}, SeO_4^{2-}, CrO_4^{2-}) the following surface species were considered: $\equiv AlHA^0$, $\equiv AlA^-$, and $\equiv AlOHA^{2-}$. In the case of SO_4^{2-}, two surface species were used: $\equiv AlHA^-$ and $\equiv AlA^{2-}$. For F^- sorption on gibbsite, three surface species were needed, that is $\equiv AlOH_2F^0$, $\equiv AlOHF^-$, and $\equiv AlF_2^-$. For $H_3BO_3^0$ and $H_3AsO_3^0$ only two surface species, $\equiv AlH_2A^0$ and $\equiv AlH_3A^-$ were used; for $H_4SiO_4^0$, only one surface species, $\equiv AlOH_2SiO_4^-$, was needed.

Surface-complexation of anions by gibbsite involves ligand-exchange reactions in which surface hydroxyl groups are replaced by the sorbing anions. An example of A^{2-} anion sorption is described with the reactions below:

$$\equiv AlOH^0 + A^{2-} + H^+ = \equiv AlA^- + H_2O \qquad K_{1A}^{app} \tag{7.1}$$

and/or

$$\equiv AlOH^0 + A^{2-} = \equiv AlOHA^{2-} \qquad K_{2A}^{app} \tag{7.2}$$

where A^{2-} represents a hypothetical divalent anion. In contrast to the need for one surface species to describe cation sorption on gibbsite, anion sorption usually requires two or three different surface species.

As in the case of protons, anions sorb onto oxide surfaces in response to chemical and coulombic forces. Therefore, a coulombic term is included in the mass law expressions for anion-sorption reactions, in order to describe the long-range electrostatic effects related to the surface charge.

$$K_A^{int} = K_A^{app}\exp(-\Delta ZF\Psi/RT) \tag{7.3}$$

where K_A^{int} is the intrinsic equilibrium constant and K_A^{app} the apparent equilibrium constant of a hypothetical anion A^{n-}, ΔZ is the net change in the charge number of the surface species, Ψ the surface potential, R the gas constant, T the absolute temperature, and F the Faraday constant.

The net charge on an oxide surface is changed if an anion is sorbed. For example, if an anion A^{2-} sorbs to an oxide and forms the charged surface complexes $\equiv AlA^-$ and $\equiv AlOHA^{2-}$, the net surface charge density is expressed by the following equation:

$$\sigma = (F/AS)[(\equiv AlOH_2^+)-(\equiv AlO^-)-(\equiv AlA^-)-2(\equiv AlOHA^{2-})] \tag{7.4}$$

A complete set of equations for the surface complexation of A^{2-} on gibbsite is given in Table 7.2 and the corresponding tableau in Table 7.3. Note that all proton-binding sites

TABLE 7.2 Anion surface complexation on gibbsite

Species

H^+, OH^-, A^{2-}, HA^-, H_2A^0, $\equiv AlOH^0$, $\equiv AlOH_2^+$, $\equiv AlO^-$, $\equiv AlA^-$, $\equiv AlOHA^{2-}$

Mass Law Equations

(OH^-)	$= (H^+)^{-1}$				$\gamma_1^{-2} \times 10^{-14}$
(H_2A^0)	$= (H^+)^2$	(A^{2-})			$\gamma_2\gamma_1^2 K_1$
(HA^-)	$= (H^+)$	(A^{2-})			$\gamma_2 K_2$
$(\equiv AlOH_2^+)$	$= (H^+)$		$(\equiv AlOH^0)$	$\exp(-F\Psi/RT)$	$\gamma_1 (K_{a1}^{int})^{-1}$
$(\equiv AlO^-)$	$= (H^+)^{-1}$		$(\equiv AlOH^0)$	$\exp(F\Psi/RT)$	$\gamma_1^{-1} K_{a2}^{int}$
$(\equiv AlA^-)$	$= (H^+)$	(A^{2-})	$(\equiv AlOH^0)$	$\exp(F\Psi/RT)$	$\gamma_1\gamma_2 K_{1A}^{int}$
$(\equiv AlOHA^{2-})$	$=$	(A^{2-})	$(\equiv AlOH^0)$	$\exp(2F\Psi/RT)$	$\gamma_1 K_{2A}^{int}$

Mole Balance Equations

TOTH	$=(H^+) - (OH^-) + (HA^-) + 2(H_2A^0) + (\equiv AlOH_2^+) - (\equiv AlO^-) + (\equiv AlA^-)$
TOTA	$=(A^{2-}) + (H_2A) + (HA^-) + (\equiv AlA^-) + (\equiv AlOHA^{2-})$
TOT(≡AlOH)	$=(\equiv AlOH_2^+) + (\equiv AlOH) + (\equiv AlO^-) + (\equiv AlA^-) + (\equiv AlOHA^{2-})$

Surface Charge

$$\sigma = (F/AS)TOTP = (F/AS)\ [(\equiv AlOH_2^+) - (\equiv AlO^-) - (\equiv AlA^-) - 2(\equiv AlOHA^{2-})]$$

Charge-Potential Relationship

$$\sigma = 0.1174c^{1/2}\sinh(Z\Psi \times 19.46)$$

TABLE 7.3 Tableau for anion surface complexation of A^{2-} anion on gibbsite

	H^+	A^{2-}	$\equiv AlOH^0$	$P = \exp(-F\Psi/RT)$
H^+	1			
OH^-	−1			
A^{2-}		1		
HA^-	1	1		
H_2A^0	2	1		
$\equiv AlOH_2^+$	1		1	1
$\equiv AlOH^0$			1	
$\equiv AlO^-$	−1		1	−1
$\equiv AlA^-$	1	1	1	−1
$\equiv AlOHA^{2-}$		1	1	−2
	TOTH	TOTA	TOT(≡AlOH)	TOTP $= \sigma(AS/F)$

are considered available for anion binding and that all sites are considered to be uniform with respect to anion binding. The surface-complexation reactions considered for the various anions examined in this study can be found at their respective sections in this chapter.

7.2 AVAILABLE SPECTROSCOPIC DATA AND USE IN MODELING

The potential to constrain the modeling of anion sorption on gibbsite using available spectroscopic data was investigated. A search of the literature revealed that some work has been done in the identification of anion surface complexes on the gibbsite surface, but spectroscopic data are limited.

Establishment of surface species in this study was inferred by available spectroscopic evidence only in a limited manner because of lack of sufficient spectroscopic data for specific anions and the uncertainty associated with the interpretation of the existing data. In this study, all anions were treated as monodentate mononuclear surface complexes with gibbsite. Different surface species were used in the case of polynuclear anions. Possible modes of ligand coordination to the gibbsite surface are shown in Figure 6.1. This section presents an overview of available spectroscopic data for the various anions studied, and discusses how these data were considered in the surface-complexation modeling for the particular anions.

7.2.1 Phosphate

Spectroscopic evidence of phosphate surface species on gibbsite is very limited. ^{31}P solid-state nuclear magnetic resonance (NMR) studies on aluminum oxides suggest that inner and outer-sphere Al–phosphate complexes and surface precipitates may form (Johnson et al., 2002). ^{31}P solid-state NMR spectra of phosphate sorption on gibbsite indicated inner-sphere complexation except at high phosphate concentration and low pH. As the pH increased, the fraction of phosphate sorbed as an inner-sphere complex increased, becoming dominant at pH >8. The Johnson et al. (2002) study suggested that monodentate complexes most possibly are formed, and that is inner-sphere complexes are bound to the singly coordinated Al–OH sites on the edges of gibbsite. Outer-sphere complexation has been found to contribute to phosphate sorption above pH 8 (van Emmerik et al., 2007). On gibbsite, phosphate can sorb in excess of the maximum monolayer sorption capacity determined by the number of singly coordinated hydroxyl groups (van Riemsdijk and Lyklema, 1980), indicative of surface precipitation.

The two-layer model used in this study does not distinguish between inner and outer-sphere surface complexes. Both types of complexes are accommodated through the surface reactions incorporated in the model. In the surface reactions for phosphate sorption on gibbsite, the simplest monodentate surface complexes capable of fitting the available sorption data were used.

7.2.2 Arsenate

Interpretation of extended X-ray absorption fine structure (EXAFS) data for arsenate sorption on gibbsite by Ladeira et al. (2001) showed that arsenate (AsO_4^{3-}) formed an inner-sphere complex on gibbsite. They found that the As(V)–Al distance of 3.19 Å is characteristic of a corner-sharing octahedral-tetrahedra arrangement resulting from a bidentate-binuclear complex (Ladeira et al., 2001). However Weerasoriya et al. (2004) noted that the findings of Ladeira et al. were confined to an experimental system at pH 5.5. Weerasooriya et al. (2004) examined possible As(V)–gibbsite surface species at pH 4.0, 7.1, and 8.5 by applying Fourier transform infrared (FTIR) spectroscopy. FTIR spectra showed that there was strong evidence of As(V)-surface direct bonding, but based solely on IR data it was difficult to assess how many As atoms were attached to the gibbsite surface. Using a molecular model calculation, they concluded that bidentate-binuclear surface species might be the most stable form of the As(V) complex with gibbsite (Weerasoriya et al., 2004). These calculations were consistent with density functional theory calculations for arsenate with aluminum oxide clusters as were done by Ladeira et al. (2001), which indicated that bidentate-binuclear species were more stable than bidentate-mononuclear, monodentate-mononuclear, or monodentate-binuclear surface species. EXAFS data for arsenate sorption on gibbsite and goethite in the presence of Cu(II) and Zn(II) revealed that arsenate formed bidentate mononuclear surface complexes in both cases (Gräfe et al., 2008). EXAFS and X-ray Absorption Near Edge Structure (XANES) studies of arsenate sorption on β-$Al(OH)_3$ at ionic strengths of 0.1 and 0.8 M and at various pH values (4, 8, and 10) indicated the formation of bidentate-binuclear surface species (Arai et al., 2001).

Interpretations of available spectroscopic data for arsenate surface complexes on gibbsite show a trend for justification of bidentate forms, but other forms are possible. The most frequently reported arsenate surface species coordination geometries using X-ray spectroscopy are the bidentate-binuclear, bidentate-mononuclear, and the monodentate-mononuclear surface complexes (Fukushi and Sverjensky, 2007).

In this study, fitting of arsenate sorption on gibbsite was done using monodentate-mononuclear surface complexes. Attempts were made to describe arsenate sorption on gibbsite using bidentate ligands but this did not improve the fitting of available sorption data.

7.2.3 Arsenite

Arsenite (AsO_3^{3-}) is a common arsenic species, along with arsenate, when As is present in aquatic systems. The stable solution species under moderately-reducing environmental conditions are H_3AsO_3, $H_2AsO_3^-$, and $HAsO_3^{2-}$. Spectroscopic data for arsenite surface species on gibbsite are scarce. Modeling calculations of proton exchange showed that As(III) is bound weakly to surface sites of gibbsite (Weerasoriya et al., 2003), implying the formation of outer-sphere complexes. Weerasoriya et al. (2003) proposed that arsenite forms bidentate-binuclear surface complexes with edge sites of gibbsite. An EXAFS and XANES study on arsenite sorption at the aluminum oxide–water interface by Arai et al. (2001) indicated that a mixture of inner- and outer-sphere As(III) complexes exist at a pH of >5.5. Outer-sphere As(III) complexes were suggested to be significant when ionic strength decreases. At a pH of ≤5.5, As(III) predominantly forms inner-sphere bidentate binuclear complexes (Arai et al., 2001).

Ligands such as AsO_3^{2-} can sorb in a bidentate or monodentate manner with a metal Al center since ≡AlOH contains oxo groups (Oliveira et al., 2006). Based on DFT calculations, in the case of As(III)–Al(III) complex the bidentate form is theromodynamically favored (Oliveira et al., 2006).

The limited spectroscopic data for arsenite sorption on gibbsite led us to use only the monodentate-mononuclear species of $\equiv AlH_2AsO_3^0$, which proved adequate to fit all available arsenite sorption data sets.

7.2.4 Molybdate

No spectroscopic data for molybdate, MoO_4^{2-}, sorption on gibbsite are available. Molybdate is a tetrahedral oxyanion, similar to phosphate and arsenate which sorbs on gibbsite and is known to compete with phosphate and arsenate for sorption sites on soil mineral surfaces (Manning and Goldberg, 1996). In this study, we used three monodentate-mononuclear species of $\equiv AlHMoO_4^0$, $\equiv AlMoO_4^-$, $\equiv AlOHMoO_4^{2-}$ which proved an adequate fit for all available surface-complexation data sets.

7.2.5 Selenate

It has been proposed that selenate (SeO_4^{2-}) and selenite (SeO_3^{2-}) sorb on hydroxylated alumina via a ligand-exchange mechanism involving the formation of mainly monodentate-mononuclear surface complexes (Rajan, 1979). The proposed mechanism has not yet been verified by direct spectroscopic evidence, however. In this study, we considered the surface-complexation species of $\equiv AlOHSeO_3^-$ to describe selenate sorption on gibbsite.

7.2.6 Chromate

No spectroscopic data for chromate CrO_4^{2-} sorption on gibbsite are available. Modeling efforts for CrO_4^{2-} sorption on hydrous oxides of Fe and Al as well as in soils suggest that Cr(VI) forms an outer-sphere complex on these surfaces. Fendorf et al. (1997) proposed that on the goethite surface CrO_4^{2-} forms monodentate, bidentate-binuclear, and bidentate-mononuclear complexes. In this study, we considered only two different monodentate surface species of chromate: $\equiv AlHCrO_4^0$ and $\equiv AlCrO_4^-$.

7.2.7 Borate

No spectroscopic data for borate, $H_3BO_3^0$, sorption on gibbsite are available. A ligand-exchange mechanism involving inner- and outer-sphere surface complexes has been proposed for B sorption on aluminum oxide minerals (Goldberg et al., 1993a; Goldberg, 2005). This study considered that borate is directly bound to the Al sites of the gibbsite surface by displacement of the reactive surface hydroxyl groups. The borate surface species, $\equiv AlH_2BO_3$ and $\equiv AlH_3BO_4^-$, enabled good agreement with the experimental data sets.

7.2.8 Sulfate

No spectroscopic data for sulfate (SO_4^{2-}) sorption on gibbsite are available. It is believed that sulfate forms outer-sphere complexes on aluminum oxides, with at least one water molecule between the sorbing ion and the surface functional group (Sposito, 1984; He et al., 1997). However, experimental evidence from studies of equilibrium sorption of sulfate on gibbsite, and generally on Al-oxides, is inconclusive (Paul et al., 2007 and the references therein). Quantum chemical calculations for the estimation of the relative Gibbs free energies for sulfate sorption on variably charged Al-(hydr)oxides suggested that the most thermodynamically favorable surface-complexation species occurring on +2 charged sites are bidentate bridging complexes, and on +1 charged sites are monodentate or bidentate complexes (Paul et al., 2007). In this study, we used two monodentate sulfate surface species: $\equiv AlSO_4^-$ and $\equiv AlOHSO_4^{2-}$.

7.2.9 Fluoride

No spectroscopic data for fluoride sorption on gibbsite are available. In this study, three fluoride surface species were required to fit available sorption data: $\equiv AlF^0$, $\equiv AlOHF^-$, and $\equiv AlF_2^-$.

7.2.10 Silicate

No spectroscopic data for silicate sorption on gibbsite are available. Under environmental conditions, fully protonated silicic acid, $H_4SiO_4^0(aq)$, is the dominant form of Si in solution. Although interactions of silicic acid with Al and Fe (hydr) oxides have been reported (Pokrovski et al., 1996; Doucet et al., 2001; Davis et al., 2002; Hiemstra et al., 2007; Houston et al., 2008), there is no evidence for the formation of surface species with gibbsite. X-ray photoelectron spectroscopy measurements of $H_4SiO_4^0$ interaction with gibbsite implied weak sorption (Alvarez et al., 1976b). On the surface of Fe (hydr)oxides, including goethite (Hiemstra et al., 2007) and ferrihydrite (Hansen et al., 1994b, 1994a), $H_4SiO_4^0$ appears to form inner-sphere complexes by exchanging ligands with surface groups of the solid (Pokrovski et al., 2003). In this study, one silicate surface species was sufficient to fit available sorption data: $\equiv AlOH_4SiO_4^-$.

7.3 PHOSPHATE

TABLE 7.4 PO_4^{3-}/gibbsite sorption data

ID	Source	Data Set
4EPO1	van Emmerik et al. (2007, Fig. 1)	$TOTPO_4^{3-} = 1.00 \times 10^{-3}$ M
4EPO2	van Emmerik et al. (2007, Fig. 1)	$TOTPO_4^{3-} = 1.00 \times 10^{-3}$ M
4EPO3	Hingston et al. (1971, Fig. 3b)	$TOTPO_4^{3-} = 1.30 \times 10^{-4}$ M
4EPO4	Hingston et al. (1971, Fig. 3b)	$TOTPO_4^{3-} = 2.60 \times 10^{-4}$ M
4EPO5	Hingston et al. (1971, Fig. 3b)	$TOTPO_4^{3-} = 6.50 \times 10^{-4}$ M
4EPO6	Manning and Goldberg (1996, Fig. 4b)	$TOTPO_4^{3-} = 1.33 \times 10^{-4}$ M
4IPO7	Parfitt et al. (1977, Fig. 1A)	$I = 0.1$ M, pH 4
4IPO8	Parfitt et al. (1977, Fig. 1A)	$I = 0.1$ M, pH 5.7
4EPO9	Okazaki et al. (1989, Fig. 4 Gi-3)	$TOTPO_4^{3-} = 3.30 \times 10^{-4}$ M
4EPO10	Essington and Anderson (2008, Fig. 3a)	$TOTPO_4^{3-} = 4.00 \times 10^{-3}$ M
4EPO11	Essington and Anderson (2008, Fig. 3a)	$TOTPO_4^{3-} = 4.00 \times 10^{-3}$ M

TABLE 7.5 PO_4^{3-}/gibbsite sorption data not used

Source	Data Set	Reason
Guan et al. (2005, Figs. 1, 2, and 3)		Amorphous aluminum hydroxide was used in the experiments
Guan et al. (2006, Figs. 1b or 2b or 3b)		Amorphous aluminum hydroxide was used in the experiments
Helyar et al. (1976a, Fig. 3)		Sorption of phosphate in solution containing divalent cations, Ca and Mg. Also, no exclusion of CO_2 was performed
Helyar et al. (1976b, Figs. 1 and 2)		Sorption of phosphate in solution containing divalent cations (Zn, Cd, Ca, Mg, Sr). Also, no exclusion of CO_2 was performed
Okazaki et al. (1989, Fig. 4 Gi-2)	$TOTPO_4^{3-} = 3.30 \times 10^{-4}$ M, Conc. (0.26 g/L), $I = 0.1$ M, $T = 23$°C	High surface area and low crystallinity of solid is indicative of amorphous aluminum hydroxide
Manning and Goldberg (1996, Fig. 4b)	$TOTPO_4^{3-} = 1.33 \times 10^{-4}$ M, $TOTAsO_4^{3-}$ or $TOTMoO_4^{2-}$ $= 1.33 \times 10^{-4}$ M, Gibbsite conc. (2.5 g/L), $I = 0.1$ M, $T = 23$°C	Competitive sorption of phosphate with arsenate or molybdate (binary anion-adsorption experiments)

TABLE 7.6 PO_4^{3-}/gibbsite surface complexation reactions

$\equiv AlOH^0 + PO_4^{3-} + 3H^+ = \equiv AlH_2PO_4^0 + H_2O$	K_1
$\equiv AlOH^0 + PO_4^{3-} + 2H^+ = \equiv AlHPO_4^- + H_2O$	K_2
$\equiv AlOH^0 + PO_4^{3-} + H^+ = \equiv AlPO_4^{2-} + H_2O$	K_3

TABLE 7.7 PO_4^{3-}/gibbsite intrinsic surface complexation constants

ID	$\log K_1^{int}$	$\sigma_{\log K}$	$\log K_2^{int}$	$\sigma_{\log K}$	$\log K_3^{int}$	$\sigma_{\log K}$	WSOS/DF
4EPO1	26.12	0.15[b]	19.41[a]	0.15[b]	11.96[a]	0.15[b]	—
4EPO2	26.27	0.15[b]	19.21[a]	0.15[b]	11.66[a]	0.15[b]	—
4EPO3	25.40[a]	0.15[b]	19.42[a]	0.15[b]	13.95[a]	0.15[b]	—
4EPO4	26.40[a]	0.15[b]	19.62[a]	0.15[b]	14.15[a]	0.15[b]	—
4EPO5	26.90[a]	0.15[b]	19.42[a]	0.15[b]	13.75[a]	0.15[b]	—
4EPO6	27.22[a]	0.15[b]	19.92[a]	0.15[b]	15.31	0.095	23.7
4IPO7	26.21	0.14	19.42[a]	0.15[b]	12.71[a]	0.15[b]	6.2
4IPO8	25.83[a]	0.15[b]	19.42[a]	0.15[b]	13.66	0.036	21.0
4EPO9	25.90[a]	0.15[b]	18.66[a]	0.15[b]	12.30[a]	0.15[b]	—
4EPO10	29.60[a]	0.15[b]	19.41[a]	0.15[b]	13.74[a]	0.15[b]	—
4EPO11	30.00[a]	0.15[b]	19.23[a]	0.15[b]	14.80[a]	0.15[b]	—

[a] Constant was fixed manually.
[b] Fixed at this value by convention.

TABLE 7.8 Best estimates for PO_4^{3-}/gibbsite surface complexation constants

Best-Estimate K	Confidence Level	$\log K^{int}(-)$	$\log K^{int}(+)$
$\log \overline{K_1^{int}} = 26.89$	0.99	26.79	26.99
	0.95	26.82	26.96
	0.90	26.83	26.95
	0.80	26.84	26.94
$\log \overline{K_2^{int}} = 19.37$	0.99	—[a]	—[a]
	0.95	—[a]	—[a]
	0.90	—[a]	—[a]
	0.80	—[a]	—[a]
$\log \overline{K_3^{int}} = 13.57$	0.99	12.95	14.19
	0.95	13.10	14.04
	0.90	13.18	13.97
	0.80	13.26	13.88

[a] Confidence limit not calculated because all K values (and hence $\sigma_{\log K}$ values) fixed manually.

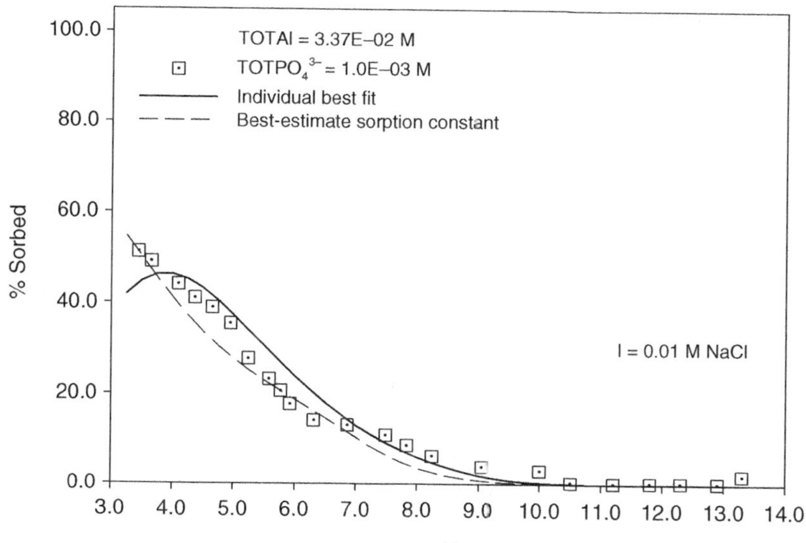

FIGURE 4EPO1

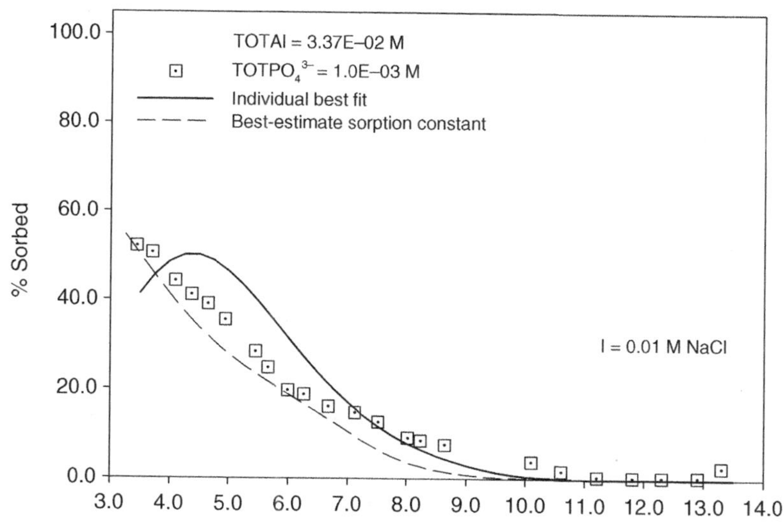

FIGURE 4EPO2

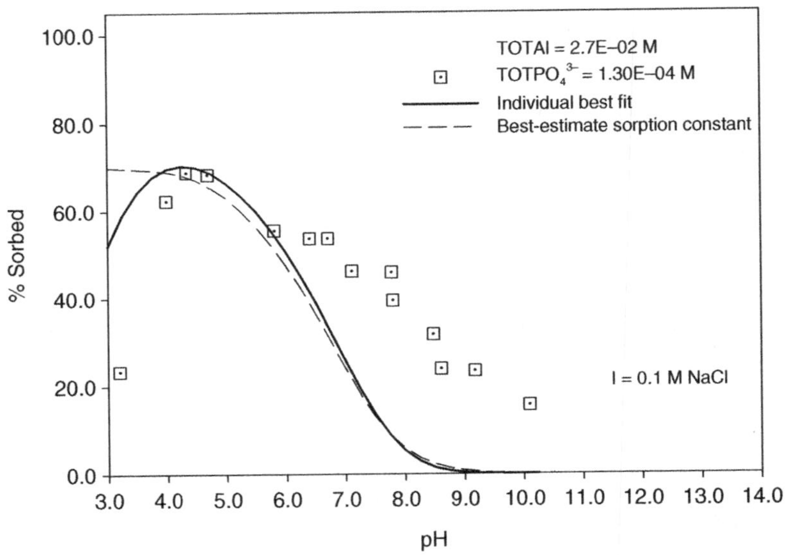

FIGURE 4EPO3

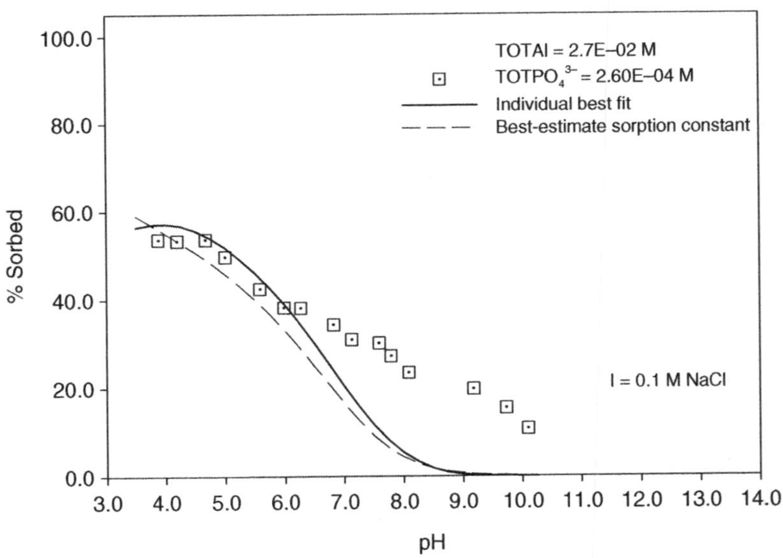

FIGURE 4EPO4

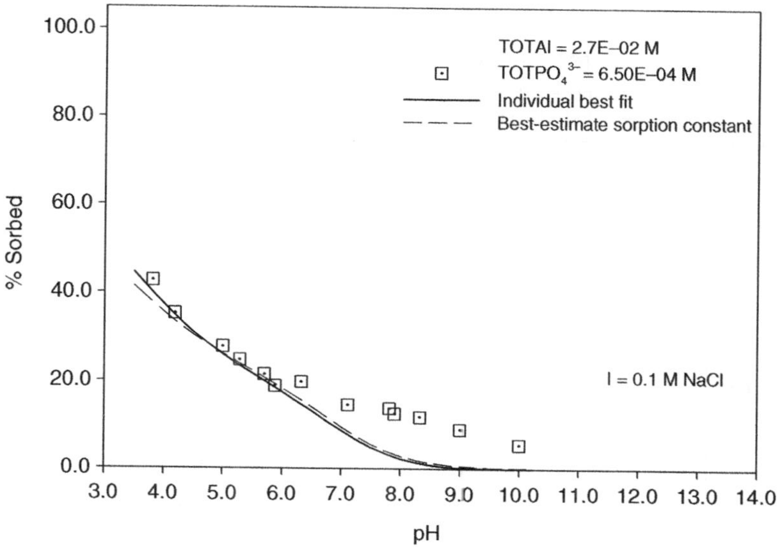

FIGURE 4EPO5

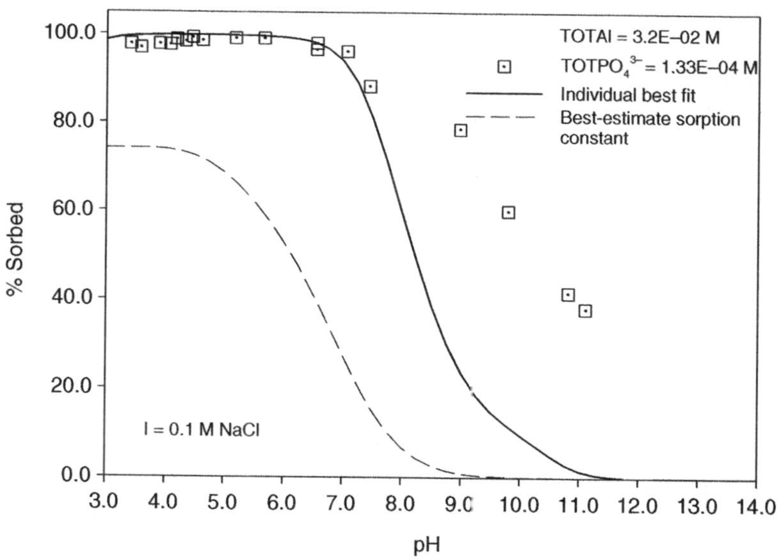

FIGURE 4EPO6

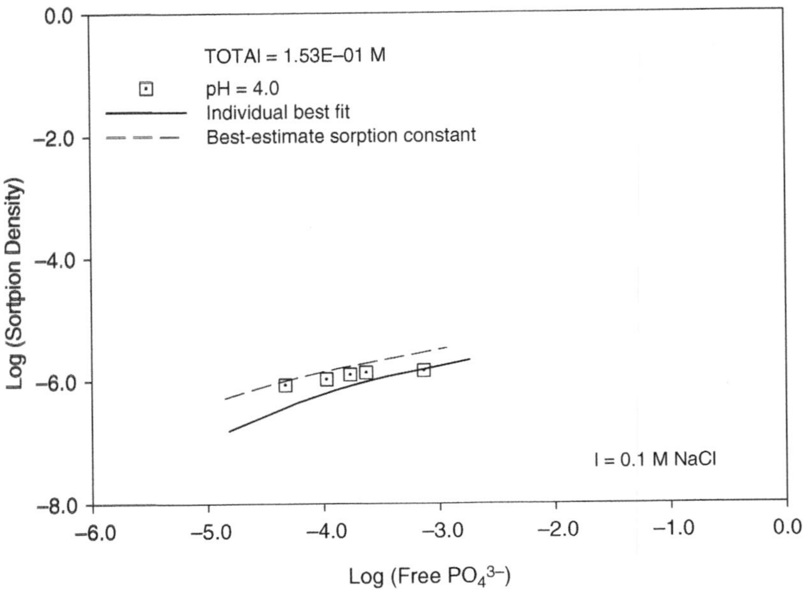

FIGURE 4IPO7

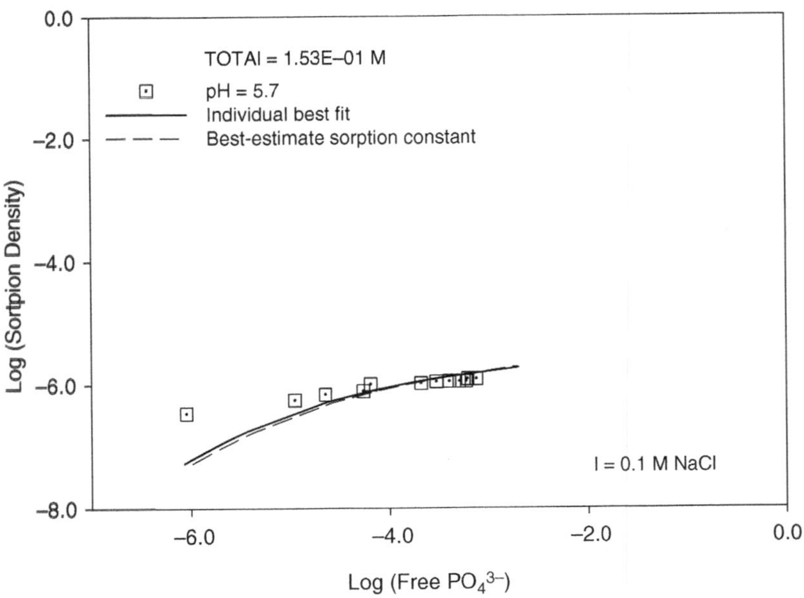

FIGURE 4IPO8

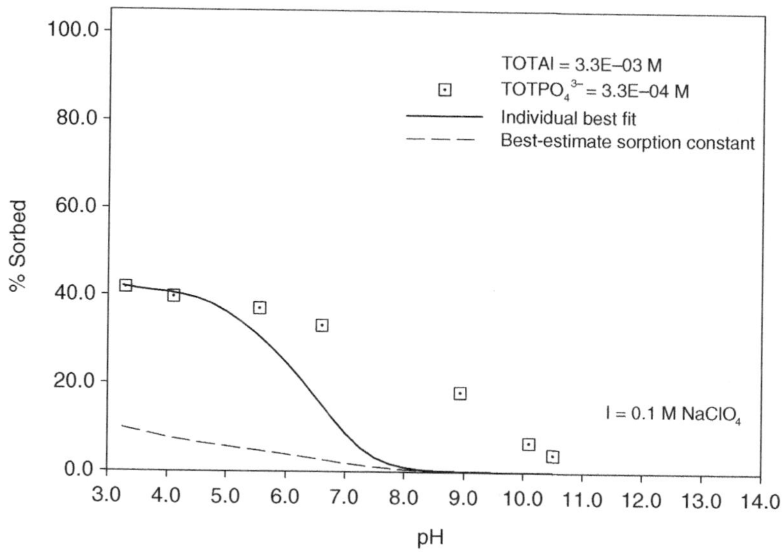

FIGURE 4EPO9

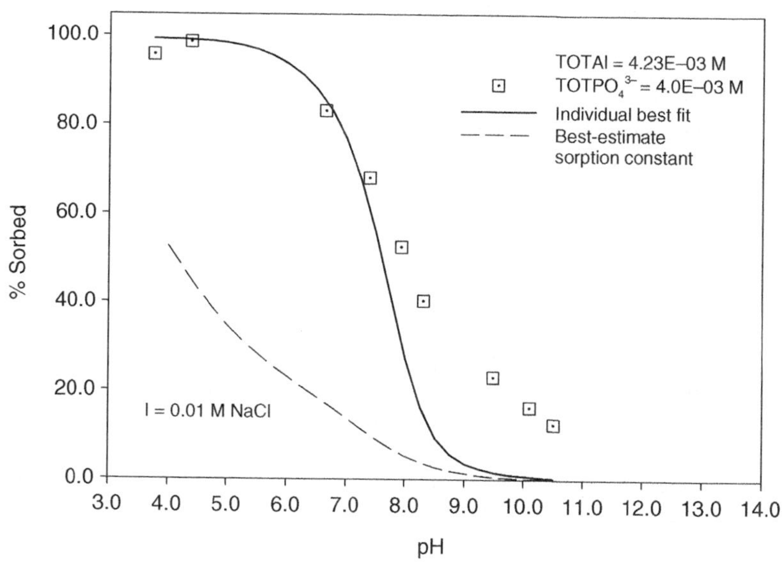

FIGURE 4EPO10

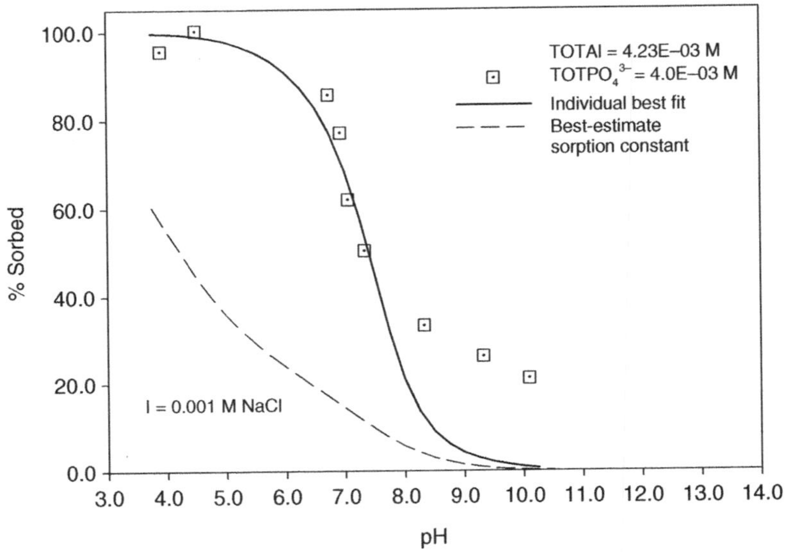

FIGURE 4EPO11

7.4 ARSENATE

TABLE 7.9 AsO_4^{3-}/gibbsite sorption data

ID	Source	Data Set
4IAS1	Liu et al. (2006, Fig. 1)	$TOTAl = 3.20 \times 10^{-2}$ M
4EAS2	Fukushi and Sverjensky (2007, Fig. 11a), original data at Manning and Goldberg (1996, Fig. 2b)	$TOTAsO_4^{3-} = 2.66 \times 10^{-4}$ M
4EAS3	Fukushi and Sverjensky (2007, Fig. 11a), original data at Manning and Goldberg (1996, Fig. 2b)	$TOTAsO_4^{3-} = 1.33 \times 10^{-4}$ M
4EAS4	Hingston et al. (1971, Fig. 3Ai)	$TOTAsO_4^{3-} = 2.70 \times 10^{-4}$ M
4EAS5	Weerasooriya et al. (2004, Fig. 3)	$TOTAsO_4^{3-} = 1.34 \times 10^{-6}$ M
4EAS6	Weerasooriya et al. (2004, Fig. 3)	$TOTAsO_4^{3-} = 1.34 \times 10^{-6}$ M
4EAS7	Weerasooriya et al. (2004, Fig. 3)	$TOTAsO_4^{3-} = 1.34 \times 10^{-6}$ M
4EAS8	Weerasooriya et al. (2004, Fig. 3)	$TOTAsO_4^{3-} = 2.67 \times 10^{-6}$ M
4EAS9	Weerasooriya et al. (2004, Fig. 3)	$TOTAsO_4^{3-} = 2.67 \times 10^{-6}$ M
4EAS10	Weerasooriya et al. (2004, Fig. 3)	$TOTAsO_4^{3-} = 2.67 \times 10^{-6}$ M
4EAS11	Weerasooriya et al. (2004, Fig. 3)	$TOTAsO_4^{3-} = 1.34 \times 10^{-5}$ M
4EAS12	Weerasooriya et al. (2004, Fig. 3)	$TOTAsO_4^{3-} = 1.34 \times 10^{-5}$ M
4EAS13	Weerasooriya et al. (2004, Fig. 3)	$TOTAsO_4^{3-} = 1.34 \times 10^{-5}$ M
4IAS14	Weerasooriya et al. (2004, Fig. 6)	$I = 0.1$ M, pH 4.30
4IAS15	Weerasooriya et al. (2004, Fig. 6)	$I = 0.1$ M, pH 7.50
4IAS16	Weerasooriya et al. (2004, Fig. 6)	$I = 0.1$ M, pH 9.20
4EAS17	Okazaki et al. (1989, Fig. 4 Gi-3)	$TOTAsO_4^{3-} = 3.3 \times 10^{-4}$ M
4EAS18	Essington and Anderson (2008, Fig. 3b)	$TOTAsO_4^{3-} = 4.0 \times 10^{-3}$ M
4EAS19	Essington and Anderson (2008, Fig. 3b)	$TOTAsO_4^{3-} = 4.0 \times 10^{-3}$ M

TABLE 7.10 AsO_4^{3-}/gibbsite sorption data not used

Source	Data Set	Reason
Liu et al. (2006, Fig. 2)		It seems that there is an experimental site limitation problem. pH_{PZC} is very low (5.06). Also low amount of sorption and no variation with pH
Xu et al. (2008, Fig. 1)		Desorption experiments. Crystalline $Al(OH)_3(s)$ verified as a mixture of gibbsite and bayerite. Competitive sorption of arsenate with phosphate
Manning and Goldberg (1996, Fig. 4a)	$TOTAsO_4^{3-} = 1.33 \times 10^{-4}$ M, $TOTPO_4^{3-} = 1.33 \times 10^{-4}$ M, Gibbsite conc. (2.5 g/L), $I = 0.1$ M, $T = 23°C$	Competitive sorption of arsenate with phosphate or molybdate (binary anion-sorption experiments).
Manning and Goldberg (1996, Fig. 4c)	$TOTAsO_4^{3-} = 1.33 \times 10^{-4}$ M, $TOTMoO_4^{2-} = 1.33 \times 10^{-4}$ M, Gibbsite conc. (2.5 g/L), $I = 0.1$ M, $T = 23°C$	Competitive sorption of arsenate with phosphate or molybdate (binary anion-sorption experiments).
Hingston et al. (1971, Fig. 3Aii)	$TOTAsO_4^{3-} = 0.27 \times 10^{-3}$ M, $TOTPO_4^{3-} = 0.13 \times 10^{-3}$ M, Gibbsite conc. (2.12 g/L)	Competitive sorption of arsenate with phosphate
Hingston et al. (1971, Fig. 3Aiii)	$TOTAsO_4^{3-} = 0.27 \times 10^{-3}$ M, $TOTPO_4^{3-} = 0.26 \times 10^{-3}$ M, Gibbsite conc. (2.12 g/L)	Competitive sorption of arsenate with phosphate
Hingston et al. (1971, Fig. 3Aiv)	$TOTAsO_4^{3-} = 0.27 \times 10^{-3}$ M, $TOTPO_4^{3-} = 0.65 \times 10^{-3}$ M, Gibbsite conc. (2.12 g/L)	Competitive sorption of arsenate with phosphate
Weerasooriya et al. (2004, Fig. 2B)		One experiment was conducted at high temperature (40°C). Ionic strength and pH of isotherms not available
Okazaki et al. (1989, Fig. 4 Gi-2, 5)	$TOTAsO_4^{3-} = 3.30 \times 10^{-4}$ M, Conc. (0.26 g/L), $I = 0.1$ M, $T = 23°C$	High surface area and low crystallinity of solid is indicative of amorphous aluminum hydroxide
Masue et al. (2009, Fig. 2a)	$TOTAl = 2.67 \times 10^{-4}$ M, I 0.1 M, pH 5	Crystalline $Al(OH)_3(s)$ verified as a mixture of gibbsite and bayerite
Masue et al. (2009, Fig. 2b)	$TOTAl = 2.67 \times 10^{-4}$ M, I 0.1 M, pH 8	Crystalline $Al(OH)_3(s)$ verified as a mixture of gibbsite and bayerite
Masue et al. (2009, Fig. 3a)	$TOTAsO_3^{3-} = 1.34 \times 10^{-5}$ M, $TOTAl = 2.67 \times 10^{-4}$ M, I 0.1 M	Crystalline $Al(OH)_3(s)$ verified as a mixture of gibbsite and bayerite

TABLE 7.11 AsO_4^{3-}/gibbsite surface complexation reactions

$\equiv AlOH^0 + AsO_4^{3-} + 3H^+ = \equiv AlH_2AsO_4^0 + H_2O$	K_1
$\equiv AlOH^0 + AsO_4^{3-} + 2H^+ = \equiv AlHAsO_4^- + H_2O$	K_2
$\equiv AlOH^0 + AsO_4^{3-} = \equiv AlOHAsO_4^{3-}$	K_4[a]

[a]The numbering of surface complexation constant follows the numbering of the generic complexation reaction constants given in Table 8.5 of Chapter 8.

TABLE 7.12 AsO_4^{3-}/gibbsite intrinsic surface complexation constants

ID	$\log K_1^{int}$	$\sigma_{\log K}$	$\log K_2^{int}$	$\sigma_{\log K}$	$\log K_4^{int}$	$\sigma_{\log K}$	WSOS/DF
4IAS1			18.00[a]	0.15[b]	4.70[a]	0.15[b]	—
4EAS2	25.86[a]	0.15[b]	19.96[a]	0.15[b]	9.39[a]	0.15[b]	—
4EAS3	25.40[a]	0.15[b]	20.31[a]	0.15[b]	8.60	0.036	—
4EAS4	17.72[a]	0.15[b]	18.11[a]	0.15[b]	5.90[a]	0.15[b]	—
4EAS5			19.81[a]	0.15[b]	6.12	0.018	51.4
4EAS6			20.40[a]	0.15[b]	5.65	0.022	182
4EAS7			25.00[a]	0.15[b]	6.55[a]	0.15[b]	—
4EAS8			15.01[a]	0.15[b]	5.80	0.011	118
4EAS9	25.00[a]	0.15[b]	24.30[a,c]	0.15[b]	6.25[a,c]	0.15[b]	—
4EAS10			24.50[a]	0.15[b]	5.75[a]	0.15[b]	—
4EAS11			20.01	0.015	6.03	0.016	110
4EAS12	25.55[a]	0.15[b]	24.30[a]	0.15[b]	6.25	0.062	1302[d]
4EAS13			24.48[a]	0.15[b]	6.84	0.046	5089[d]
4IAS14	30.07[a]	0.15[b]	24.21[a]	0.15[b]	6.19[a]	0.15[b]	—
4IAS15	30.07[a]	0.15[b]	24.31	0.025	6.19[a]	0.15[b]	13.5
4IAS16	30.07[a]	0.15[b]	23.80[a]	0.15[b]	6.19[a]	0.15[b]	—
4EAS17	27.68[a]	0.15[b]	23.00[a]	0.15[b]	5.56[a]	0.15[b]	—
4EAS18	30.05[a]	0.15[b]	25.00[a]	0.15[b]	12.40[a]	0.15[b]	—
4EAS19	28.90[a]	0.15[b]	22.97[a]	0.15[b]	9.86[a]	0.15[b]	—

[a]Constant was fixed manually.
[b]Fixed by convention.
[c]Based on findings of 3EAS12.
[d]Although FITEQL overweighed some data points in these two data sets resulting in a very high WSOS/DF, the fit is good.

TABLE 7.13 Best estimates for AsO_4^{3-}/gibbsite surface complexation constants

Best-Estimate K	Confidence Level	$\log K^{int}(-)$	$\log K^{int}(+)$
$\log \overline{K_1^{int}} = 26.94$	0.99	—[a]	—[a]
	0.95	—[a]	—[a]
	0.90	—[a]	—[a]
	0.80	—[a]	—[a]
$\log \overline{K_2^{int}} = 21.79$	0.99	20.78	22.80
	0.95	21.02	22.56
	0.90	21.14	22.44
	0.80	21.29	22.29
$\log \overline{K_4^{int}} = 6.39$	0.99	6.14	6.64
	0.95	6.19	6.58
	0.90	6.23	6.55
	0.80	6.26	6.51

[a]Confidence limit not calculated because all K values (and hence $\sigma_{\log K}$ values) fixed manually.

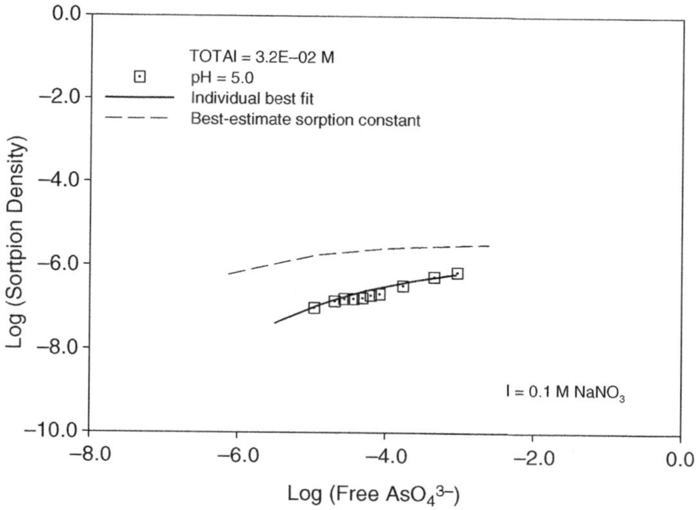

FIGURE 4IAS1

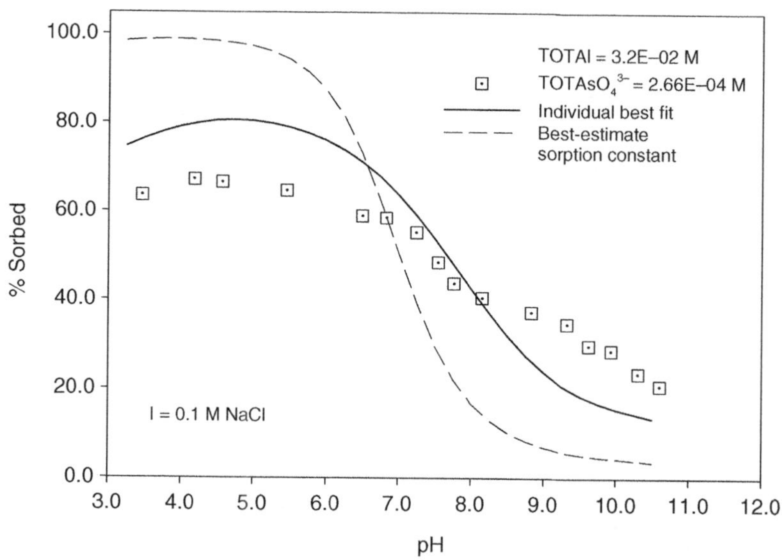

FIGURE 4IAS2

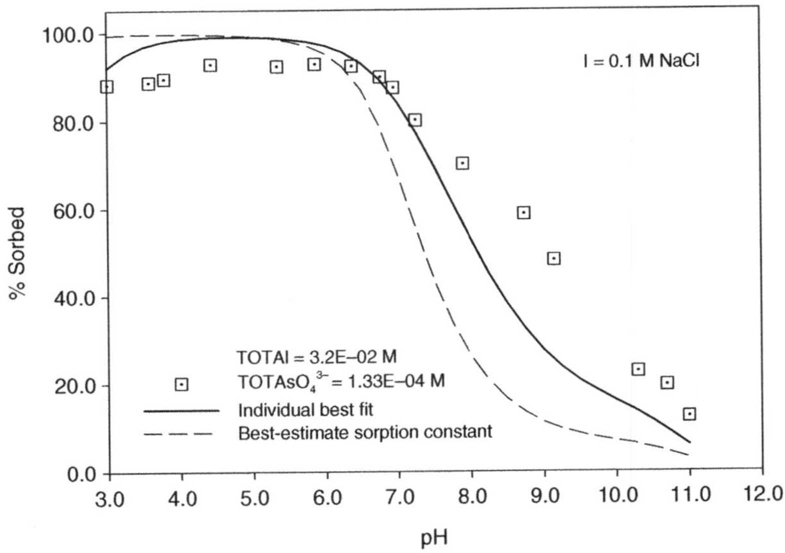

FIGURE 4IAS3

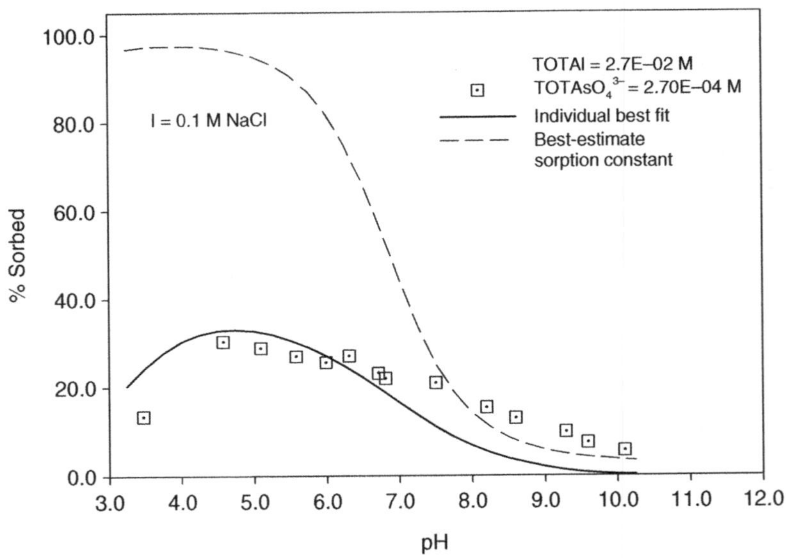

FIGURE 4IAS4

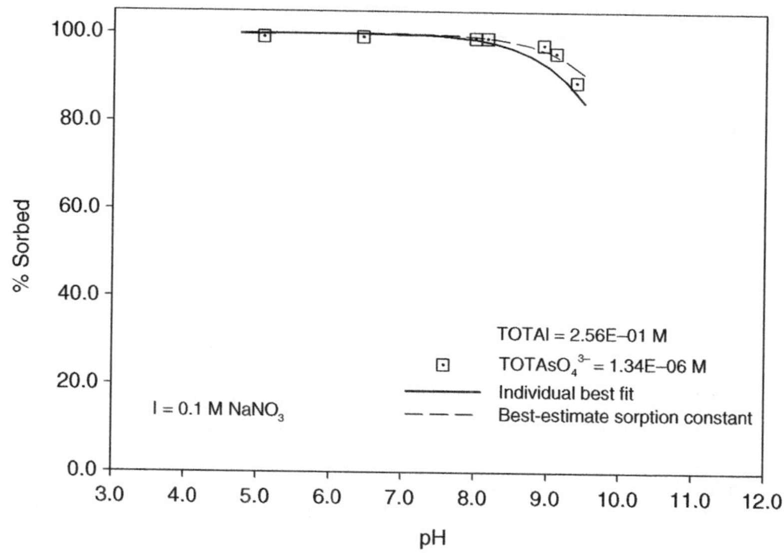

FIGURE 4IAS5

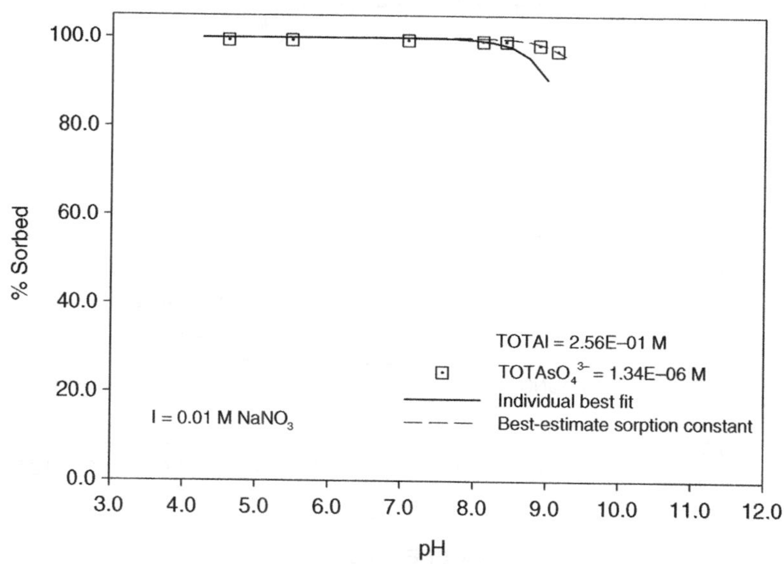

FIGURE 4IAS6

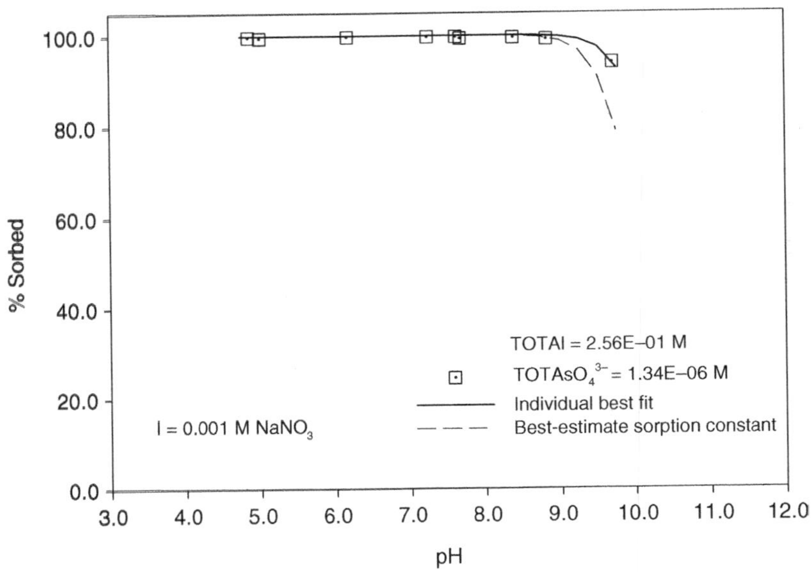

FIGURE 4IAS7

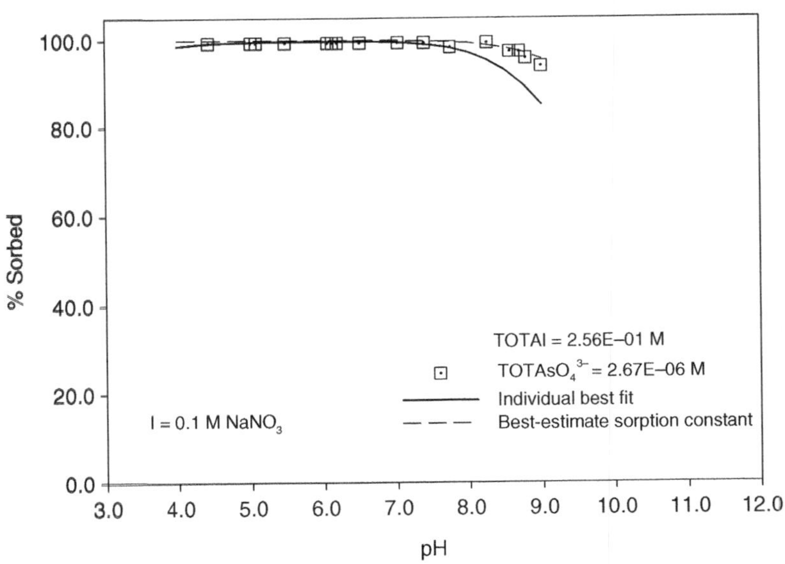

FIGURE 4IAS8

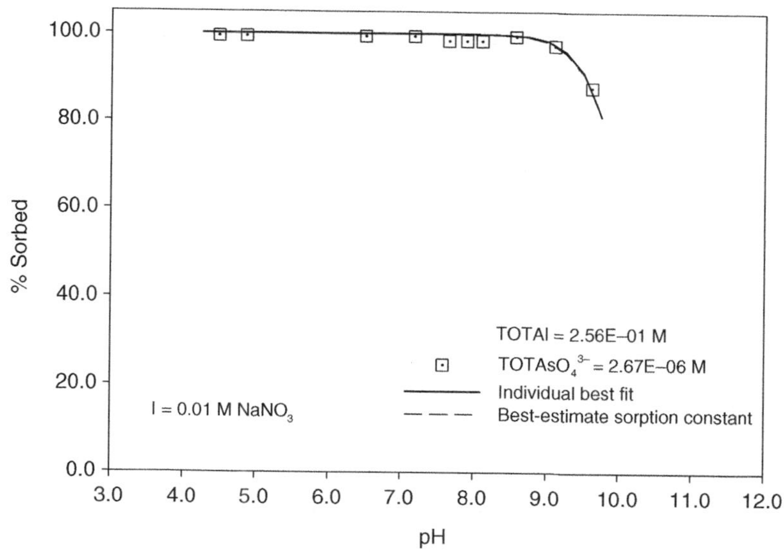

FIGURE 4IAS9

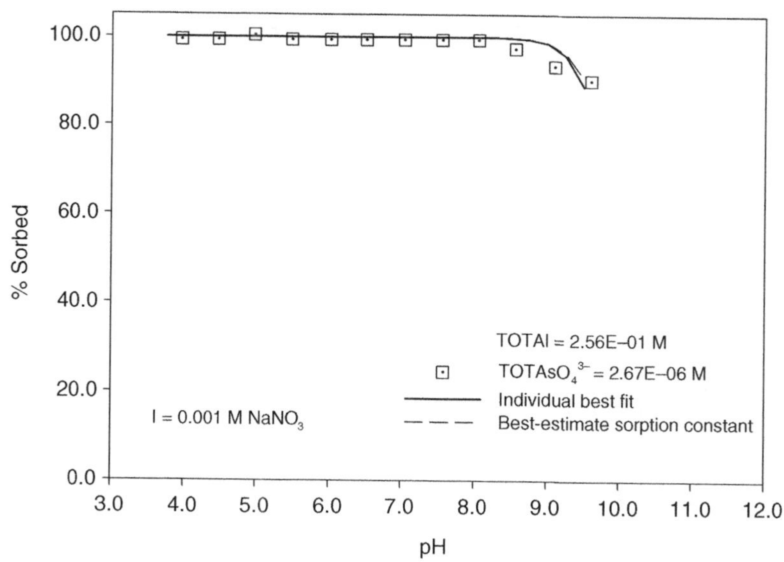

FIGURE 4IAS10

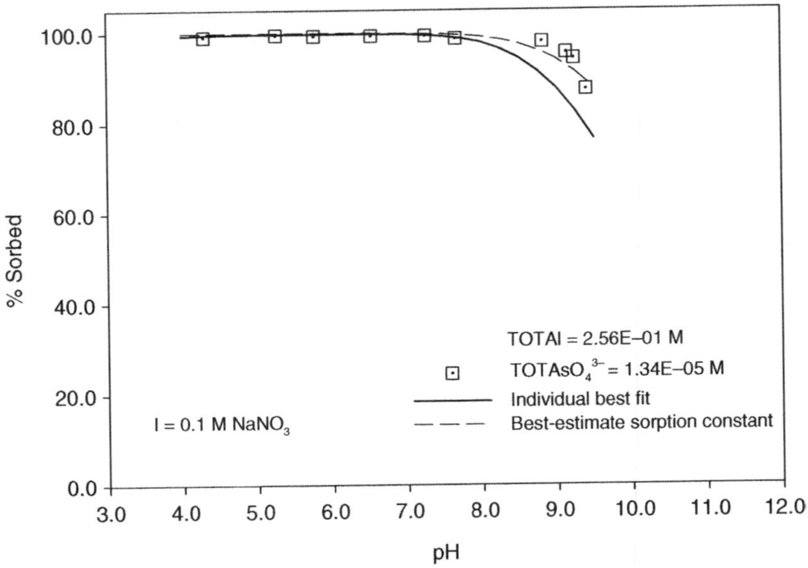

FIGURE 4IAS11

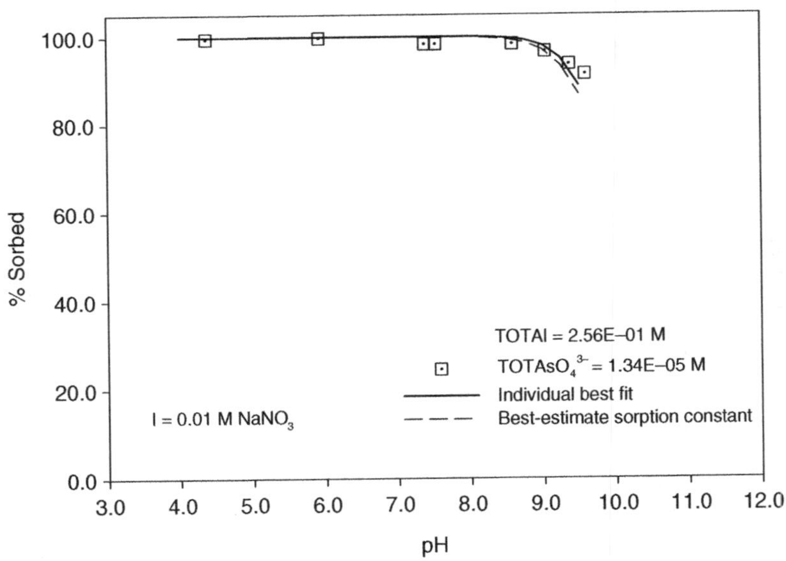

FIGURE 4IAS12

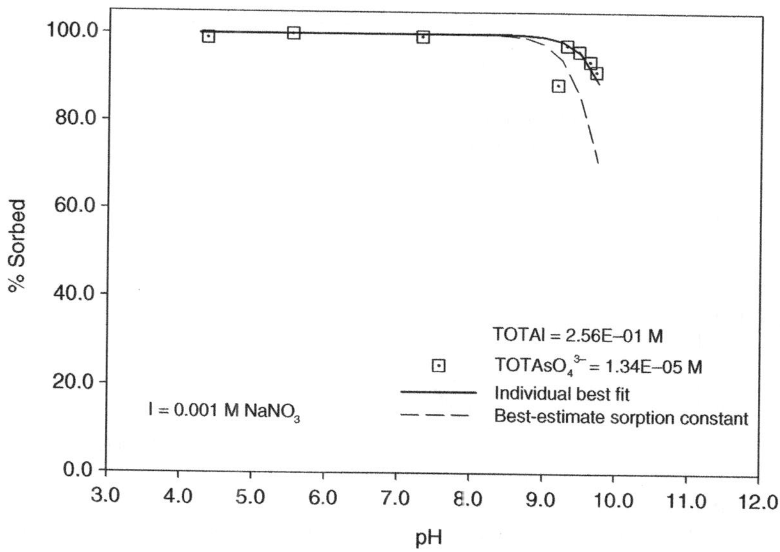

FIGURE 4IAS13

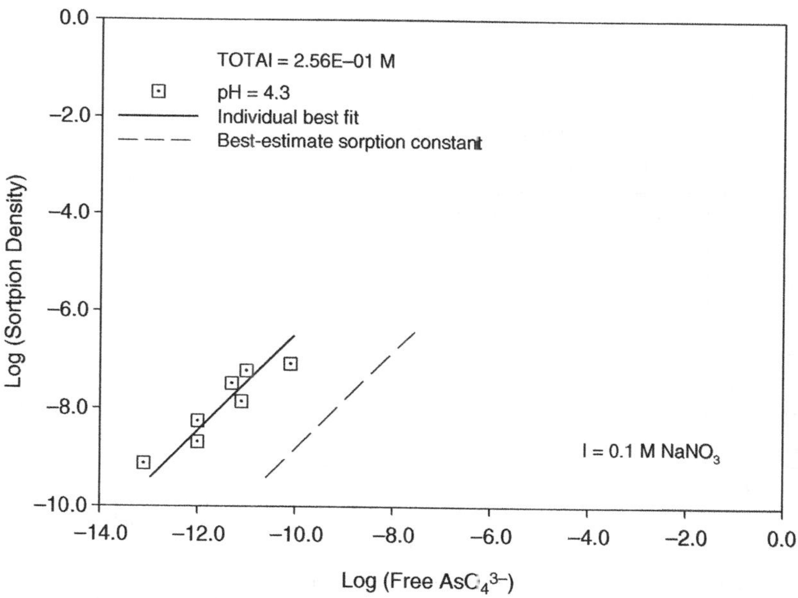

FIGURE 4IAS14

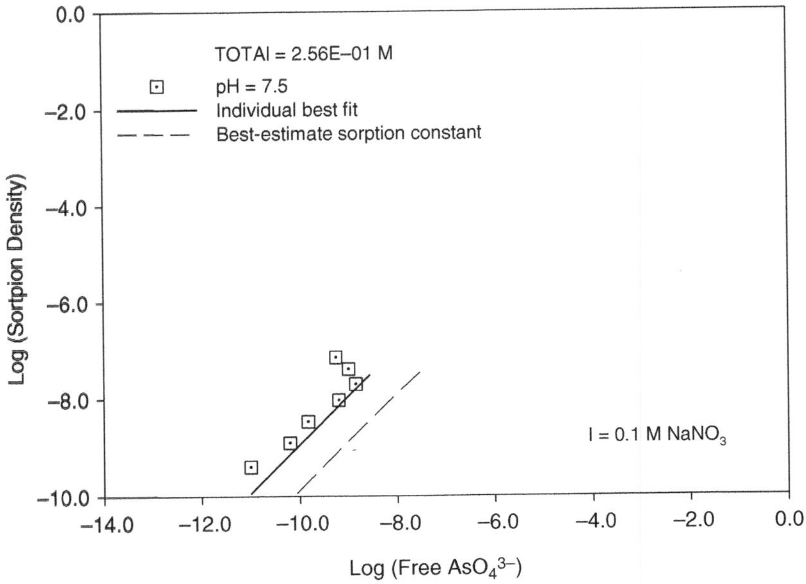

FIGURE 4IAS15

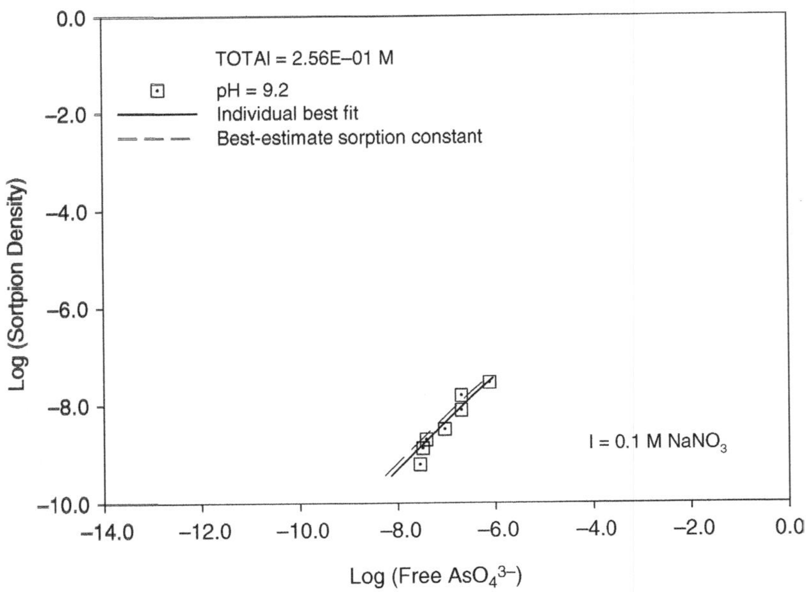

FIGURE 4IAS16

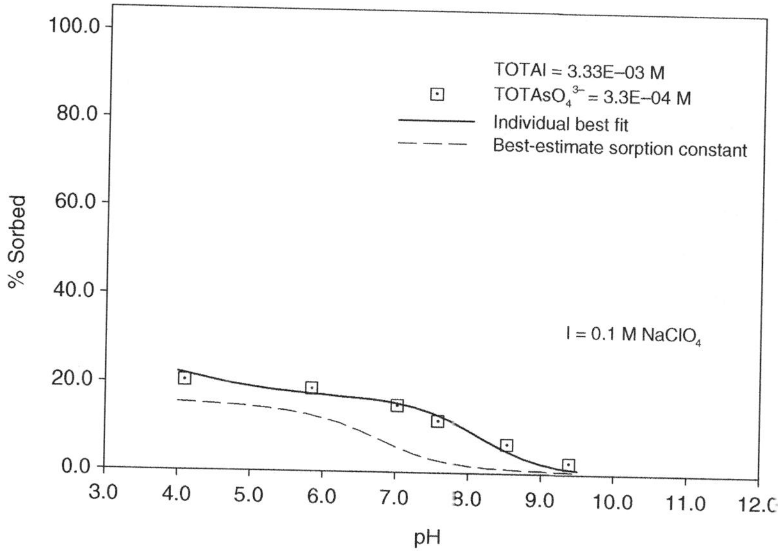

FIGURE 4IAS17

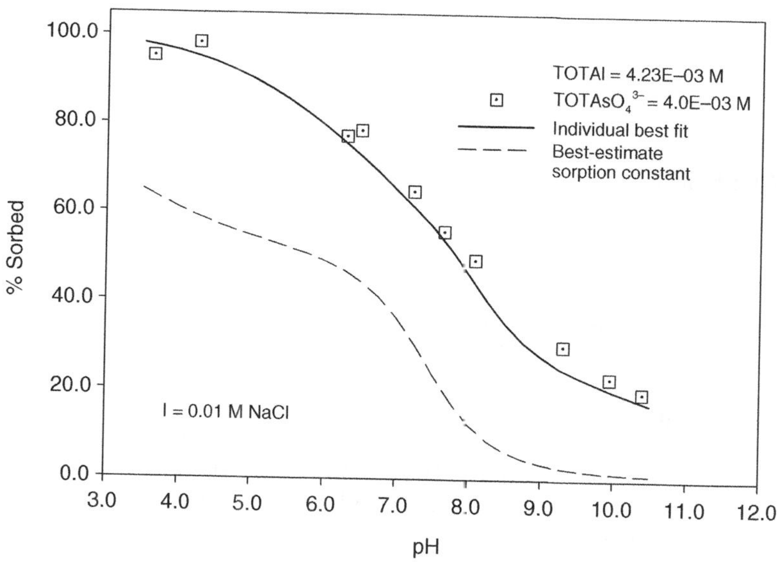

FIGURE 4IAS18

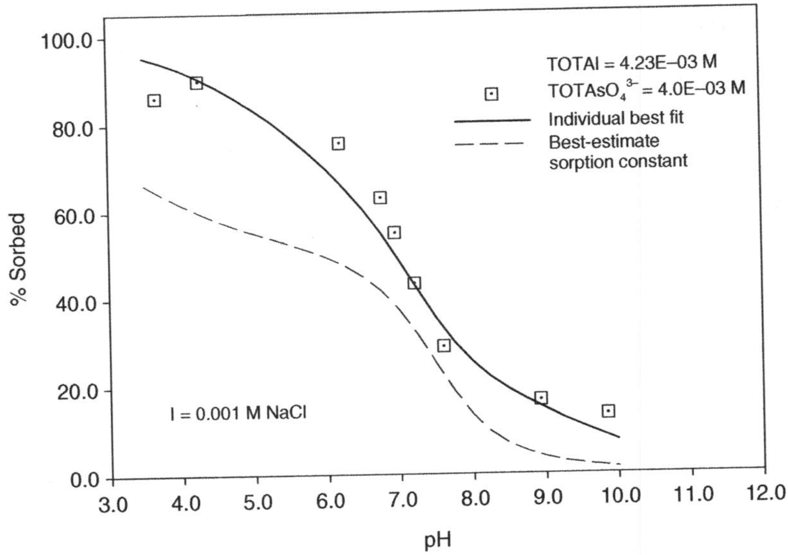

FIGURE 4IAS19

7.5 ARSENITE

TABLE 7.14 AsO_3^{3-}/gibbsite sorption data

ID	Source	Data Set
4EAT1	Weerasooriya et al. (2003, Fig. 5A)	$TOTAsO_3^{3-} = 2.67 \times 10^{-6}$ M
4EAT2	Weerasooriya et al. (2003, Fig. 5A)	$TOTAsO_3^{3-} = 2.67 \times 10^{-6}$ M
4EAT3	Weerasooriya et al. (2003, Fig. 5A)	$TOTAsO_3^{3-} = 2.67 \times 10^{-6}$ M
4EAT4	Weerasooriya et al. (2003, Fig. 5B)	$TOTAsO_3^{3-} = 1.35 \times 10^{-5}$ M
4EAT5	Weerasooriya et al. (2003, Fig. 5B)	$TOTAsO_3^{3-} = 1.35 \times 10^{-5}$ M
4EAT6	Weerasooriya et al. (2003, Fig. 5B)	$TOTAsO_3^{3-} = 1.35 \times 10^{-5}$ M
4IAT7	Weerasooriya et al. (2003, Fig. 3)	$I = 0.01$ M, pH 4.0
4IAT8	Weerasooriya et al. (2003, Fig. 3)	$I = 0.01$ M, pH 7.5
4IAT9	Weerasooriya et al. (2003, Fig. 3)	$I = 0.01$ M, pH 8.2

TABLE 7.15 AsO_3^{3-}/gibbsite sorption data not used

Source	Data Set	Reason
Masue et al. (2009, Fig. 2c)	TOTAl = 2.67×10^{-4} M, I 0.1 M, pH 5	Crystalline $Al(OH)_3(s)$ verified as a mixture of gibbsite and bayerite
Masue et al. (2009, Fig. 2d)	TOTAl = 2.67×10^{-4} M, I 0.1 M, pH 8	Crystalline $Al(OH)_3(s)$ verified as a mixture of gibbsite and bayerite
Masue et al. (2009, Fig. 3b)	$TOTAsO_3^{3-} = 1.34 \times 10^{-5}$ M, TOTAl = 2.67×10^{-4} M, I 0.1 M	Crystalline $Al(OH)_3(s)$ verified as a mixture of gibbsite and bayerite

TABLE 7.16 AsO_3^{3-}/gibbsite surface complexation reaction

$\equiv AlOH^0 + H_3AsO_3^0 = \equiv AlH_2AsO_3^0 + H_2O$	K_1

TABLE 7.17 AsO_3^{3-}/gibbsite intrinsic surface complexation constants

ID	$\log K_1^{int}$	$\sigma_{\log K}$	WSOS/DF
4EAT1	1.70	0.024	2.2
4EAT2	1.72	0.027	1.1
4EAT3	1.73	0.024	1.4
4EAT4	1.79	0.026	2.6
4EAT5	1.81	0.029	3.9
4EAT6	1.79	0.026	1.5
4IAT7	1.73	0.025	0.7
4IAT8	1.77	0.019	0.2
4IAT9	1.96	0.017	8.3

TABLE 7.18 Best estimates for AsO_3^{3-}/gibbsite surface complexation constants

Best-Estimate K	Confidence Level	$\log K_1^{int}(-)$	$\log K_1^{int}(+)$
$\overline{\log K_1^{int}} = 1.78$	0.99	1.75	1.82
	0.95	1.76	1.81
	0.90	1.77	1.80
	0.80	1.77	1.80

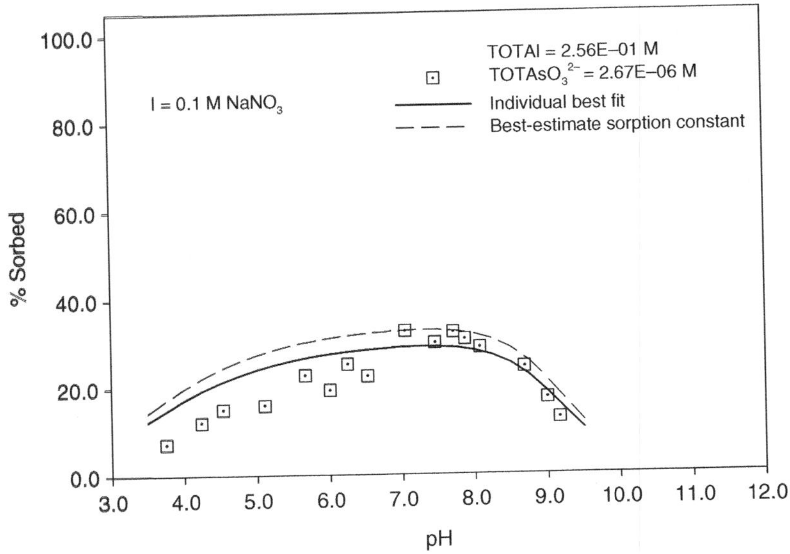

FIGURE 4EAT1

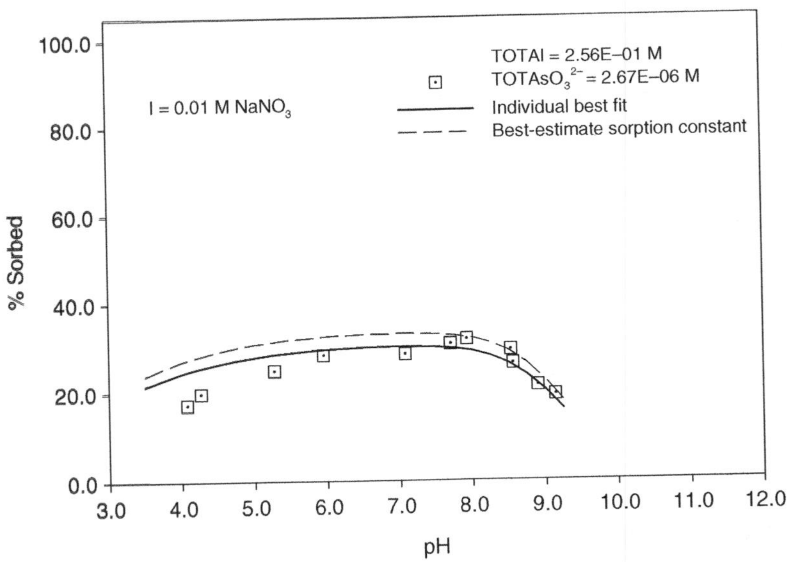

FIGURE 4EAT2

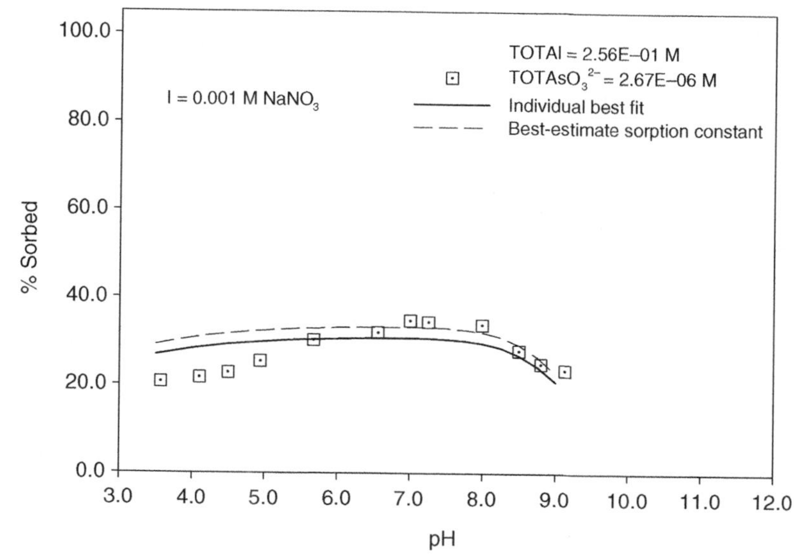

FIGURE 4EAT3

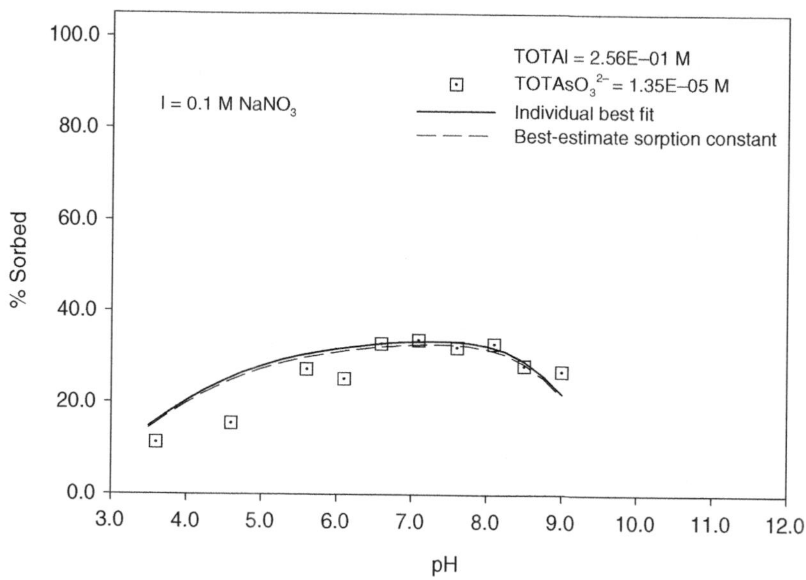

FIGURE 4EAT4

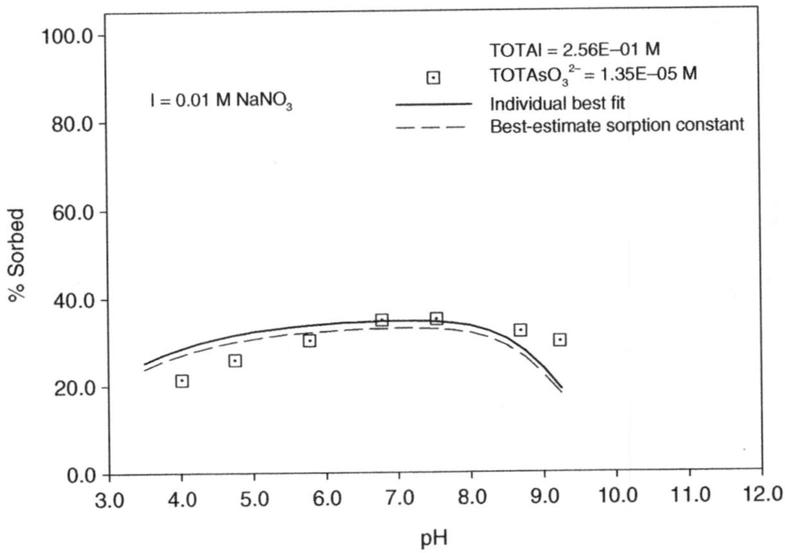

FIGURE 4EAT5

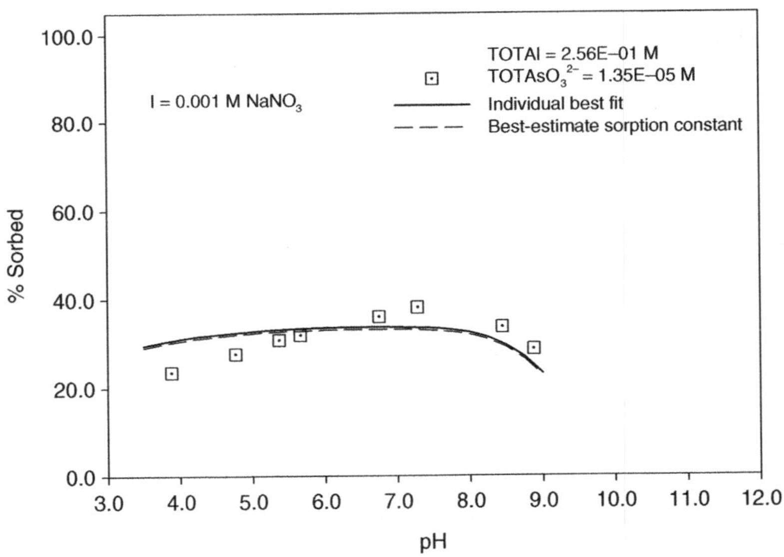

FIGURE 4EAT6

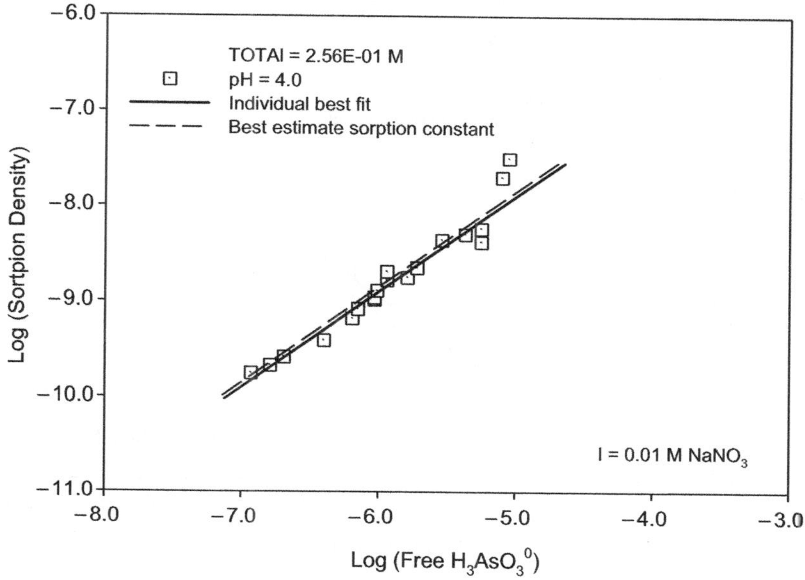

FIGURE 4IAT7

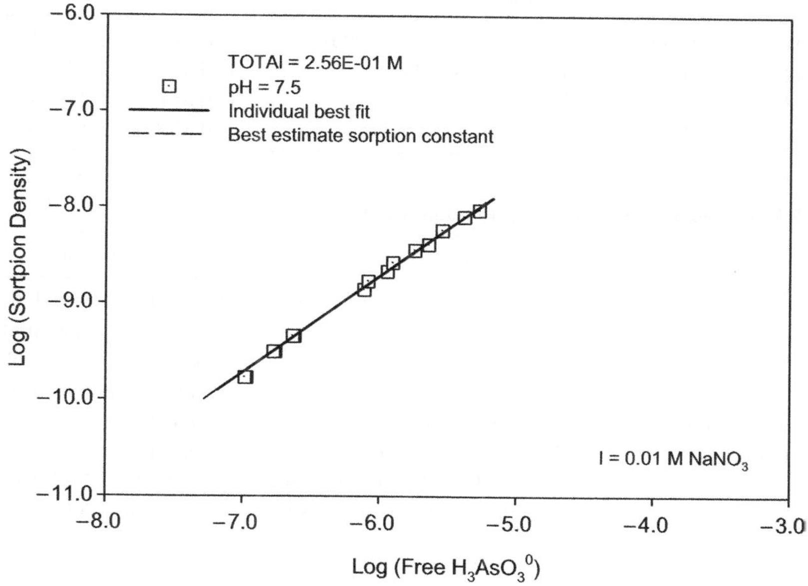

FIGURE 4IAT8

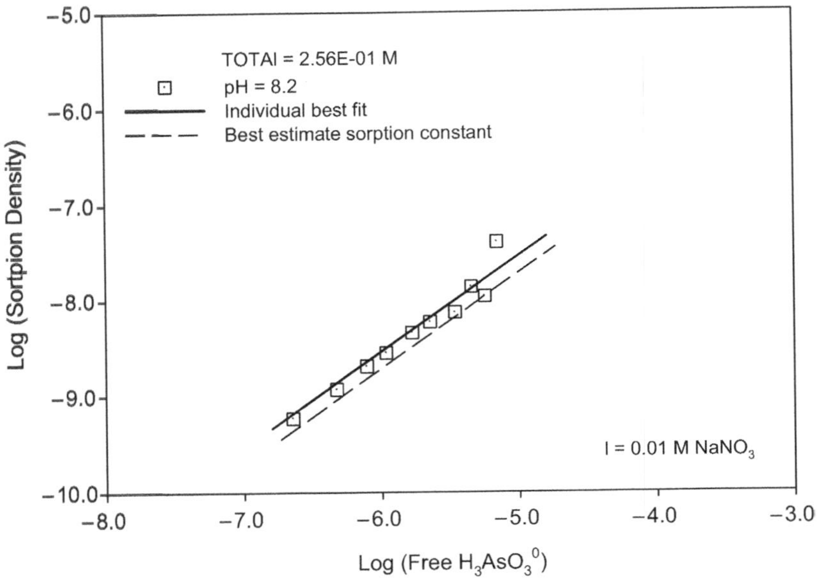

FIGURE 4IAT9

7.6 MOLYBDATE

TABLE 7.19 MoO_4^{2-}/gibbsite sorption data

ID	Source	Data Set
4EMO1	Manning and Goldberg (1996, Fig 4D)	$TOTMoO_4 = 1.33 \times 10^{-4}$ M
4EMO2	Goldberg et al. (1996, Fig 5B)	$TOTMoO_4 = 2.92 \times 10^{-4}$ M
4EMO3	Hingston et al. (1972, Fig. 1G)	$TOTMoO_4 = 1.7 \times 10^{-3}$ M
4EMO4	Okazaki et al. (1989, Fig. 4 Gi-3)	$TOTMoO_4 = 3.3 \times 10^{-4}$ M

TABLE 7.20 MoO_4^{2-}/gibbsite sorption data not used

Source	Data Set	Reason
Okazaki et al. (1989, Fig. 4 Gi-2)	$TOTMoO_4^{2-} = 3.3 \times 10^{-4}$ M	High surface area, low degree of crystallinity indicative of amorphous aluminum hydroxide
Manning and Goldberg (1996, Fig. 4d)	$TOTMoO_4^{2-} = 1.33 \times 10^{-4}$ M, $TOTAsO_4^{3-}$ or $TOTPO_4^{3-} = 1.33 \times 10^{-4}$ M, Gibbsite conc. (2.5 g/L), $I = 0.1$ M, $T = 23$°C	Competitive sorption of phosphate with arsenate or molybdate (binary anion-sorption experiments)

TABLE 7.21 MoO_4^{2-}/gibbsite surface complexation reactions

$\equiv AlOH^0 + MoO_4^{2-} + 2H^+ = \equiv AlHMoO_4 + H_2O$	K_1
$\equiv AlOH^0 + MoO_4^{2-} + H^+ = \equiv AlMoO_4^- + H_2O$	K_2
$\equiv AlOH^0 + MoO_4^{2-} = \equiv AlOHMoO_4^{2-}$	K_3

TABLE 7.22 MoO_4^{2-}/gibbsite intrinsic surface complexation constants

ID	$\log K_1^{int}$	$\sigma_{\log K}$	$\log K_2^{int}$	$\sigma_{\log K}$	$\log K_3^{int}$	$\sigma_{\log K}$	WSOS/DF
4EMO1			0.69	0.020	0.90[a]	0.15[b]	11.6
4EMO2	15.01	0.082	10.20	0.064	2.84[a]	0.15[b]	7.7
4EMO3	13.81[a]	0.15[b]	8.45	0.108			2.3
4EMO4	16.62	0.52	9.99	0.922			2.0

[a]Constant was fixed manually.
[b]Fixed at this value by convention.

TABLE 7.23 Best estimates for MoO_4^{2-}/gibbsite surface complexation constants

Best-Estimate K	Confidence Level	$\log K^{int}(-)$	$\log K^{int}(+)$
$\log \overline{K_1^{int}} = 14.77$	0.99	12.25	17.29
	0.95	12.86	16.69
	0.90	13.16	16.38
	0.80	13.52	16.03
$\log \overline{K_2^{int}} = 3.72$	0.99	3.13	4.32
	0.95	3.27	4.17
	0.90	3.35	4.10
	0.80	3.43	4.02
$\log \overline{K_3^{int}} = 1.87$	0.99	1.40[a]	2.34[a]
	0.95	1.51[a]	2.23[a]
	0.90	1.57[a]	2.17[a]
	0.80	1.63[a]	2.10[a]

[a]Surface-complexation constants were estimated based on two data sets.

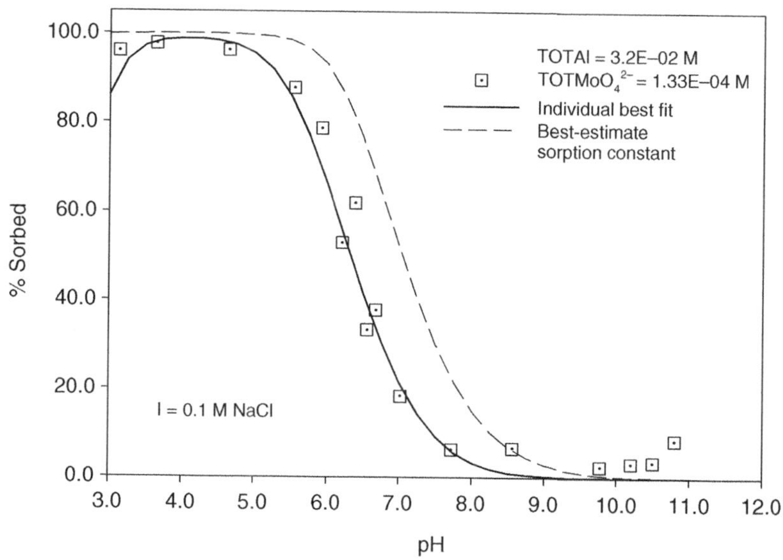

FIGURE 4EMO1

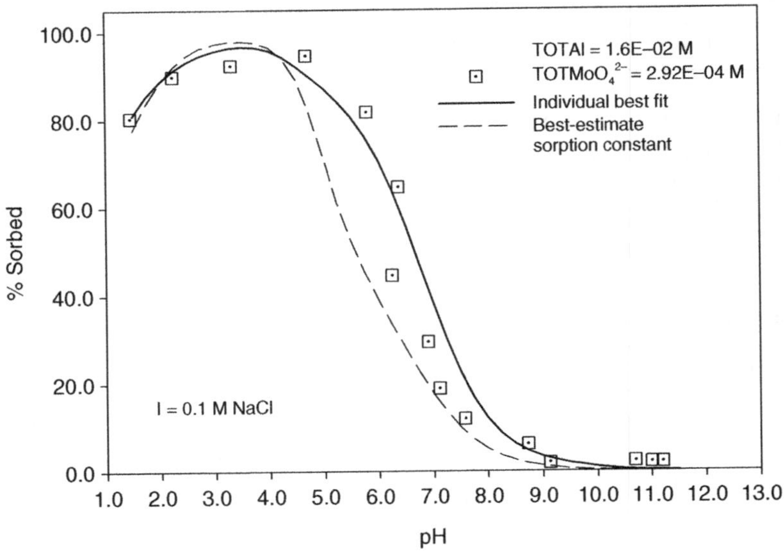

FIGURE 4EMO2

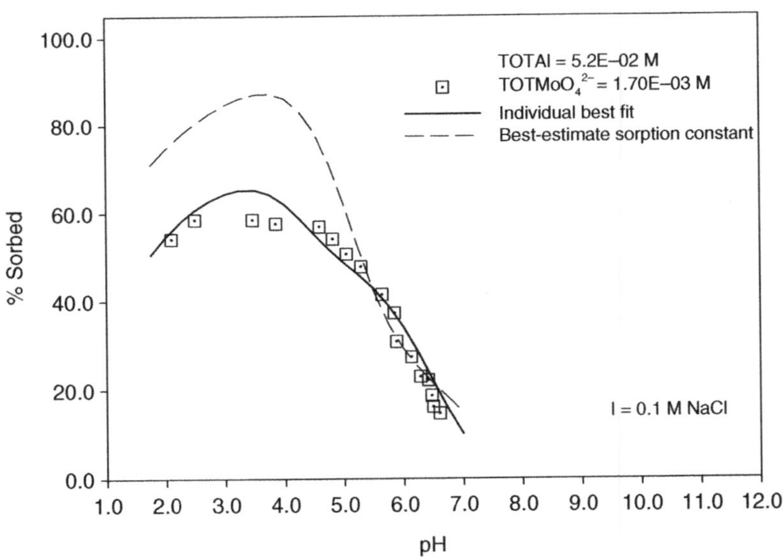

FIGURE 4EMO3

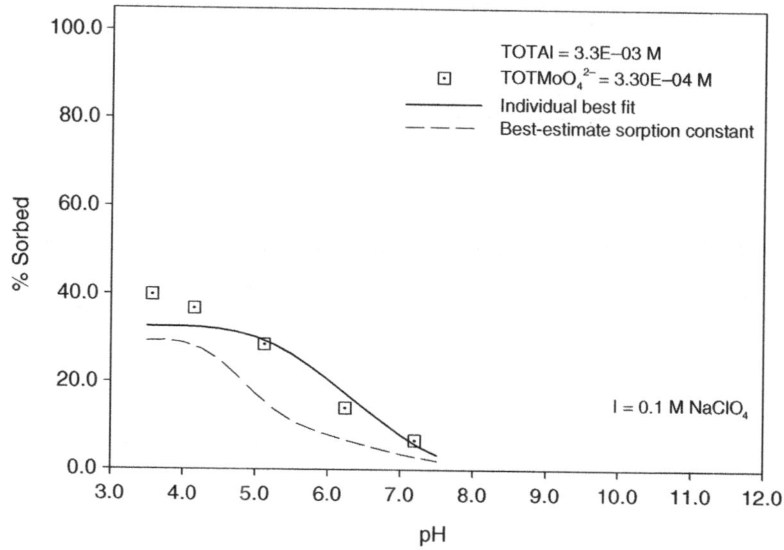

FIGURE 4EMO4

7.7 SELENATE

TABLE 7.24 SeO_4^{2-}/gibbsite sorption data

ID	Source	Data Set
4ESE1	Hingston et al. (1972, Fig. 1E)	$TOTSeO_4^{2-} = 1.00 \times 10^{-3}$ M

TABLE 7.25 SeO_4^{2-}/gibbsite sorption data not used

Source	Data Set	Reason
Peak (2006, Fig. 8)	$TOTSeO_4^{2-} = 1.0 \times 10^{-4}$ M, $I = 0.1$ and 0.001 M, HAO conc. 0.5 g/L	The hydrous aluminum oxide (HAO) was not verified as gibbsite. High surface area (BET 267 m^2/g) and inhibition of conversion from Al_2O_3 to $Al(OH)_3$ verifies non-gibbsite solid

TABLE 7.26 SeO_4^{2-}/gibbsite surface complexation reaction

$\equiv AlOH^0 + SeO_4^{2-} = \equiv AlOHSeO_4^{2-}$	$K_3{}^a$

[a]The numbering of surface complexation constant follows the numbering of the generic complexation reaction constants given in Table 8.5 of Chapter 8.

TABLE 7.27 SeO_4^{2-}/gibbsite intrinsic surface complexation constants

ID	$\log K_3^{int}$	$\sigma_{\log K}$	WSOS/DF
4ESE1	4.53	0.18	2.5

TABLE 7.28 Best estimates for SeO_4^{2-}/gibbsite surface complexation constants

Best-Estimate K	Confidence Level	$\log K^{int}(-)$	$\log K^{int}(+)$
$\log \overline{K_3^{int}} = 4.53$	0.99		
	0.95		
	0.90		
	0.80		

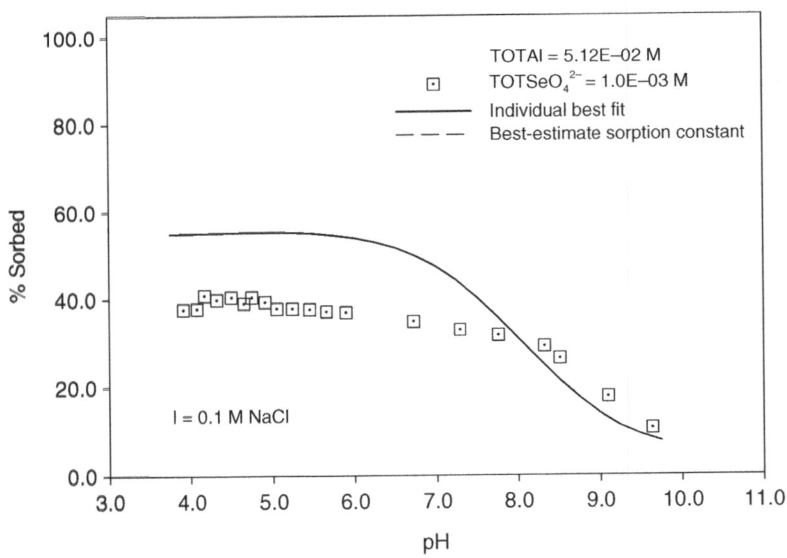

FIGURE 4ESE1

7.8 CHROMATE

TABLE 7.29 CrO_4^{2-}/gibbsite sorption data

ID	Source	Data Set
4ECH1	Okazaki et al. (1989, Fig. 4 Gi-3)	$TOTCrO_4^{2-} = 3.3 \times 10^{-4}$ M

TABLE 7.30 CrO_4^{2-}/gibbsite sorption data not used

Source	Data Set	Reason
Okazaki et al. (1989, Fig. 4 Gi-2)	$TOTCrO_4^{2-} = 3.3 \times 10^{-4}$ M	High surface area, low degree of crystallinity indicative of amorphous aluminum hydroxide
Music (1986, Fig. 1)	$TOTAl = 10^{-2}$ M, $TOTCrO_4^{2-}$ = traces, $t = 30$ min	The $Al(OH)_3$ was freshly precipitated. There is no evidence that it was gibbsite
Music (1986, Fig. 2)	$TOTAl = 10^{-2}$ M, pH = 6.5, I = unknown	The $Al(OH)_3$ was an amorphous form
Music (1986, Fig. 5)	$TOTAl = 10^{-2}$ M, $TOTCrO_4^{2-} = 10^{-4}$ M, $TOTCl^- = 0$, 0.25, 0.5, 1 M, $t = 30$ min	The $Al(OH)_3$ was an amorphous form. There was also competitive sorption with chloride
Music (1986, Fig. 6)	$TOTAl = 10^{-2}$ M, $TOTCrO_4^{2-} = 10^{-4}$ M, $TOTMoO_4^{2-} = 0$, 2×10^{-4}, 5×10^{-4}, and 10^{-3} M, $t = 30$ min	The $Al(OH)_3$ was an amorphous form. There was also competitive sorption with molybdate

TABLE 7.31 CrO_4^{2-}/gibbsite surface complexation reactions

$\equiv AlOH^0 + CrO_4^{2-} + 2H^+ = \equiv AlHCrO_4^0 + H_2O$	K_1
$\equiv AlOH^0 + CrO_4^{2-} + H^+ = \equiv AlCrO_4^- + H_2O$	K_2

TABLE 7.32 CrO_4^{2-}/gibbsite intrinsic surface complexation constants

ID	$\log K_1^{int}$	$\sigma_{\log K}$	$\log K_2^{int}$	$\sigma_{\log K}$	WSOS/DF
4ECH1	13.40[a]	0.15[b]	9.50[a]	0.15[b]	—

[a] Constant was fixed manually.
[b] Fixed at this value by convention.

TABLE 7.33 Best estimates for CrO_4^{2-}/gibbsite surface complexation constants

Best-Estimate K	Confidence Level	$\log K^{int}(-)$	$\log K^{int}(+)$
$\log\overline{K_1^{int}} = 13.40$	0.99		
	0.95		
	0.90		
	0.80		
$\log\overline{K_2^{int}} = 9.50$	0.99		
	0.95		
	0.90		
	0.80		

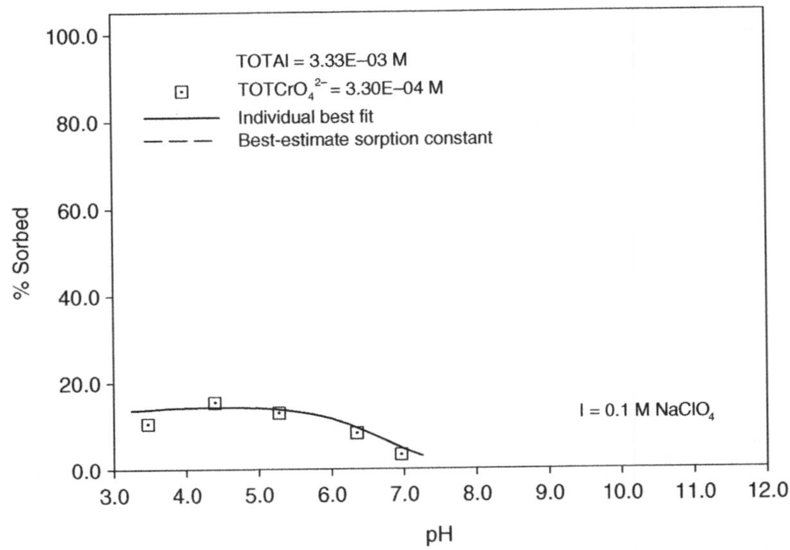

FIGURE 4ECH1

7.9 BORATE

TABLE 7.34 H_3BO_3/gibbsite sorption data

ID	Source	Data Set
4EBO1	Goldberg (2005, Fig 3); Goldberg et al. (1993a, Fig. 2b)	$TOTB = 4.62 \times 10^{-4}$ M
4EBO2	Goldberg (2005, Fig 3); Goldberg et al. (1993a, Fig. 2b)	$TOTB = 4.62 \times 10^{-4}$ M
4EBO3	Goldberg (2005, Fig 3); Goldberg et al. (1993a, Fig. 2b)	$TOTB = 4.62 \times 10^{-4}$ M
4EBO4	Goldberg et al. (1993b, Fig. 1A)	$TOTB = 4.62 \times 10^{-4}$ M

TABLE 7.35 H_3BO_3/gibbsite sorption data not used

Source	Data Set	Reason
Goldberg (2005, Fig. 3)	TOTB $= 4.62 \times 10^{-4}$ M, Gibbsite conc. (25 g/L) $I = 1$ M	High ionic strength. Davies equation cannot be applied
Goldberg et al. (1993b, Fig. 1A)	TOTB $= 4.62 \times 10^{-3}$ M, Gibbsite conc. (25 g/L) $I = 0.1$ M, $T = 10°C$	Low temperature
Goldberg et al. (1993b, Fig. 1A)	TOTB $= 4.62 \times 10^{-3}$ M, Gibbsite conc. (25 g/L) $I = 0.1$ M, $T = 40°C$	High temperature

TABLE 7.36 H_3BO_3/gibbsite surface complexation reactions

$\equiv AlOH^0 + H_3BO_3{}^0 = \equiv AlH_2BO_3{}^0 + H_2O$	K_1
$\equiv AlOH^0 + H_3BO_3{}^0 = \equiv AlH_3BO_4{}^- + H^+$	K_2

TABLE 7.37 H_3BO_3/gibbsite intrinsic surface complexation constants

ID	$\log K_1^{int}$	$\sigma_{\log K}$	$\log K_2^{int}$	$\sigma_{\log K}$	WSOS/DF
4EBO1	1.80[a]	0.15[b]	−7.60[a]	0.15[b]	—
4EBO2	1.82[a]	0.15[b]	−7.70[a]	0.15[b]	—
4EBO3	1.86[a]	0.15[b]	−7.82[a]	0.15[b]	—
4EBO4	0.80[a]	0.15[b]	−7.40[a]	0.15[b]	—

[a] Constant was fixed manually.
[b] Fixed at this value by convention.

TABLE 7.38 Best estimates for H_3BO_3/gibbsite surface complexation constants

Best-Estimate K	Confidence Level	$\log K^{int}(-)$	$\log K^{int}(+)$
$\overline{\log K_1^{int}} = 1.57$	0.99	_[a]	_[a]
	0.95	_[a]	_[a]
	0.90	_[a]	_[a]
	0.80	_[a]	_[a]
$\overline{\log K_2^{int}} = -7.63$	0.99	_[a]	_[a]
	0.95	_[a]	_[a]
	0.90	_[a]	_[a]
	0.80	_[a]	_[a]

[a] Confidence limit not calculated because all K values (and hence $\sigma_{\log K}$ values) fixed manually.

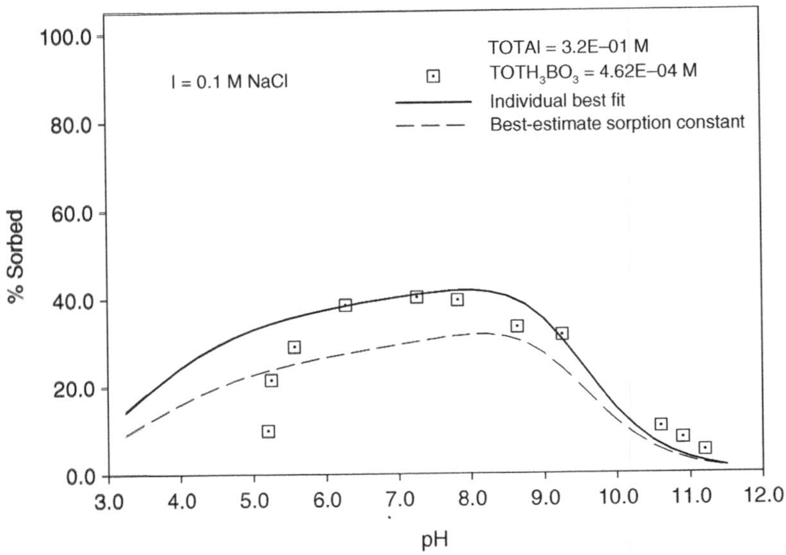

FIGURE 4EBO1

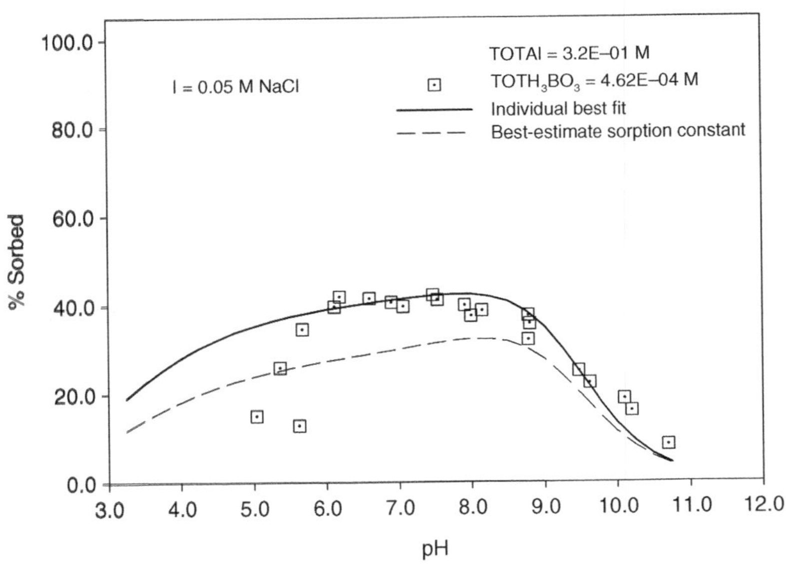

FIGURE 4EBO2

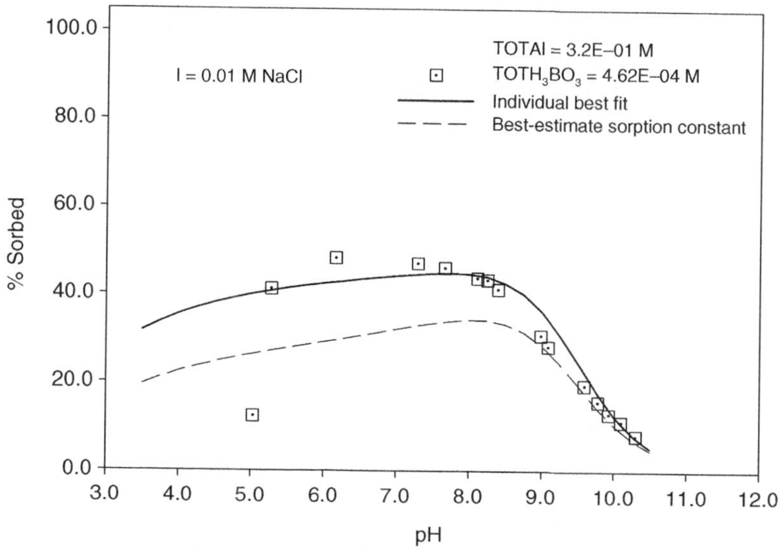

FIGURE 4EBO3

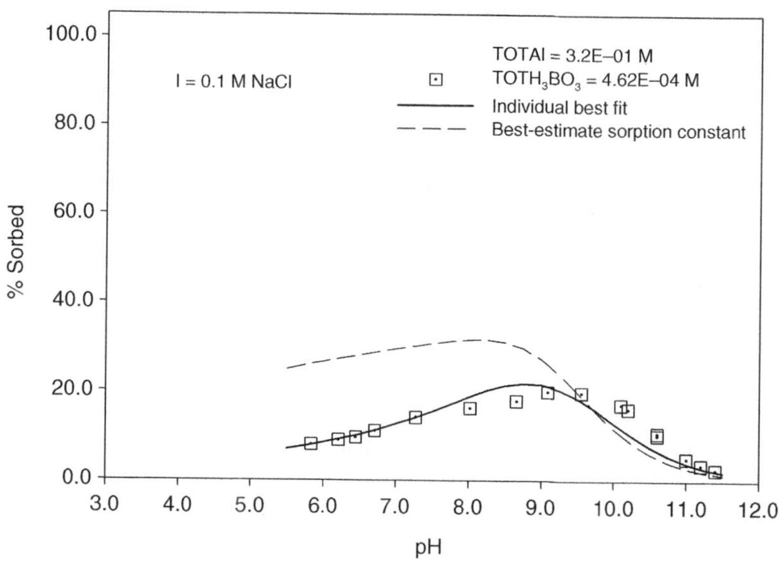

FIGURE 4EBO4

7.10 SULFATE

TABLE 7.39 SO_4^{2-}/gibbsite sorption data

ID	Source	Data Set
4ESO1	Hingston et al. (1972, Fig. 1C)	$TOTSO_4^{2-} = 1.70 \times 10^{-3}$ M
4ESO2	Essington and Anderson (2008, Fig. 3c)	$TOTSO_4^{2-} = 4.00 \times 10^{-3}$ M
4ESO3	Essington and Anderson (2008, Fig. 3c)	$TOTSO_4^{2-} = 4.00 \times 10^{-3}$ M

TABLE 7.40 SO_4^{2-}/gibbsite sorption data not used

Source	Data Set	Reason
Meng and Letterman (1993b, Fig. 8)	$TOTSO_4^{2-} = 1.0 \times 10^{-4}$ M	The characterization of $Al(OH)_3$ used in this study was reported in Meng and Letterman (1993a), where TEM micrographs verified that the $Al(OH)_3$ precipitate was an amorphous form
Meng and Letterman (1993b, Fig. 9)	$TOTSO_4^{2-} = 1.0 \times 10^{-4}$ M	

TABLE 7.41 SO_4^{2-}/gibbsite surface complexation reactions

$\equiv AlOH^0 + SO_4^{2-} + H^+ = \equiv AlSO_4^- + H_2O$	K_2
$\equiv AlOH^0 + SO_4^{2-} = \equiv AlOHSO_4^{2-}$	K_3

TABLE 7.42 SO_4^{2-}/gibbsite intrinsic surface complexation constants

ID	$\log K_2^{int}$	$\sigma_{\log K}$	$\log K_3^{int}$	$\sigma_{\log K}$	WSOS/DF
4ESO1	−0.45[a]	0.15[b]	−1.92	0.22	0.4
4ESO2	−0.45[a]	0.15[b]	0.94[a]	0.15[b]	—
4ESO3	−0.45[a]	0.15[b]	2.42[a]	0.15[b]	—

[a] Constant was fixed manually.

[b] Fixed at this value by convention.

TABLE 7.43 Best estimates for SO_4^{2-}/gibbsite surface complexation constants

Best-Estimate K	Confidence Level	$\log K^{int}(-)$	$\log K^{int}(+)$
$\log\overline{K_2^{int}} = -0.45$	0.99	_[a]	_[a]
	0.95	_[a]	_[a]
	0.90	_[a]	_[a]
	0.80	_[a]	_[a]
$\log\overline{K_3^{int}} = 1.19$	0.99	0.71	1.67
	0.95	0.82	1.55
	0.90	0.88	1.49
	0.80	0.95	1.43

[a]Confidence limit not calculated because all K values (and hence $\sigma_{\log K}$ values) fixed manually.

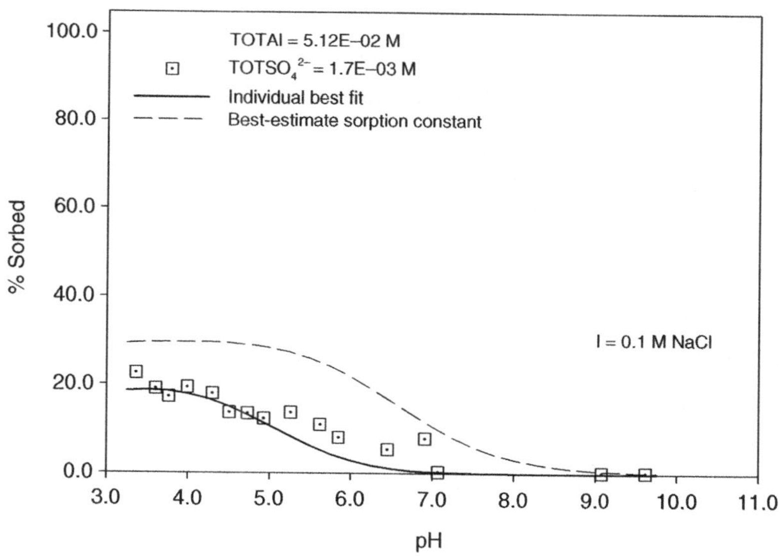

FIGURE 4ESO1

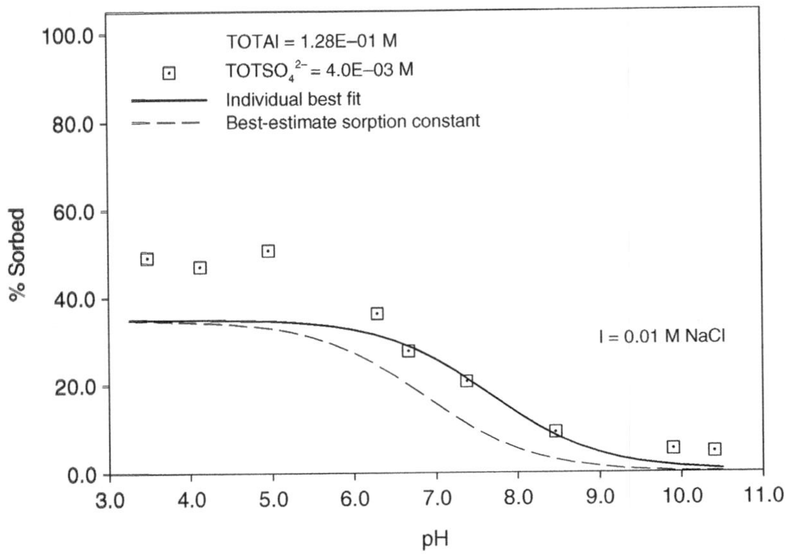

FIGURE 4ESO2

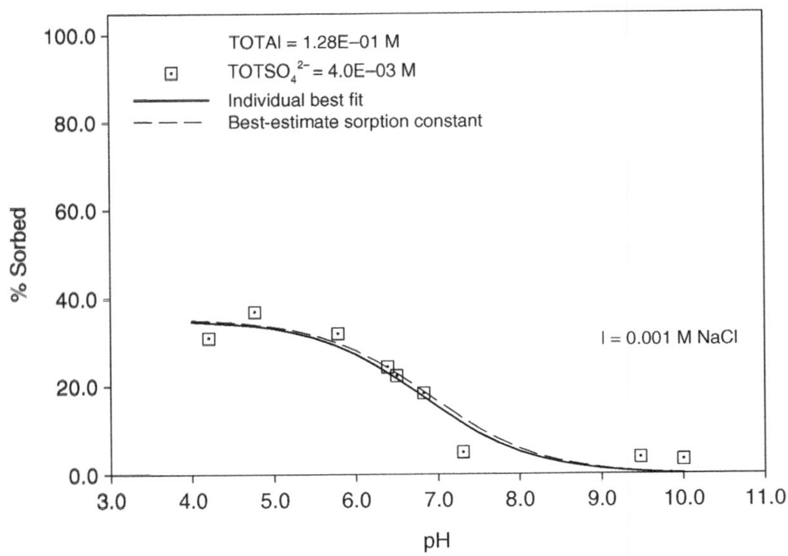

FIGURE 4ESO3

7.11 FLUORIDE

TABLE 7.44 F^-/gibbsite sorption data

ID	Source	Data Set
4EF1	Hingston et al. (1972, Fig. 1a)	TOTF = 4.00×10^{-3} M

TABLE 7.45 F^-/gibbsite sorption data not used

Source	Data Set	Reason
Farrah et al. (1987, Fig. 4b)	TOTF = 6.0×10^{-4} M, Equilibration period = 1, 2, 7, 14 days TOTF = 1.0×10^{-3} M, Equilibration period = 1, 2, 7, 14 days TOTF = 3.0×10^{-4} M, Equilibration period = 1, 2, 7 days TOTF = 1.0×10^{-4} M, Equilibration period = 1, 2, 7 days	No report of the solution ionic strength

TABLE 7.46 F^-/gibbsite surface complexation reactions

$\equiv AlOH^0 + F^- + H^+ = \equiv AlF^0 + H_2O$	K_1
$\equiv AlOH^0 + F^- = \equiv AlOHF^-$	K_2
$\equiv AlOH^0 + 2F^- + H^+ = \equiv AlF_2{}^- + H_2O$	K_3

TABLE 7.47 F^-/gibbsite intrinsic surface complexation constants

ID	$\log K_1^{int}$	$\sigma_{\log K}$	$\log K_2^{int}$	$\sigma_{\log K}$	$\log K_3^{int}$	$\sigma_{\log K}$	WSOS/DF
4EF1	8.78[a]	0.15[b]	2.88[a]	0.15[b]	11.94[a]	0.15[b]	—

[a] Constant was fixed manually.
[b] Fixed at this value by convention.

TABLE 7.48 Best estimates for F^-/gibbsite surface complexation constants

Best-Estimate K	Confidence Level	$\log K^{int}(-)$	$\log K^{int}(+)$
$\log \overline{K_1^{int}} = 8.78$	0.99		
	0.95		
	0.90		
	0.80		
$\log \overline{K_2^{int}} = 2.88$	0.99		
	0.95		
	0.90		
	0.80		
$\log \overline{K_3^{int}} = 11.94$	0.99		
	0.95		
	0.90		
	0.80		

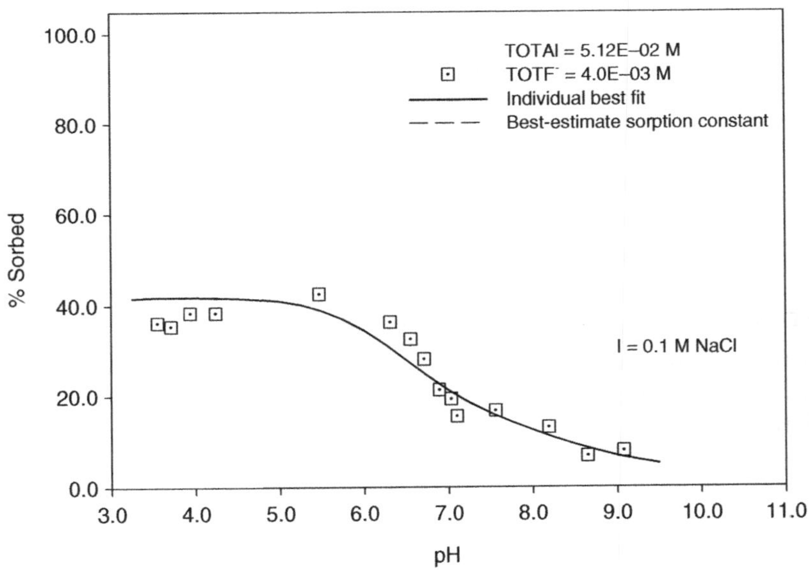

FIGURE 4EF1

7.12 SILICATE

TABLE 7.49 H_4SiO_4/gibbsite sorption data

ID	Source	Data Set
4ESI1	Jepson et al. (1976, Fig. 1)	$TOTH_4SiO_4 = 1.6 \times 10^{-3}$ M

TABLE 7.50 H_4SiO_4/gibbsite sorption data not used

Source	Data Set	Reason
Adu-Wusu and Wilcox (1991b, Fig. 1)	Various sorption isotherms at T = 30°C, 40°C, 50°C, 60°C	High temperature of the experiments (>25°C)
Jepson et al. (1976, Fig. 2)	$I = 0.02$ M, pH = 4.0 $I = 0.02$ M, pH = 9.0	Potential precipitation of $SiO_2(s)$

TABLE 7.51 H_4SiO_4/gibbsite surface complexation reaction

$\equiv AlOH^0 + H_4SiO_4{}^0 = \equiv AlOH_2SiO_4{}^- + H^+$	K_2[a]

[a]The numbering of surface complexation constant follows the numbering of the generic complexation reaction constants given in Table 8.5 of Chapter 8.

TABLE 7.52 H_4SiO_4/gibbsite intrinsic surface complexation constants

ID	$\log K_2^{int}$	$\sigma_{\log K}$	WSOS/DF
4ESI1	−4.16[a]	0.15[b]	—

[a]Constant was fixed manually.
[b]Fixed at this value by convention.

TABLE 7.53 Best estimates for H_4SiO_4/gibbsite surface complexation constants

Best-Estimate K	Confidence Level	$\log K_1^{int}(-)$	$\log K_1^{int}(+)$
$\overline{\log K_2^{int}} = -4.16$	0.99		
	0.95		
	0.90		
	0.80		

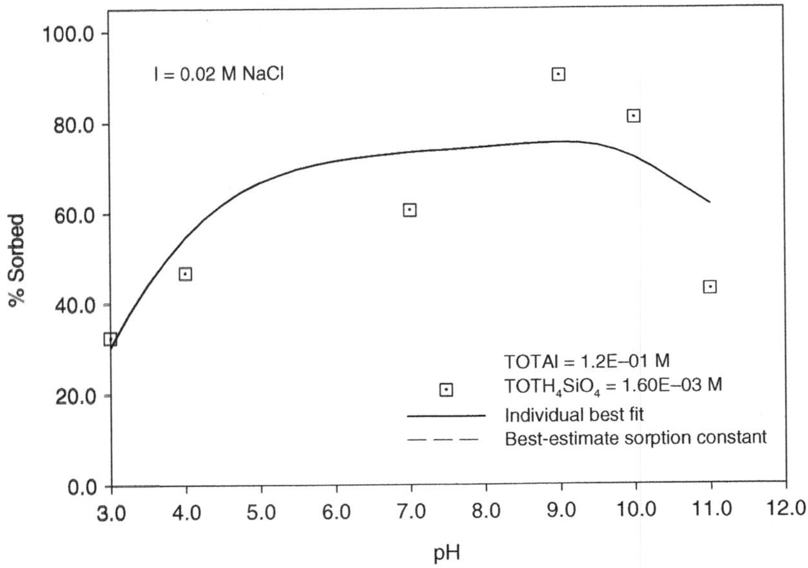

FIGURE 4ESI1

8

COHERENCE AND EXTRAPOLATION OF THE RESULTS

In this chapter, the results obtained from modeling available data for cation and anion sorption on gibbsite with the generalized two-layer model are examined for internal consistency. The results are also used in linear free energy relationships (LFERs) to estimate surface-complexation constants for ions for which limited or no sorption data exists. In addition, LFERs are developed and used to compare the surface-complexation constants obtained for ion sorption on gibbsite with those obtained previously with the generalized two-layer model for hydrous ferric oxide (HFO), goethite, and manganese dioxide.

8.1 CATION SORPTION ON GIBBSITE

The best-estimate sorption constants for cations are presented in Table 8.1. Their corresponding reactions are also provided. The individual best fits showed that no more than two surface species were required to describe the sorption for any cation except for UO_2^{2+}, for which three surface species were needed. The surface species $\equiv AlOM^+$ was sufficient for most cations. For UO_2^{2+}, the additional species $\equiv AlOUO_2(OH)_3^{2-}$ and $\equiv AlO(UO_2)_3(OH)_5^0$ were needed. Similarly, for modeling mercury and thorium-sorption data additional surface species were necessary, namely $\equiv AlOH\text{–}HgOHCl$ and $\equiv AlOTh(OH)^{2+}$ respectively.

For cation sorption on gibbsite, only a single type of surface site had to be considered, as was also the case with goethite (Mathur and Dzombak, 2006). However, in previous studies of cation surface complexation with HFO (Dzombak

Surface Complexation Modeling: Gibbsite By Athanasios K. Karamalidis and David A. Dzombak

TABLE 8.1 Surface complexation reactions and constants for some cations on gibbsite

Ion	$\log K_1^{int}$	$\log K_2^{int}$	$\log K_3^{int}$	$\log K_4^{int}$	$\log K_5^{int}$	$\log K_6^{int}$
Cu^{2+}	0.25 ± 0.08	—	—	—	—	—
Pb^{2+}	0.37 ± 0.03	—	—	—	—	—
Co^{2+}	-2.52 ± 0.05	—	—	—	—	—
Cd^{2+}	-2.73 ± 0.07	—	—	—	—	—
Mn^{2+}	$-5.49 \pm ?$	—	—	—	—	—
Fe^{2+}	$-3.77 \pm ?$	—	—	—	—	—
Ca^{2+}	-10.49 ± 0.04	—	—	—	—	—
Zn^{2+}	-0.96 ± 0.2	—	—	—	—	—
Hg^{2+}	3.17^a					
UO_2^{2+}	1.22 ± 0.17	—	22.0 ± 0.01	15.62 ± 1.40	—	—
Th^{4+}	—	—	—	—	10.83 ± 0.24	3.40 ± 0.18

Reactions

$\equiv AlOH^0 + M^{2+}$	$= \equiv AlOM^+ + H^+$	K_1
$\equiv AlOH^0 + M^{2+} + 2H_2O$	$= \equiv AlOM(OH)_2^- + 3H^+$	K_2
$\equiv AlOH^0 + M^{2+} + 3H_2O$	$= \equiv AlOM(OH)_3^{2-} + 4H^+$	K_3
$\equiv AlOH^0 + 3M^{2+} + 5H_2O$	$= \equiv AlOM_3(OH)_5^0 + 6H^+$	K_4
$\equiv AlOH^0 + M^{4+}$	$= \equiv AlOM^{3+} + H^+$	K_5
$\equiv AlOH^0 + M^{4+} + H_2O$	$= \equiv AlOM(OH)^{2+} + 2H^+$	K_6

Note: The reported uncertainties correspond to 95% confidence limit; T = 20–30°C; I = 0; dash indicates insufficient data to determine constants or nonexisting reaction for the specific cation.

? indicates that constants were derived from one data set, so uncertainties could not be calculated.

[a] Surface-complexation constant for Hg^{2+} was calculated from the surface-complexation constant of 9.34 ± 0.14 which corresponds to the surface-complexation reaction with $Hg(OH)_2^0$. Modeling was performed based on the surface-complexation reactions for $Hg(OH)_2^0$ (see Chapter 6, Table 6.42).

and Morel, 1990) and with hydrous manganese oxide (HMO) (birnessite, vernadite, and δ-MnO_2) (Tonkin et al., 2004) two site types were needed: a small concentration of high-affinity sites and a large concentration of low-affinity sites. As was the case with goethite, the more ordered crystalline structure of gibbsite appears to present less diversity of binding sites. However, aluminum hydroxide polymorphism causes variations in gibbsite crystals and some diversity in binding sites among gibbsite crystals is expected to occur.

Due to the similarity of cation surface-complexation reactions on oxides and hydrolysis reactions in solution, the energy of interaction of metal ions with hydroxyl ions in solution can be correlated with the energy of interaction of metal ions with surface hydroxyl groups (Schindler et al., 1976; Dzombak and Morel, 1990; Mathur and Dzombak, 2006). Such a relationship between the free energies of formation of aqueous complexes and thermodynamic properties of the metal ions is commonly termed a LFER.

In Figure 8.1, the cation surface-complexation constants for the formation of the species $\equiv AlOM^+$ are plotted against the first hydrolysis constants of the divalent cations studied. Similar plots for other surface species are not presented because other species were used for only a few cations and the results are not sufficient for LFER development. The LFER plot presented in Figure 8.1 can be used to predict surface-complexation constants for divalent cations for which limited or no sorption data exist. Predictions of surface-complexation constants are presented in Table 8.2.

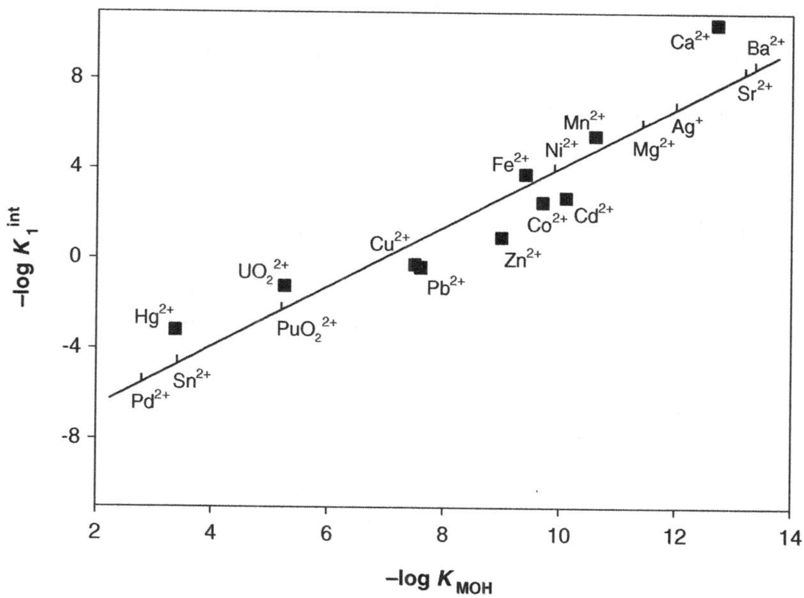

FIGURE 8.1 Correlation of $\log K_1^{int}$ with the first hydrolysis constant for various divalent cations (see Table 8.1 for the surface reactions corresponding to K_1^{int}). The fitted line has a slope of 1.32, a y-intercept of −9.19, and is the result of a linear least squares regression with $R^2 = 0.84$ (95 percent confidence limit). Filled boxes (■) are derived, while vertical marks (I) are estimated surface complexation constants.

TABLE 8.2 Derived and estimated surface complexation constants for sorption of some cations on gibbsite

Ion	$-\log K_{MOH}$ [a]	$\log K_1^{int}$	$\log K_2^{int}$	$\log K_3^{int}$	$\log K_4^{int}$
Ba^{2+}	13.36	−8.50 [b]	—	—	—
Sr^{2+}	13.18	−8.26 [b]	—	—	—
Ca^{2+}	12.69	−10.49	—	—	—
Mg^{2+}	11.42	−5.93 [b]	—	—	—
Ag^{+}	11.99	−6.70 [b,c]	—	—	—
Mn^{2+}	10.59	−5.49	—	—	—
Co^{2+}	9.69	−2.52	—	—	—
Ni^{2+}	9.89	−3.92 [b]	—	—	—
Cd^{2+}	10.09	−2.73	—	—	—
Zn^{2+}	8.99	−0.96	—	—	—
Cu^{2+}	7.49	0.25	—	—	—
Pb^{2+}	7.59	0.37	—	—	—
UO_2^{2+}	5.25	1.22	—	−22.00	−15.62
PuO_2^{2+}	5.20	2.31 [b]	—	—	—
Hg^{2+}	3.36	3.17	—	—	—
Sn^{2+}	3.39	4.69 [b]	—	—	—
Pd^{2+}	2.79	5.50 [b]	—	—	—
Fe^{2+}	9.39	−3.77	—	—	—

Note: This represents a completion and extension of Table 8.1; see Table 8.1 for corresponding reactions. Dash indicates reactions that were not considered because they were not necessary for fitting the data sets of the respective cation.

[a] From Smith et al. (2003; NIST 46.7). All reactions are written as: $M^{2+} + H_2O = MOH^+ + H^+$

[b] Estimated using the LFER in Figure 8.1

[c] Surface reactions for monovalent Ag^+ ion: $\equiv AlOH^0 + Ag^+ = \equiv AlOAg^0 + H^+ \quad K_1$
$\equiv AlOH^0 + Ag^+ = \equiv AlOHAg^0 \quad K_2$

A second LFER approach for estimating surface-complexation constants is based on the dependency of both surface complexation and hydrolysis constants on ion properties such as charge and size. Brown et al. (1985) determined that the effective nuclear charge of a cation can be related to the hydrolysis constant via:

$$\log\beta_{11} = \text{intercept} + \text{slope}\left[g_1\left(\frac{z}{r^2} + g_2\right)\right] \tag{8.1}$$

where β_{11} is the first hydrolysis constant for the cation, z the formal charge of the cation, r the ionic radius, and g_1 and g_2 are defined as

$$g_1 = (1+2S+D)(z+2) \tag{8.2}$$

$$g_2 = g(n)(z-1) + 0.1d(n-3)^2(1-S) \tag{8.3}$$

S is equal to 0 if there are no s electrons in the outer shell of the ion and equal to 1 if such electrons are present; D is the number of d electrons in the neutral metal; n is the principal quantum number of the outermost shell of the ion; $g(n)$ is equal to 0 when n is 1 and equal to 1 when n is greater than 1; and d is the number of d electrons possessed by the ion. Values of g_1, g_2, and r for cations considered in this study are reported in Table 8.3.

Equation 8.1 can be extended to surface-complexation reactions and to $\log K_1^{\text{int}}$.

$$\log K_1^{\text{int}} = \text{intercept} + \text{slope}\left[g_1\left(\frac{z}{r^2} + g_2\right)\right] \tag{8.4}$$

Table 8.2 presents $\log K_1^{\text{int}}$ values for the cations modeled in this study along with corresponding g_1, g_2, and r values. In Figure 8.2, $\log K_1^{\text{int}}$ values are plotted against the $[g_1((z/r^2) + g_2)]$ values for the various cations. The resulting linear best fit has an intercept of -10.11 and slope equal to 0.19. Thus, Equation (8.4) becomes:

$$\log K_1^{\text{int}} = -10.11 + 0.19\left[g_1\left(\frac{z}{r^2} + g_2\right)\right] \tag{8.5}$$

The linear relationship presented in Figure 8.2 and Equation 8.5 can be used to predict surface-complexation constants for cations for which limited or no sorption data exist. These results are presented in Table 8.4. The constants predicted for Ba^{2+}, Sr^{2+}, Mg^{2+}, and Ag^{+} are very similar to those predicted using the correlation with the

TABLE 8.3 Values for surface complexation constants, and g_1, g_2, and r for some cations

Ion	$\log K_1^{\text{int}}$	g_1	g_2	r (Å)
Cu^{2+}	0.25	8	1	0.62
Pb^{2+}	0.37	16	1	0.94
Co^{2+}	−2.52	8	1	0.74
Cd^{2+}	−2.73	8	2	0.97
Mn^{2+}	−5.49	8	1	0.82
Fe^{2+}	−3.77	8	1	0.77
Ca^{2+}	−10.49	4	1	1.00
Zn^{2+}	−0.96	8	1	0.74
Hg^{2+}	3.17	8	5	0.69
UO_2^{2+}	1.22	8	1	0.87[a]
Th^{4+}	—	12	3	1.00

Note: Values for $\log K_1^{\text{int}}$ are from Table 8.1. Values for g_1, g_2, and r are obtained from Brown et al. (1985).
[a]For consistency with the rest of values reported in this table taken from Brown et al. (1985), this ionic radius corresponds to a uranyl ion with maximum equatorial coordination number of six, with an overall coordination number eight around uranium(VI). There is also evidence of an equatorial coordination number eight for the UO_2^{2+} ion with an overall coordination number ten around uranium(VI) in the solid state (Ahuja et al., 1980). If this is the case, then an ionic radius of 1.00 should be used instead.

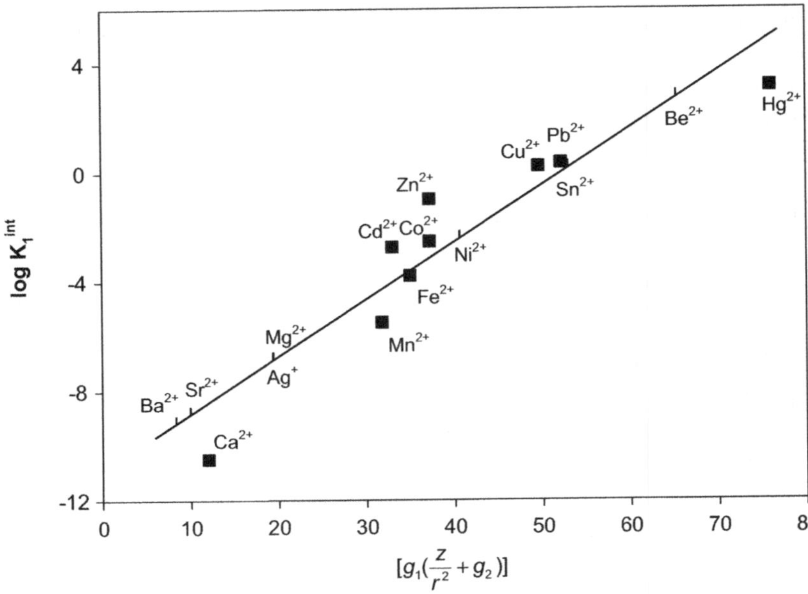

FIGURE 8.2 Correlation of $\log K_1^{int}$ with the $[g_1(z/r^2 + g_2)]$ for various divalent cations (see Table 8.3 for values of g_1, g_2, and r). The fitted line has a slope of 0.21, a y-intercept of -10.91, and is the result of a linear least squares regression with $R^2 = 0.87$ (95 percent confidence limit). Filled boxes (■) are derived, while vertical marks (I) are estimated surface complexation constants. The $\log K_1^{int}$ versus $[g_1(z/r^2 + g_2)]$ for uranyl (UO_2^{2+}) is not shown here. If it is included in the plot, then the fitted line would have a slope of 0.18, a y-intercept of -9.38, and $R^2 = 0.65$ (95 percent confidence limit).

first hydrolysis constants (Fig. 8.1), while for Ni^{2+} and Sn^{2+} some differences were observed.

8.2 ANION SORPTION ON GIBBSITE

The best-estimate surface-complexation constants for anions are presented in Tables 8.5 and 8.6. The generalized forms of the reactions used to describe anion sorption on gibbsite are presented in the same tables.

Anion sorption occurs over a wide range of pH. Most of the anion data were fit considering multiple surface species, reflecting the multiple protonated states of surface species over the pH range also. The necessity of using several different surface species is consistent with previous modeling of anion sorption on HFO, goethite, and other oxides. For simplicity, only 1 : 1 surface species were considered, although bidentate surface species in the sorption of some anions have been reported in the literature (see the discussion in Chapter 7).

LFERs can be developed to relate the tendency of anions to react with protons and with surface metal atoms to form surface complexes,that is, the relationship of the logarithms of the surface-complexation constants and the pK_a value(s) for the

TABLE 8.4 Derived and estimated surface complexation constants for sorption of some cations on gibbsite

Ion	g_1	g_2	r(Å)	$g_1([z/r^2]+g_2)$	$\log K_1^{int}$
Ba^{2+}	4	1	1.36	8.32	−8.50
Sr^{2+}	4	1	1.16	9.96	−8.19
Ca^{2+}	4	1	1.00	12.00	−10.49[a]
Mg^{2+}	4	1	0.72	19.43	−6.36
Ag^{+}	6	1	0.67	19.36	−6.48
Mn^{2+}	8	1	0.82	31.80	−5.49[a]
Co^{2+}	8	1	0.74	37.20	−2.52[a]
Ni^{2+}	8	1	0.70	40.65	−2.25
Cd^{2+}	8	2	0.97	33.00	−2.73[a]
Zn^{2+}	8	1	0.74	37.20	−0.96[a]
Fe^{2+}	8	1	0.77	35.00	−3.77[a]
Cu^{2+}	8	1	0.62	49.64	0.25[a]
Pb^{2+}	16	1	0.94	52.20	0.37[a]
UO_2^{2+}	8	1	0.87	29.13	1.22[a]
Be^{2+}	4	0	0.35	65.32	2.51
Hg^{2+}	8	5	0.69	76.00	3.17[a]
Sn^{2+}	16	1	0.93	52.99	0.12
Th^{4+}	12	3	1.00	84.00	—

[a] Values for $\log K_1^{int}$ are from Table 8.1. Values for g_1, g_2, and r are obtained from Brown et al. (1985), except for UO_2^{2+} the values for which are calculated in this study.

deprotonation of the sorbing anion. In Figures 8.3 and 8.4, this linear relationship is depicted by the plots of LFERs obtained for log K_1^{int} and log K_2^{int} versus pK_a for the second deprotonation constants of the anions evaluated in this study. LFERs could not be obtained for log K_3^{int} and log K_4^{int} due to the small number of anions for which these constants were determined.

The LFERs in Figures 8.3 and 8.4 can be used to predict surface-complexation constants for anions for which limited or no data exist. Derived and estimated surface-complexation constants for various anions on gibbsite are presented in Table 8.7.

In Table 8.7, predicted surface-complexation constants are presented only for those anions for which there were no available data to fit surface-complexation reactions and constants. The accuracy of these predictions is unknown. Experience with similar predictions made for HFO (Dzombak and Morel, 1990) has indicated that LFER-predicted surface-complexation constants overpredict sorption for some anions, such as cyanide (CN^-) (Dzombak et al., 2006). In contrast to the LFER predictions made for HFO (Dzombak and Morel, 1990) this study did not predict surface-complexation constants for anions for which there were available data, that is, no predictions were made for constants associated with surface-complexation reactions for which the data available were not sufficient to fit reactions and constants that could be relevant. The addition of a predicted surface-complexation reaction and constant could affect the fit of available data. The predicted constants should be used with caution, considering the uncertainty about their accuracy.

TABLE 8.5 Surface complexation constants for some anions on gibbsite

Ion	$-\log K_a$ [a]	$\log K_1^{int}$	$\log K_2^{int}$	$\log K_3^{int}$	$\log K_4^{int}$
AsO_4^{3-}	11.80	26.26	21.60	—	6.24
SO_4^{2-}	1.99	—	-0.45	1.19	—
SeO_4^{2-}	1.70	—	—	4.53	—
CrO_4^{2-}	6.51	13.40	9.50	—	—
MoO_4^{2-}	4.24	14.77	3.72	1.87	—
$H_2SiO_4^{2-}$	13.20	—	18.88[b]	—	—
PO_4^{3-}	12.37	26.31	19.48	13.56	—
F^-	3.18	8.78	2.88	11.94	—

Reactions

$\equiv AlOH^0 + A^{3-} + 3H^+$	$= \equiv AlH_2A^0\equiv + H_2O$	K_1
$\equiv AlOH^0 + A^{3-} + 2H^+$	$= \equiv AlHA^- + H_2O$	K_2
$\equiv AlOH^0 + A^{3-} + H^+$	$= \equiv AlA^{2-} + H_2O$	K_3
$\equiv AlOH^0 + A^{3-}$	$= \equiv AlOHA^{3-}$	K_4
$\equiv AlOH^0 + A^{2-} + 2H^+$	$= \equiv AlHA^0 + H_2O$	K_1
$\equiv AlOH^0 + A^{2-} + H^+$	$= \equiv AlA^- + H_2O$	K_2
$\equiv AlOH^0 + A^{2-}$	$= \equiv AlOHA^{2-}$	K_3
$\equiv AlOH^0 + A^{2-}$	$= \equiv AlOA^{3-} + H^+$	K_4
$\equiv AlOH^0 + A^- + H^+$	$= \equiv AlOH_2A^0$	K_1
$\equiv AlOH^0 + A^-$	$= \equiv AlOHA^-$	K_2
$\equiv AlOH^0 + 2A^-$	$= \equiv AlA_2^- + H_2O$	K_2'

Note: The reported uncertainties correspond to 95% confidence limit; $T = 20$–$30°C$; $I = 0$; dash indicates insufficient data to determine constants or nonexisting reaction for the specific anion.
[a]From Smith et al. (2003; NIST 46.7). Corresponds to the last deprotonation reaction, that is, pK_{a2} for divalent ions and pK_{a3} for trivalent ions.
[b]The reported value was calculated from the reactions: $\equiv AlOH^0 + H_4SiO_4^0 = \equiv AlOH_4SiO_4^- + H^+$, $\log K_2^{int} = -4.16$, and $H_2SiO_4^{2-} + 2H^+ = H_4SiO_4^0$, $\log K = 23.04$.

TABLE 8.6 Surface complexation constants for H_3A^0 type anions on gibbsite

Ion	$-\log K_a$ [a]	$\log K_1^{int}$	$\log K_2^{int}$
H_3AsO_3	9.17	1.78	—
H_3BO_3	9.24	1.57	-7.63

Reactions

$\equiv AlOH^0 + H_3A^0$	$= \equiv AlH_2A^0 + H_2O$	K_1
$\equiv AlOH^0 + H_3A^0$	$= \equiv AlOH_3A^- + H^+$	K_2

Note: The reported uncertainties correspond to 95% confidence limit; $T = 20$–$30°C$; $I = 0$; dash indicated insufficient data to determine constants or nonexisting reaction for the specific anion.
[a]From Smith et al. (2003; NIST 46.7). Corresponds to the first deprotonation reaction.

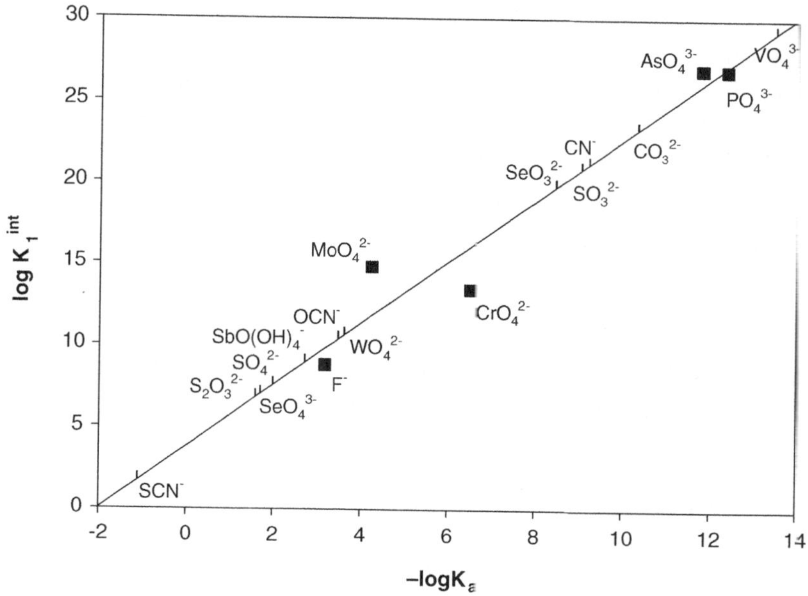

FIGURE 8.3 Correlation of logK_1^{int} with the second deprotonation constant for various anions (see Table 8.5 for the surface reactions corresponding to K_1^{int}). The fitted line has a slope of 1.886, a y-intercept of 3.784, and is the result of a linear least squares regression with $R^2 = 0.94$ (95 percent confidence limit). Filled boxes (■) are derived, while vertical marks (I) are estimated surface complexation constants.

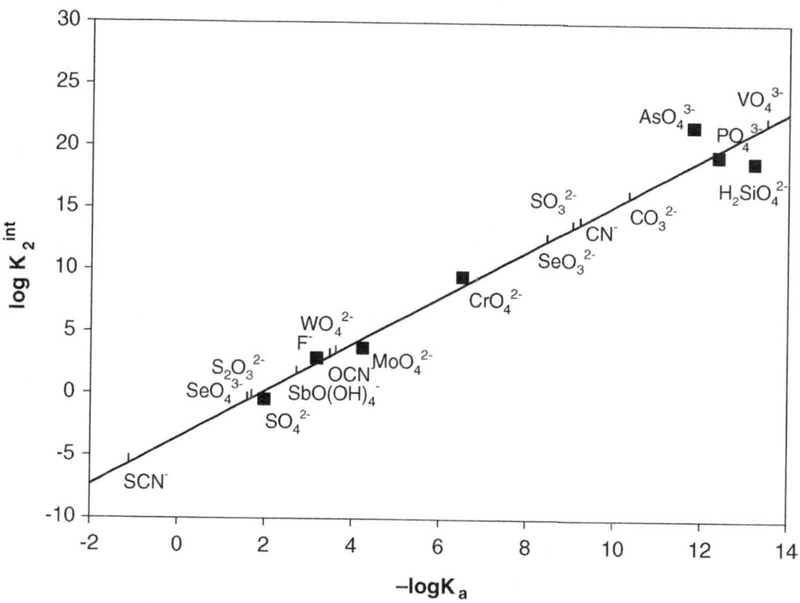

FIGURE 8.4 Correlation of logK_2^{int} with the second deprotonation constant for various anions (see Table 8.5 for the surface reactions corresponding to K_2^{int}). The fitted line has a slope of 1.890, a y-intercept of −3.575, and is the result of a linear least squares regression with $R^2 = 0.97$ (95 percent confidence limit). Filled boxes (■) are derived, while vertical marks (I) are estimated surface complexation constants.

From the knowledge of the equilibrium constants for the formation of AlA in solution, where A is an anion, surface-complexation constants can in principle be estimated for the formation of $\equiv$AlA by means of the LFER in Figure 8.5. In this study, the limited surface-complexation reactions with analogous solution reactions, all presented in Table 8.8, were not sufficient to allow predictions. However, the plot of log K^{int} versus log K for the three matching reactions yields a good correlation which is another indication of the internal consistency of the data and modeling methodology used in this study.

8.3 COMPARISON OF GIBBSITE SURFACE-COMPLEXATION CONSTANTS WITH THOSE OF GOETHITE, HYDROUS FERRIC OXIDE, AND HYDROUS MANGANESE OXIDE

In this section, we examine the hypothesis that surface hydroxyl groups on common hydrous metal oxides such as goethite, HFO, gibbsite exhibit similar reactivity and surface-complexation constants for the same sorbing ions when data interpretation is done in a consistent manner. The surface reactivity of hydrous metal oxides is determined by their reactive surface hydroxyl groups, and it might be expected that the energy of reaction of a particular ion with a surface hydroxyl site would be consistent, especially for oxides of the same metal, provided that differences in surface parameters such as surface area and surface site density are treated consistently and taken into account (Mathur and Dzombak, 2006). Here, we compare the surface-complexation constants for ion sorption on gibbsite, goethite, HFO, and HMO. All surface-complexation constants were obtained using the generalized two-layer model, and consistent surface properties were used in the modeling of ion sorption on each solid. Tonkin et al. (2004) compiled data for the sorption of divalent metal cations on HMO, and fitted the data with the generalized two-layer model following the procedures of Dzombak and Morel (1990). Data for the synthetic HMO phases birnessite, vernasite, and δ-MnO_2 were compiled and considered together.

A comparison of cation surface-complexation constants obtained for gibbsite, goethite, HFO, and HMO is presented in Table 8.9. All surface-complexation constants for goethite are from Mathur and Dzombak (2006), for HFO are from Dzombak and Morel (1990), and for HMO are from Tonkin et al. (2004). Mathur and Dzombak (2006) used one type of binding site to describe cation sorption on goethite, while Dzombak and Morel (1990) and Tonkin et al. (2004) used two site types for HFO and HMO, respectively: strong (high-affinity) and weak (low-affinity) sites. In Table 8.9, only the surface-complexation constants obtained for sorption on weak sites were used for the HFO and HMO minerals, for comparison with the cation surface-complexation constants for gibbsite and goethite.

The cation surface-complexation constants for gibbsite, goethite, and HFO in Table 8.9 show consistency with each other, while those for HMO differ from the other solids. This consistency among the first three minerals and the inconsistency with HMO can be seen graphically by plotting the surface-complexation constants (log K_1^{int}) for goethite, HFO, and HMO versus those of gibbsite (Fig. 8.6). Figures 8.6a and

TABLE 8.7 Derived and estimated surface complexation constants for sorption of some anions on gibbsite

Ion	$-\log K_a$[a]	$\log K_1^{\text{int}}$	$\log K_2^{\text{int}}$	$\log K_3^{\text{int}}$	$\log K_4^{\text{int}}$
AsO_4^{3-}	11.8	26.94	21.79		6.39
H_3AsO_3	9.17	1.78			
CO_3^{2-}	10.33	21.48[c]	15.93[d]		
SO_4^{2-}	1.99		−0.45	1.19	
SO_3^{2-}	9.04	20.07[c]	13.50[d]		
SeO_4^{2-}	1.70			4.53	
SeO_3^{2-}	8.46	19.44[c]	12.40[d]		
H_4SiO_4	9.84		−4.16		
$S_2O_3^{2-}$	1.60	11.94[c]	−0.55[d]		
CrO_4^{2-}	6.51	13.40	9.50		
WO_4^{2-}	3.62	14.15[c]	3.26[d]		
MoO_4^{2-}	4.24	14.77	3.72	1.87	
VO_4^{3-}	13.49[b]	28.52[c]	23.41[d]		
$SbO(OH)_4^-$	2.72[b]	13.16[c]	1.56[d]		
CNO^-	3.48	13.99[c]	3.00[d]		
CN^-	9.21	20.26[c]	13.82[d]		
SCN^-	−1.10	8.99[c]	−5.65[d]		
F^-	3.18	8.78	2.88	11.94	
PO_4^{3-}	12.37	26.89	19.37	13.57	
H_3BO_3	9.23	1.57	−7.63		

Note: This represents a completion and extension of Table 8.5; see Table 8.5 for corresponding reactions.
[a]From Smith et al. (2003) (NIST 46.7). Corresponds to the last deprotonation reaction, that is, pK_{a2} for divalent ions and pK_{a3} for trivalent ions.
[b]From Smith and Martell (1976), Cruywagen (2000) (for vanadate).
[c]Estimated using the LFER in Figure 8.3.
[d]Estimated using the LFER in Figure 8.4.

8.6b show linear relationships with relatively high R^2 values, while the plot for HMO versus gibbsite surface-complexation constants in Figure 8.6c shows a poor correlation ($R^2 = 0.13$).

The weaker correlation exhibited by the surface-complexation constants obtained for HMO relative to those for HFO, goethite, and gibbsite may relate to the difficulty of achieving uniform composition and properties of HMO in synthesizing it for sorption experiments which translates to difficulty in compiling sorption data obtained with solids of consistent properties. Tonkin et al. (2004) acknowledged this in their work. Because of the multiple oxidation states of Mn under environmental conditions (Mn^{2+}, Mn^{3+}, Mn^{4+}), several different forms of manganese oxides and hydroxides occur under environmental conditions (Murray, 1974; Hem, 1985). The numerous HMO phases found in nature commonly have the added complexity of multiple oxidation states (Foster, 2003) and form more complex tunnel and layer structures than iron or aluminum (hydr)oxides (Manceau and Combes, 1988). In the database-development effort of Tonkin et al. (2004) for HMO, which comprised data for three different types of synthesized manganese oxide, it was difficult for the

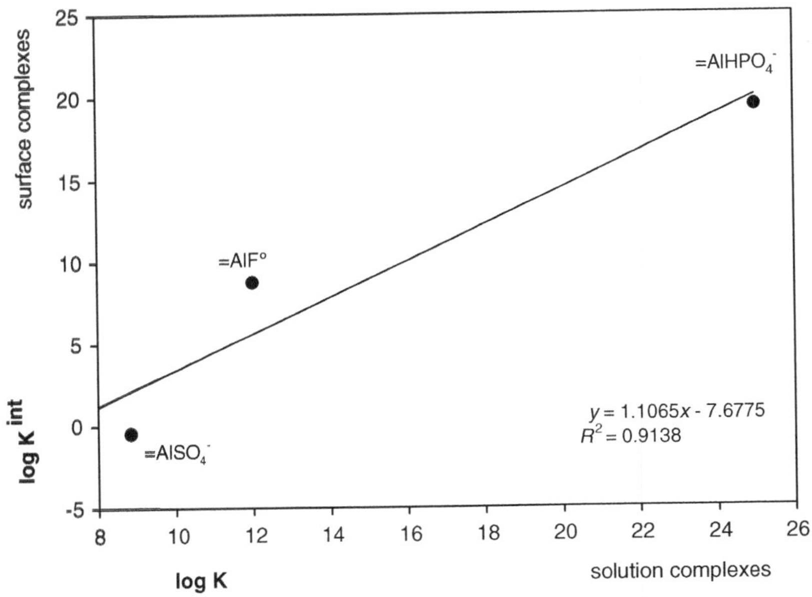

FIGURE 8.5 Surface complex formation constants (log K^{int}) versus log K for analogous solution reactions. The reactions are presented in Table 8.8.The fitted line has a slope of 1.1065, a y-intercept of −7.6775, and is the result of a linear least squares regression with $R^2 = 0.91$ (95 percent confidence limit).

authors to ensure that all sorption data sets evaluated were obtained with manganese oxide solids possessing the same basic surface properties of specific surface area and site density. Nevertheless, the database is a valuable product and the cation surface-complexation constants obtained for HMO are consistent in order of affinity with those obtained for HFO, goethite, and gibbsite.

The intrinsic surface-complexation constants derived for metal cation sorption on gibbsite, goethite, and HFO are plotted against their respective first hydrolysis constants in Figure 8.7. As evident in the plot, and as also indicated by Figures 8.6a and 8.6b, the cation surface-complexation constants derived for the three oxides are quite consistent. Thus, Figure 8.7 supports the hypothesis that

TABLE 8.8 Surface reactions and analogous solution reactions for selected anions

Reaction	log K	log K^{int}
$AlOH^{2+} + PO_4^{3-} + 2H^+ = AlHPO_4^+ + H_2O$	25.007	
$\equiv AlOH^0 + PO_4^{3-} + 2H^+ = \equiv AlHPO_4^-$		19.37 ± ?
$AlOH^{2+} + SO_4^{2-} + H^+ = AlSO_4^+ + H_2O$	8.837	
$\equiv AlOH^0 + SO_4^{2-} + H^+ = \equiv AlSO_4^- + H_2O$		−0.45 ± ?
$AlOH^{2+} + F^- + H^+ = AlF^{2+} + H_2O$	12.007	
$\equiv AlOH^0 + F^- + H^+ = \equiv AlF^0$		8.78 ± ?

? indicates that constants were derived from very few datasets, so uncertainties could not be calculated.

TABLE 8.9 Comparison of intrinsic surface complexation constants for cation sorption on goethite, hydrous ferric oxide (HFO), and hydrous manganese oxide (HMO)

Ion	$-\log K_{MOH}$ [a]	$\log K_1^{int}$(Gibbsite)	$\log K_1^{int}$(Goethite)	$\log K_1^{int}$(HFO)	$\log K_1^{int}$(HMO)
Cu^{2+}	7.49	0.25	1.39	—	0.85
Cd^{2+}	10.09	−2.73	−1.96	−2.90	−2.4
Zn^{2+}	8.99	−0.96	−0.96	−1.99	−0.01
Pb^{2+}	7.59	0.37	0.44	—	—
Co^{2+}	9.69	−2.52	−0.79	−3.01	1.0
Ni^{2+}	9.89	— [b]	−1.96	—	−0.48
Hg^{2+}	3.36	3.17	3.84	6.45	—
Mg^{2+}	11.42	—	−3.02	—	−2.4
Ca^{2+}	12.69	−10.49	−6.48	−5.85	−1.5
Sr^{2+}	13.18	—	—	−6.58	−1.6
Mn^{2+}	10.59	−5.49	—	—	1.2
Fe^{2+}	9.39	−3.77	—	—	—

[a] Hydrolysis constants are from Smith et al. (NIST 46.6, 2001; NIST 46.7, 2003) for hydrolysis reactions of the form $M^{2+} + H_2O = MOH^+ + H^+$, where M^{2+} is a divalent metal ion.
[b] — indicates that data were not sufficient or did not support extraction of a constant.

the intrinsic constants for ion sorption on hydrous metal oxides are approximately the same, at least for iron and aluminum oxides. Additional testing of the hypothesis is needed, of course, but the consistency reflected in Figure 8.7 nevertheless has implications for the practical application of the surface-complexation modeling. The consistency of the cation surface-complexation constants for HFO, goethite, and gibbsite can be explained in part by structural similarities. The common aluminum (hydr)oxide phases gibbsite, boehmite, and diaspore contain similar linkages and have structures similar to the HFO because the Al(III) atom has the same charge and a nearly identical ionic radius as the Fe(III) atom (Waychunas, 1991; Foster, 2003). Also, goethite is formed by close-packed oxygen atoms in a hexagonal pattern (Tan, 1998), a structure similar to that of gibbsite.

The comparison of the intrinsic surface-complexation constants for gibbsite, goethite, and HFO was performed for anions as well. Surface-complexation constants for anion sorption on HMO were not developed by Tonkin et al. (2004). Table 8.10 presents the comparison of the intrinsic anion surface-complexation constants for gibbsite, goethite, and HFO. Here, only the first three intrinsic surface-complexation constants ($\log K_1^{int}$, $\log K_2^{int}$, and $\log K_3^{int}$) are compared. The constants presented are only for those anions that are common among the three different databases. A plot for $\log K_4^{int}$ was not prepared because of the small number of anions for which the reaction was needed. Figures 8.8, 8.9, and 8.10 show the correlations of each one of the three intrinsic anion surface-complexation constants for gibbsite, goethite, and HFO with the second deprotonation constant for the various anions.

The surface-complexation constants for gibbsite, goethite, and HFO in Table 8.10 and in Figures 8.8, 8.9, and 8.10 show a good consistency, as also observed with cation surface-complexation constants determined for these oxides. This consistency tends

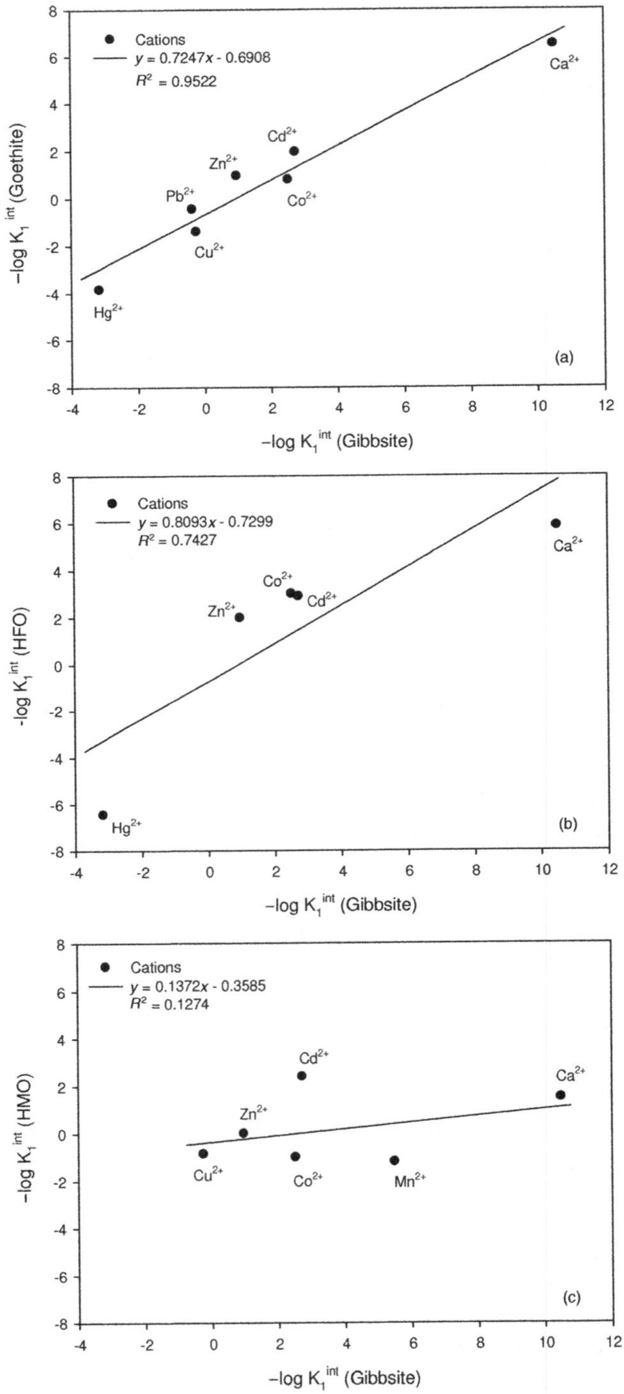

FIGURE 8.6 Correlation of each $\log K_1^{int}$ of goethite, HFO, and HMO with $\log K_1^{int}$ of gibbsite for various cations. (a) Goethite versus gibbsite. The fitted line has a slope of 0.72, a y-intercept of -0.69, and is the result of a linear least squares regression with $R^2 = 0.95$ (95 percent confidence limit). (b) HFO versus gibbsite. Slope $= 0.81$, y-intercept $= -0.73$ and $R^2 = 0.74$. (c) HMO versus gibbsite. Slope $= 0.137$, y-intercept $= -0.36$, and $R^2 = 0.13$.

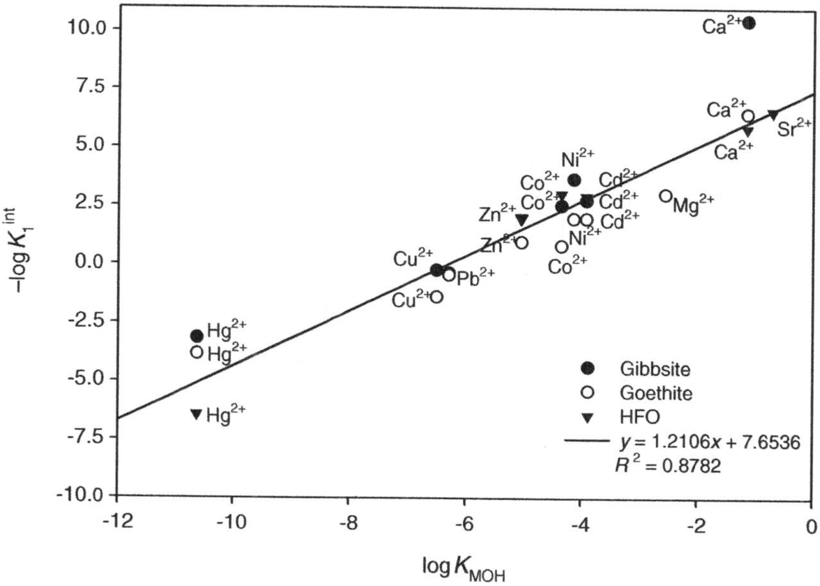

FIGURE 8.7 Correlation of $\log K_1^{int}$ for cation sorption on gibbsite, goethite, and HFO with the first hydrolysis constant for various cations (see Table 8.9 for details). The fitted line has a slope of 1.21, a *y*-intercept of 7.65, and is the result of a linear least squares regression with $R^2 = 0.88$ (95 percent confidence limit).

to support the hypothesis that the intrinsic surface-complexation constants are approximately the same for hydrous metal oxides with similar structure, but further database development is needed to test this hypothesis.

8.4 SUMMARY

In this work, a surface-complexation database for gibbsite was developed to extend the available surface-complexation databases for modeling ion sorption on common hydrous metal oxides. Previously, surface-complexation databases have been developed with the same procedures and generalized two-layer model for HFO (Dzombak and Morel, 1990), goethite (Mathur, 1995; Mathur and Dzombak, 2006), and HMO (Tonkin et al., 2004). For the construction of the gibbsite database, available acid–base and ion-sorption data were fitted with the generalized two-layer model using the same methodology employed in the development of the databases for HFO, goethite, and HMO. Optimal surface-complexation reactions and equilibrium constants for fitting available gibbsite-sorption data were determined.

To evaluate the internal consistency of the surface-complexation constants extracted from gibbsite sorption data, and to estimate constants for ions for which limited or no sorption data exist, LFERs were developed. LFERs were also used to compare the surface-complexation constants obtained for ion sorption on gibbsite

TABLE 8.10 Comparison of intrinsic surface complexation constants for the sorption of anions on gibbsite, goethite, and hydrous ferric oxide

Ion	$-\log K_a$ [a]	$\log K_1^{int}$			$\log K_2^{int}$			$\log K_3^{int}$		
		Gibbsite	Goethite	HFO	Gibbsite	Goethite	HFO	Gibbsite	Goethite	HFO
AsO_4^{3-}	11.80	26.94	—	29.31	21.79	—	23.51	—	—	—
SO_4^{2-}	1.99	— [b]	12.89	—	−0.45	6.74	7.78	1.19	—	0.79
SeO_4^{2-}	1.70	—	11.65	—	—	6.54	7.73	4.53	—	0.80
CrO_4^{2-}	6.51	13.40	17.11	—	9.50	11.17	10.85	—	4.05	—
PO_4^{3-}	12.37	26.89	30.72	31.29	19.37	24.91	25.39	13.57	19.65	17.72
F^-	3.18	8.78	9.20	—	2.88	1.59	—	11.94	—	—

Note: Goethite constants are taken from Mathur and Dzombak (2006) and HFO constants are for sorption on weak binding sites and are taken from Dzombak and Morel (1990).

[a] From Smith et al. (2003; NIST 46.7). Corresponds to the last deprotonation reaction, that is, pK_{a2} for divalent ions and pK_{a3} for trivalent ions.

[b] — indicates that data were not sufficient or did not support extraction of a constant.

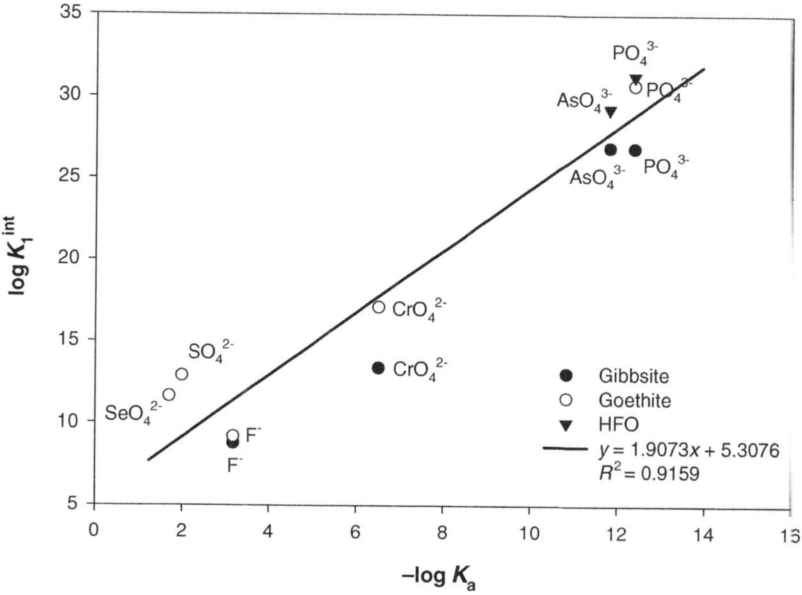

FIGURE 8.8 Correlation of log K_1^{int} for anion sorption on gibbsite, goethite, and HFO with the second deprotonation constant for various anions (see Table 8.10 and Table 8.5 for the surface reactions and constants corresponding to K_1^{int}). The fitted line has a slope of 1.91, a y-intercept of 5.31, and is the results of a linear least squares regression with $R^2 = 0.92$.

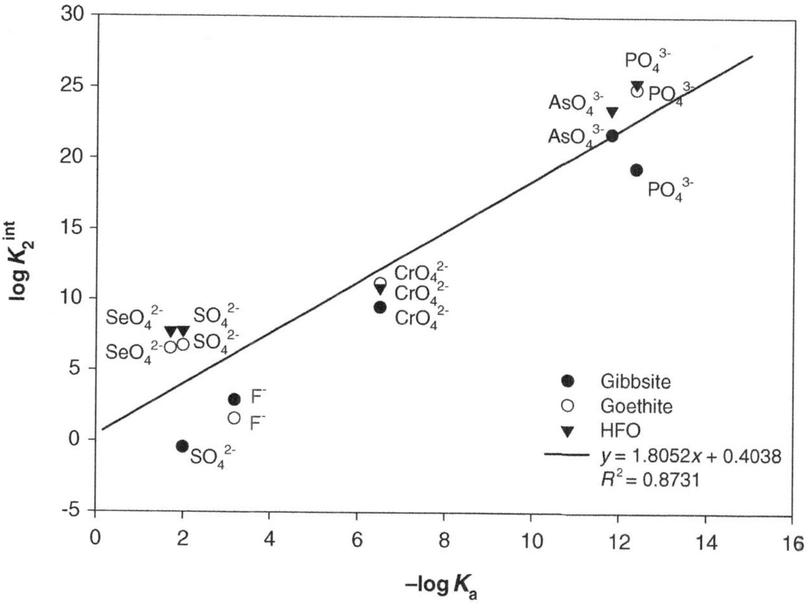

FIGURE 8.9 Correlation of log K_2^{int} for anion sorption on gibbsite, goethite, and HFO with the second deprotonation constant for various anions (see Table 8.10 and Table 8.5 for the surface reactions and constants corresponding to K_2^{int}). The fitted line has a slope of 1.80, a y-intercept of 0.40, and is the result of a linear least squares regression with $R^2 = 0.87$.

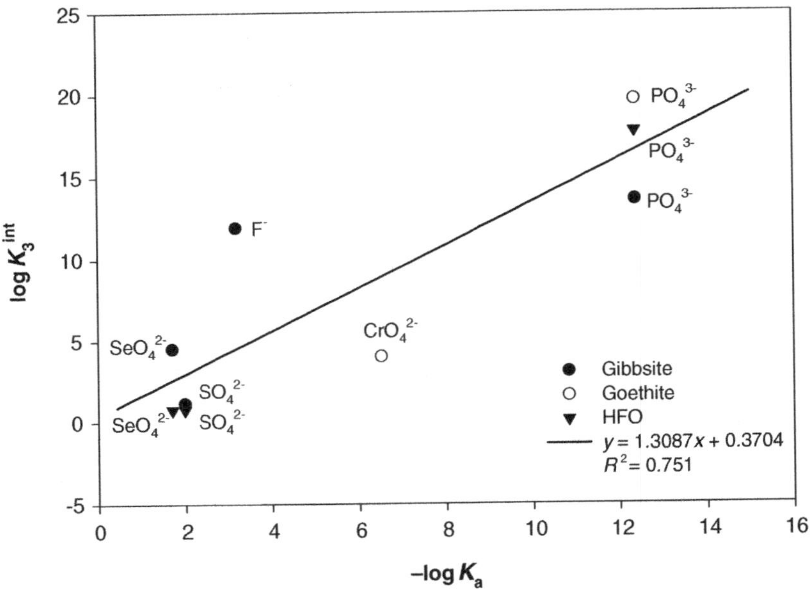

FIGURE 8.10 Correlation of log K_3^{int} for anion sorption on gibbsite, goethite, and HFO with the second deprotonation constant for various anions (see Table 8.10 and Table 8.5 for the surface reactions and constants corresponding to K_3^{int}). The fitted line has a slope of 1.31, a *y*-intercept of 0.37, and is the result of a linear least squares regression with $R^2 = 0.75$.

with those obtained previously with the generalized two-layer model for HFO, goethite, and HMO.

LFERs were successfully applied to correlate the surface-complexation constants for gibbsite with equilibrium constants for related solution reactions. Plots of cation surface-complexation constants against the first hydrolysis constants for the various cations resulted in log-linear relationships. These plots allowed the prediction of surface-complexation constants for Ba^{2+}, Mg^{2+}, Ni^{2+}, Ag^{+}, Sr^{2+}, PuO_2^{2+}, Pd^{2+}, and Sn^{2+}.

LFERs were also developed for anion surface-complexation constants plotted against their second deprotonation constants, and reasonably good correlations were again obtained. Predictions of surface-complexation constants for several anions for which limited or no sorption data exist was achieved using the LFERs. Anions for which surface-complexation constants were estimated were: CO_3^{2-}, SO_3^{2-}, SeO_4^{2-}, SeO_3^{2-}, $SbO(OH)_4^{-}$, WO_4^{2-}, CNO^{-}, CN^{-}, and SCN^{-}.

Finally, a comparison of intrinsic surface-complexation constants for similar surface reactions of ions on gibbsite, goethite, HFO, and HMO was performed. It may be hypothesized that the intrinsic surface-complexation constants obtained with the same model and in a consistent manner for ion sorption on hydrous metal oxides are approximately the same. Put another way, the hypothesis is that the reactivity of surface hydroxyl groups with particular sorbing ions is similar for the common hydrous metal oxides. Intrinsic constants for surface complexation of cations and anions on gibbsite, goethite, and HFO were close to each other and quite consistent, but the ion surface-complexation constants for HMO were not as consistent with those of the other three solids, perhaps reflecting differences in the reactivity of ≡MnOH surface groups relative to those associated with iron and aluminum oxides. Further database development is necessary to test the hypothesis. Nevertheless, the surface-complexation databases developed with the generalized two-layer model for HFO, goethite, HMO, and gibbsite and the similarity of the surface-complexation constants extracted from the data are useful for the practical application of surface-complexation modeling.

REFERENCES

Addai-Mensah, J.; Dawe, J.; Hayes, R.; Prestidge, C.; Ralston, J. (1998) The unusual colloid stability of gibbsite at high pH. *J. Colloid Interface Sci.* 203: 115–121.

Adu-Wusu, K.; Wilcox, W.R. (1991a) Kinetics of silicate reaction with gibbsite. *J. Colloid Interface Sci.* 143(1): 127–138.

Adu-Wusu, K.; Wilcox, W.R. (1991b) Sorption and desorption isotherms for silicate on gibbsite. *J. Colloid Interface Sci.* 143(1): 139–145.

Ahuja, I.S.; Singh, R.; Singh, R. (1980) Infrared spectral evidence for equatorially eight-coordinate $[UO_2^{2+}]$. *Transition Met. Chem.* 5: 303–305.

Akitt, J.W.; Greenwood, N.N.; Khandelwal, B.L.; Lester, G.D. (1972) ^{27}Al nuclear magnetic resonance studies of the hydrolysis and polymerisation of the hexa-aquo-aluminum(III) cation. *J. Chem. Soc., Dalton. Trans.* 604–610.

Aldcroft, D.; Bye, G.C. (1967) Formation and thermal decomposition of nordstrandite. *Sci. Ceram.* 3: 75–94.

Aldcroft, D.; Bye, G.C.; Hughes, C.A. (1969) Crystallization process in aluminum hydroxide gels. IV. Factors influencing the formation of the crystalline trihydroxides. *J. Appl. Chem.* 19: 167–172.

Alloway, B.J. (1995) *Heavy Metals in Soil*, 2nd ed., Blackie Academic and Professional, Glasgow, UK.

Alvarez, E.; Perez, A.; Calvo, R. (1993) Aluminium speciation in surface waters and soil solutions in areas of sulphide mineralization in Galicia (N. W. Spain). *Sci. Total Environ.* 133: 17–37.

Alvarez, R.; Cramer, R.E.; Silva, J.A. (1976a) Laser Raman spectroscopy: a technique for studying adsorption on aluminum sesquioxide, gibbsite. *Soil Sci. Soc. Am. J.* 40: 317–319.

Alvarez, R.; Fadley, C.S.; Silva, J.A.; Uehara, G. (1976b) A study of silicate adsorption on gibbsite ($Al(OH)_3$) by X-ray photoelectron spectroscopy (XPS). *Soil Sci. Soc. Am. J.* 40: 615–617.

Alwitt, R.S. (1976) The aluminum–water system. In: J.W. Diggleand K.V. Ashik, editors, *Oxides and Oxide Films*, vol. 4, Marcel Dekker, New York.

Anderson, P.R.; Benjamin, M.M. (1985) Effects of silicon on the crystallization and adsorption properties of ferric oxides. *Environ. Sci. Technol.* 19(11): 1048–1053.

Surface Complexation Modeling: Gibbsite By Athanasios K. Karamalidis and David A. Dzombak

Anthony, J.W.; Bideaux, R.A.; Bladh, K.W.; Nichols, M.C. (1997) Halides, hydroxides, oxides. In: *Handbook of Mineralogy*, vol III, Mineralogical Society of America.

Arai, Y.; Elzinga, E.J.; Sparks, D.L. (2001) X-ray absorption spectroscopic investigation of arsenite and arsenate adsorption at the aluminum oxide–water interface. *J. Colloid Interface Sci.* 235: 80–88.

Authier-Martin, M.; Forté, G.; Ostap, S.; See, J. (2001) The mineralogy of bauxite for producing smelter-grade alumina. *JOM* 53(12): 36–40.

Ayoob, S.; Gupta, A.K.; Bhat, V.T. (2008) A conceptual overview on sustainable technologies for the defluoridation of drinking water. *Crit. Rev. Environ. Sci. Technol.* 38(6): 401–470.

Bardossy, G. (1982) *Karst Bauxites: Bauxite Deposits on Carbonate Rocks*, Elsevier, p. 441.

Bargar, J.R.; Brown, G.E., Jr.; Parks, G.A. (1997) Surface complexation of Pb(II) at oxide–water interfaces. I. XAFS and bond-valence determination of mononuclear and polynuclear Pb(II) sorption products on aluminum oxides. *Geochim. Cosmochim. Acta* 61(13): 2617–2637.

Bargar, J.R.; Reitmeyer, R.; Davis, J.A. (2000) Spectroscopic confirmation of uranium (VI)-carbonato complexes on hematite. *Environ. Sci. Technol.* 33: 2481–2484.

Barnishel, R.I.; Rich, C.I. (1963) Gibbsite formation from aluminum-interlayer in montmorillonite. *Soil Sci. Soc. Am. J.* 27: 632–635.

Barnishel, R.I.; Rich, C.I. (1965) Gibbsite, bayerite, and nordstrandite formation as affected by anions, pH, and mineral surfaces. *Soil Sci. Soc. Proc.* 531–534.

Baroni, T.C.; Griffin, B.J.; Browne, J.R.; Lincoln, F.J. (2000) Correlation between charge contrast imaging and the distribution of some trace level impurities in gibbsite. *Microsc. Microanal.* 6: 49–58.

Barron, V.; Torrent, J. (1996) Surface hydroxyl configuration of various crystal faces of hematite and goethite. *J. Colloid Interface Sci.* 177: 407–410.

Baumann, N.; Brendler, V.; Arnold, T.; Geipel, G.; Bernhard, G. (2005) Uranyl sorption onto gibbsite studied by time-resolved laser-induced fluorescence spectroscopy (TRLFS). *J. Colloid Interface Sci.* 290: 318–324.

Beattie, J.K.; Cleaver, J.K.; Waite, T.D. (1996) Anomalous aggregation behavior of aluminum oxyhydroxides. *Colloids Surf. A* 111: 131–138.

Bénézeth, P.; Palmer, D.A.; Wesolowski, D.J. (2008) Dissolution/precipitation kinetics of boehmite and gibbsite: application of a pH relaxation technique to study near equilibrium rates. *Geochim. Cosmochim. Acta* 72(10): 2429–2453.

Berkowitz, J.; Anderson, M.A.; Graham, R.C. (2005) Laboratory investigation of aluminum solubility and solid-phase properties following alum treatment of lake waters. *Water Res.* 39: 3918–3928.

Bertsch, P.M.; Parker, D.R. (1996). Aqueous polynuclear aluminum species. In: G. Sposito, editor, *The Environmental Chemistry of Aluminum*, 2nd ed., CRC Press Inc., Boca Raton, FL.

Beyrouty, C.A.; van Scoyoc, G.E.; Feldkamp, J.R. (1984) Evidence supporting specific adsorption of boron on synthetic aluminum hydroxides. *Soil Sci. Soc. Am. J.* 48(2): 284–287.

Bhattacharyya, K.G.; Gupta, S.S. (2008a) Kaolinite and montmorillonite as adsorbents for Fe(III), Co(II) and Ni(II) in aqueous medium. *Appl. Clay Sci.* 41: 1–9.

Bhattacharyya, K.G.; Gupta, S.S. (2008b) Adsorption of a few heavy metals on natural and modified kaolinite and montmorillonite: a review. *Adv. Colloid Interface Sci.* 140: 114–131.

Bi, S.; Wang, C.; Cao, Q.; Zhang, C. (2004) Studies on the mechanism of hydrolysis and polymerization of aluminum salts in aqueous solution: correlations between the "Core-links" model and "Cage-like" Keggin-Al_{13} model. *Coord. Chem. Rev.* 248: 441–455.

Bickmore, B.R.; Nagy, K.L.; Sandlin, P.E.; Crater, T.S. (2002) Quantifying surface areas of clays by atomic force microscopy. *Am. Miner.* 87: 780–783.

Bickmore, B.R.; Tadanier, C.J.; Rosso, K.M.; Monn, W.D.; Eggett, D.L. (2004) Bond-valence methods for pK_a prediction: critical reanalysis and a new approach. *Geochim. Cosmochim. Acta* 68(9): 2025–2042.

Bird, M.I.; Longstaffe, F.J.; Fyfe, W.S.; Tazaki, K.; Chivas, A.R. (1994) Oxygen-isotope fractionation in gibbsite: synthesis experiments versus natural samples. *Geochim. Cosmochim. Acta* 58(23): 5267–5277.

Bloom, P.R. (1983) The kinetic of gibbsite dissolution in nitric acid. *Soil Sci. Soc. Am. J.* 47(1): 164–168.

Bloom, P.R.; Erich, M.S. (1987) Effect of solution composition on the rate and mechanism of gibbsite dissolution in acid solutions. *Soil Sci. Soc. Am. J.* 51(5): 1131–1136.

Bolt, G.H.; Van Riemsdijk, W.H. (1982) In: G.H. Bolt, editor, *Soil Chemistry. B. Physico-Chemical Models*, 2nd ed., Elsevier, Amsterdam, pp. 459–504.

Bosmans, H.J. (1970) Unit cell and crystal structure of nordstrandite, $Al(OH)_3$. *Acta Crystallogr. Sect. B: Struct. Sci.* B26: 649–652.

Bottero, J.Y.; Axelos, M.; Tchoubar, D.; Cases, J.M.; Fripiat, J.J.; Fiessinger, F. (1987) Mechanism of formation of aluminum trihydroxide from Keggin Al_{13} polymer. *J. Colloid Interface Sci.* 117: 47–57.

Bottero, J.Y.; Cases, J.M.; Fiessinger, F.; Poirier, J.E. (1980) Studies of hydrolyzed aluminum chloride solutions. 1. Nature of aluminum species and composition of aqueous solutions. *J. Phys. Chem.* 84: 2933–2939.

Bowen, H.J.M. (1979) *Environmental Chemistry of the Elements*, Academic Press, London.

Brady, N.C.; Weil, R.R. (2002) *The Nature and Properties of Soils*, 13th ed. Prentice Hall, Upper Saddle River, NJ.

Brady, P.V. (1994) Alumina surface chemistry at 25, 40, and 60°C. *Geochim. Cosmochim. Acta* 58(3): 1213–1217.

Brattebo, H.; Odegaard, H. (1986) Phosphorus removal by granular activated alumina. *Water Res.* 20(8): 977–986.

Brosset, C.; Biedermann, G.; Sillén, L.G. (1954) Studies on the hydrolysis of metal ions. XI. The aluminum ion, Al^{3+}. *Acta Chem. Scand.* 8: 1917–1926.

Brown, P.L.; Sylva, R.N.; Ellis, J. (1985) An equation for predicting the formation constants of hydroxo-metal complexes. *J. Chem. Soc., Dalton Trans.* 723–730.

Brunauer, S.; Emmett, P.H.; Teller, E. (1938) Adsorption of gases in multimolecular layers. *J. Am. Chem. Soc.* 60: 309–319.

Bye, G.C.; Robinson, J.G. (1964) Crystallization processes in aluminum hydroxide gels. *Kolloid Z. Z. Polym.* 198(1–2): 53–60.

Calvet, E.; Boivinet, P.; Noel, M.; Thidon, H.; Maillard, A.; Tertian, R. (1953) Contribution a l'etude des gels d'alumine. *Bull. Soc. Chim.* 99–108.

Carrier, X.; Marceau, E; Lambert, J.-F.; Che, M. (2007) Transformations of γ-alumina in aqueous suspensions. 1. Alumina chemical weathering studied as a function of pH. *J. Colloid Interface Sci.* 308: 429–437.

Cate, R.B.; McCracken, R.J. (1972) Gibbsite in western N. *Carolina. Southeastern Geol.* 14: 107–112.

Cesteros, Y.; Salagre, P.; Medina, F.; Sueiras, J.E. (1999) Several factors affecting faster rates of gibbsite formation. *Chem. Mater.* 11: 123–129.

Cesteros, Y.; Salagre, P.; Medina, F.; Sueiras, J.E. (2001) A new route to the synthesis of fine-grain gibbsite. *Chem. Mater.* 13: 2595–2600.

Chang, H.-S.; Korshin, G.V.; Wang, Z.; Zachara, J.M. (2006) Adsorption of uranyl on gibbsite: a time-resolved laser-induced fluorescence spectroscopy study. *Environ. Sci. Technol.* 40: 1244–1249.

Chang, T.W.; Wang, M.K.; Jang, L.Y. (2005) An extended X-ray absorption spectroscopy study of copper(II) sorption by oxides. *Geoderma* 129: 211–218.

Chao, G.Y.; Baker, J.; Sabina, A.P.; Roberts, A.C. (1985) Doyleite, a new polymorph of $Al(OH)_3$, and its relationship to bayerite, gibbsite and nordstrandite. *Can. Miner.* 23: 21–28.

Chisholm-Brause, C.J.; Berg, J.M.; Little, K.M.; Matzner, R.A.; Morris, D.E. (2004) Uranyl sorption by smectites: spectroscopic assessment of thermodynamic modeling. *J. Colloid Interface Sci.* 277: 366–382.

Chowdhury, Z.K.; Amy, G.L.; Bales, R.C. (1991) Coagulation of submicron colloids in water treatment by incorporation into aluminum hydroxide floc. *Environ. Sci. Technol.* 25(10): 1766–1773.

Chvedov, D.; Logan, E.L.B. (2004) Surface charge properties of oxides and hydroxides formed on metal substrates determined by contact angle titration. *Colloids Surf. A* 240(1–3): 211–223.

Clark, G.R.; Rodgers, K.A.; Henderson, G.S. (1998) The crystal chemistry of doyleite, $Al(OH)_3$. *Z. Kristallogr.* 213(2): 96–100.

Clifford, D.A. (1999) Ion exchange and inorganic adsorption. In: *Water Quality and Treatment*, 5th ed., McGraw-Hill, Chapter 9.

Colombo, C.; Ricciardella, M.; Di Cerce, A.; Maiuro, L.; Violante, A. (2004) Effect of tannate, pH, sample preparation, ageing and temperature on the formation and nature of Al oxyhydroxides. *Clays Clay Miner.* 52(6): 721–733.

Cronan, C.S.; Schofield, C.L. (1979) Aluminum leaching response to acid precipitation: effects on high-elevation watersheds in the northeast. *Science* 204: 304–306.

Cruywagen, J.J. (2000) Protonation, oligomerization, and condensation reactions of vanadate(V), molybdate(VI), and tungstate(VI). *Adv. Inorg. Chem.* 49: 127–182.

Davis, C.C.; Chen, H.-W.; Edwards, M. (2002) Modeling silica sorption to iron hydroxide. *Environ. Sci. Technol.* 36(4): 582–587.

Dayton, E.A.; Basta, N.T. (2005) A method for determining the phosphorus sorption capacity and amorphous aluminum of aluminum-based drinking water treatment residuals. *J. Environ. Qual.* 34: 1112–1118.

De-Bashan, L.E.; Bashan, Y. (2004) Recent advances in removing phosphorus from wastewater and its future use as fertilizer (1997–2003). *Water Res.* 38: 4222–4246.

De Boer, J.H.; Fortuin, J.M.H.; Lippens, B.C.; Meijs, W.H. (1963) Study of the nature of surfaces with polar molecules. II. The adsorption of water on aluminas. *J. Catal.* 2(1): 1–7.

Dejou, J.; Guyot, J.; Pedro, G.; Chaumont, C.; Antoine, H. (1970) Sur la présence de gibbsite dans les formations superficielles développés en Haute Vienne sur micaschiste francs. *C. R. Acad. Sci. Paris* 270(Série D): 287–289.

Demichelis, R.; Catti; M. Dovesi, R. (2009) Structure and stability of the $Al(OH)_3$ polymorphs doyleite and nordstrandite: a quantum mechanical *ab initio* study with the CRYSTAL06 code. *J. Phys. Chem. C* 113: 6785–6791.

Dietzel, M.; Böhme, G. (2005) The dissolution rates of gibbsite in the presence of chloride, nitrate, silica, sulfate, and citrate in open and closed systems at 20°C. *Geochem. Cosmochim. Acta* 69(5): 1199–1211.

Digne, M.; Sautet, P.; Raybaud, P.; Toulhoat, H.; Artacho, E. (2002) Structure and stability of aluminum hydroxides: a theoretical study. *J. Phys. Chem.* 106: 5155–5162.

Dixon, J.B.; Weed S.B.; Kittrick, J.A.; Milford, M.H.; White, J.L. (1989) *Minerals in Soil Environments*, 2nd ed. ASA-CSSA-SSSA, Madison, WI.

Dollimore, D.; Spooner, P.; Turner, A. (1976) The BET method of analysis of gas adsorption data and its relevance to the calculation of surface areas. *Surf. Technol.* 4: 121–160.

Doucet, F.; Schneider, C.; Bones, S.J.; Kretchmer, A.; Moss, I.; Tekely, P.; Exley, C. (2001) The formation of hydroxyaluminosilicates of geochemical and biological significance. *Geochem. Cosmochim. Acta* 65(15): 2461–2467.

Driscoll, C.T.; Postek, K.M. (1996) The chemistry of aluminum in surface waters. In: G. Sposito, editor, *The Environmental Chemistry of Aluminum*, 2nd ed., Lewis Publishers, New York, pp. 363–418.

Dubbin, W.E.; Sposito, G. (2005) Copper-glyphosate sorption to microcrystalline gibbsite in the presence of soluble Keggin Al_{13} polymers. *Environ. Sci. Technol.* 39: 2509–2514.

Dubbin, W.E.; Sposito, G.; Zavarin, M. (2000) X-ray absorption spectroscopic study of Cu-glyphosate adsorbed by microcrystalline gibbsite. *Soil Sci.* 165: 699–707.

Dyer, C.; Hendra, P.J.; Forsling, W.; Ranheimer, M. (1993) Surface hydration of aqueous γ-Al_2O_3 studied by Fourier transform Raman and infrared spectroscopy. I. Initial results. *Spectrochim. Acta* 49A: 691–705.

Dzombak, D.A.; Ghosh, R.S.; Young, T.C. (2006) Physical–chemical properties and reactivity of cyanide in water and soil. In: *Cyanide in Water and Soil: Chemistry, Risk, and Management*, CRC/Taylor & Francis Group, Boca Raton, FL, Chapter 5, pp. 57–92.

Dzombak, D.A.; Morel, F.M.M. (1990) *Surface Complexation Modeling: Hydrous Ferric Oxide*, Wiley, New York.

Essington, M.E.; Anderson, R.M. (2008) Competitive adsorption of 2-ketogluconate and inorganic ligands onto gibbsite and kaolinite. *Soil Sci. Soc. Am. J.* 72(3): 595–604.

Eswaran, H.; Stoops, G.; Sys, C. (1977) The micromorphology of gibbsite forms in soils. *J. Soil Sci.* 28: 136–143.

Exley, C.; Schneider, C.; Doucet, F.J. (2002) The reaction of aluminium with silicic acid in acidic solution: an important mechanism in controlling the biological availability of aluminium. *Coord. Chem. Rev.* 228: 127–135.

FAO (1998) *World Reference Base for Soil Resources*, World Soil Resources Reports, vol. 84, Food and Agricultural Organization of the United Nations, Rome.

Farrah, H.; Slavek, J.; Pickering, W.F. (1987) Fluoride interactions with hydrous aluminum oxides and alumina. *Aust. J. Soil Res.* 25: 55–69.

Feldkamp, J.R.; Shah, D.N.; Meyer, S.L.; White, J.L.; Hem, S.L. (1981) Effect of adsorbed carbonate on surface-charge characteristics and physical properties of aluminum hydroxide gel. *J. Pharm. Sci.* 70(6): 638–640.

Fendorf, S.; Eick, M.J.; Grossl, P.; Sparks, D.L. (1997) Arsenate and chromate retention mechanisms on goethite. 1. Surface structure. *Environ. Sci. Technol.* 31(2): 315–320.

Fendorf, S.E.; Sparks, D.L. (1996) X-ray absorption fine structure spectroscopy. In: A.L. Page, editor, *Methods of Soil Analysis. Part 3. Chemical Methods*, Soil Science Society of America, Madison, WI, pp. 337–416.

Foster, A.L. (2003) Spectroscopic investigations of arsenic species in solid phases. In: A.H. Welch and K.G. Stollenwerk, editors, *Arsenic in Ground Water*, Kluwer Academic Publishers, Dordrecht, Chapter 2.

Foster, A.L.; Brown, G.E., Jr. ; Parks, G.A. (1998) X-ray absorption fine-structure spectroscopy study of photocatalyzed, heterogeneous As(III) oxidation on kaolin and anatase. *Environ. Sci. Technol.* 32: 1444–1452.

Franks, G.V.; Gan, Y. (2007) Charging behavior at the alumina-water interface and implications for creaming processing. *J. Am. Ceram. Soc.* 90(11): 3373–3388.

Freeze, R.A.; Cherry, J.A. (1979) *Groundwater*, Prentice Hall, Englewood Cliffs, NJ.

Frempong, E.M.; Yanful, E.K. (2006) Chemical and mineralogical transformations in three tropical soils due to permeation with acid mine drainage. *Bull. Eng. Geol. Environ.* 65: 253–271.

Frenzel, J.; Oliveirab, A.F.; Duarteb, H.A.; Heinea, T.; Seifert, G. (2005) Structural and electronic properties of bulk gibbsite and gibbsite surfaces. *Z. Anorg. Allg. Chem.* 631: 1267–1271.

Froideval, A.; Del Nero, M.; Gaillard, C.; Barillon, R.; Rossini, I.; Hazemann, J.L. (2006) Uranyl sorption species at low coverage on Al-hydroxide: TRLFS and XAFS studies. *Geochim. Cosmochim. Acta* 70: 5270–5284.

Fukushi, K.; Sverjensky, D.A. (2007) A predictive model (ETLM) for arsenate adsorption and surface speciation on oxides consistent with spectroscopic and theoretical molecular evidence. *Geochim. Cosmochim. Acta* 71: 3717–3745.

Fukushi, K.; Tsukimura, K.; Yamada, H. (2006) Surface acidity of amorphous aluminum hydroxide. *Acta Geol. Sin.* 80(2): 206–211.

Funaki, K.; Shimizu, Y. (1960) On the study of hydrous aluminum and alumina (Part 1). *Electrochemistry* 28: 302–312 (in Japanese).

Galas, Z.J.; Janiak, S.; Mista, W.; Wrzyszcz, J.; Zawadzki, M. (1993) Morphological and phase changes of transition aluminas during their rehydration. *J. Mater. Sci.* 28: 2075–2078.

Gale, J.D.; Rohl, A.L.; Milman, V.; Warren, M.C. (2001) An *ab initio* study of the structure and properties of aluminum hydroxide: gibbsite and bayerite. *J. Phys. Chem. B* 105: 10236–10242.

Gan, Y.; Franks, G.V. (2006) Charging behavior of the gibbsite basal (001) surface in NaCl solution investigated by AFM colloidal probe technique. *Langmuir* 22: 6087–6092.

Ganor, J.; Mogollón, J.L.; Lasaga, A.G. (1999) Kinetics of gibbsite dissolution under low ionic strength conditions. *Geochim. Cosmochim. Acta* 63(11–12): 1635–1651.

Gardner, L.R. (1972) Conditions for direct formation of gibbsite from K-feldspar: further discussion. *Am. Miner.* 57: 294–300.

Gastuche, M.C.; Herbillon, A. (1962) Étude des gels d'alumine: cristallisation en milieu désionisé. *Bull. Soc. Chim. Fr.* 1404–1412.

Gensemer, R.W.; Playle, R.C. (1999) The bioavailability and toxicity of aluminum in aquatic environments. *Crit. Rev. Environ. Sci. Technol.* 29(4): 315–450.

Georgantas, D.A.; Grigoropoulou, H.P. (2007) Orthophosphate and metaphosphate ion removal from aqueous solution using alum and aluminum hydroxide. *J. Colloid Interface Sci.* 315: 70–79.

Gessa, C.; De Cherchi, M.L.; Melis, P.; Micera, G.; Strinna Erre, L. (1984) Anion-induced metal binding in amorphous aluminium hydroxide. *Colloids Surf.* 11: 109–117.

Ghorai, S.; Pant, K.K. (2004) Investigations on the column performance of fluoride adsorption by activated alumina in a fixed-bed. *Chem. Eng. J.* 98: 165–173.

Ghorai, S.; Pant, K.K. (2005) Equilibrium, kinetics and breakthrough studies for adsorption of fluoride on activated alumina. *Sep. Purif. Technol.* 42: 265–271.

Giessibl, F.J. (2003) Advances in atomic force microscopy. *Rev. Mod. Phys.* 75(3): 949–983.

Ginsberg, H.; Hüttig, W.; Strunk-Lichtenberg, G. (1957) *Anorg. Allg. Chem.* 293: 33–46 204–213.

Girvin, D.C.; Gassman, P.L.; Bolton, H., Jr. (1993) Adsorption of aqueous cobalt ethylenediaminetetraacetate by δ-Al_2O_3. *Soil Sci. Soc. Am. J.* 57(1): 47–57.

Girvin, D.C.; Gassman, P.L.; Bolton, H., Jr. (1996) Adsorption of nitrilotriacetate (NTA), Co and CoNTA by gibbsite. *Clays Clay Miner.* 44(6): 757–768.

Goldberg, S. (1992) Use of surface complexation models in soil chemical systems. *Adv. Agron.* 47: 233–329.

Goldberg, S. (2002) Competitive adsorption of arsenate and arsenite on oxides and clay minerals. *Soil Sci. Soc. Am. J.* 66(2): 413–421.

Goldberg, S. (2005) Inconsistency in the triple layer model description of ionic strength dependent boron adsorption. *J. Colloid Interface Sci.* 285: 509–517.

Goldberg, S.; Davis, J.A.; Hem, J.D. (1996a) The surface chemistry of aluminum oxides and hydroxides. In: G. Sposito, editor, *The Environmental Chemistry of Aluminum*, 2nd ed., Lewis Publishers, New York, pp. 271–331.

Goldberg, S.; Forster, H.S.; Godfrey, C.L. (1996b) Molybdenum adsorption on oxides, clay minerals, and soils. *Soil Sci. Soc. Am. J.* 60(2): 425–432.

Goldberg, S.; Forster, H.S.; Heick, E.L. (1993a) Boron adsorption mechanisms on oxides, clay minerals, and soils inferred from ionic strength effects. *Soil Sci. Soc. Am. J.* 57: 704–708.

Goldberg, S.; Forster, H.S.; Heick, E.L. (1993b) Temperature effects on boron adsorption by reference minerals and soils. *Soil Sci. Soc. Am. J.* 156(5): 316–321.

Goldberg, S.; Glaubig, R.A. (1988) Boron and silicon adsorption on an aluminum oxide. *Soil Sci. Soc. Am. J.* 52(1): 87–91.

Gorman-Lewis, D.J.; Fein, J.B. (2004) Experimental study of the adsorption of an ionic liquid onto bacterial and mineral surfaces. *Environ. Sci. Technol.* 38: 2491–2495.

Gräfe, M.; Singh, B.; Balasubramanian, M. (2007) Surface speciation of Cd(II) and Pb(II) on kaolinite by XAFS spectroscopy. *J. Colloid Interface Sci.* 315: 21–32.

Gräfe, M.; Tappero, R.V.; Marcus, M.A.; Sparks, D.L. (2008) Arsenic speciation in multiple metal environments. I. Bulk-XAFS spectroscopy of model and mixed compounds. *J. Colloid Interface Sci.* 320: 383–399.

Gu, C.; Karthikeyan, K.G. (2005) Interaction of tetracycline with aluminum and iron hydrous oxides. *Environ. Sci. Technol.* 39: 2660–2667.

Guan, X.-H.; Chen, G.-H.; Shang, C. (2005) Competitive adsorption between orthophosphate and other phosphates on aluminum hydroxide. *Soil Sci.* 170(5): 340–349.

Guan, X.-H.; Chen, G.-H.; Shang, C. (2007) ATR-FTIR and XPS study on the structure of complexes formed upon the adsorption of simple organic acids on aluminum hydroxide. *J. Environ. Sci.* 19: 438–443.

Guan, X.-H.; Shang, C.; Chen, G.-H. (2006) Competitive adsorption of organic matter with phosphate on aluminum hydroxide. *J. Colloid Interface Sci.* 296: 51–58.

Gupta, S.S.; Bhattacharyya, K.G. (2006) Adsorption of Ni(II) on clays. *J. Colloid Interface Sci.* 295: 21–32.

Gupta, S.S.; Bhattacharyya, K.G. (2008) Immobilization of Pb(II), Cd(II) and Ni(II) ions on kaolinite and montmorillonite surfaces from aqueous medium. *J. Environ. Manag.* 87: 46–58.

Gustafsson, J.P. (2009) *Visual MINTEQ*, version 2.61 KTH Royal Institute of Technology, Stockholm, Sweden.

Hansen, H.C.B.; Raben-Lange, B.; Raulund-Rasmussen, K.; Borggaard, O.K. (1994a) Monosilicate adsorption by ferrihydrite and goethite at pH 3–6. *Soil Sci.* 158(1): 40–46.

Hansen, H.C.B.; Wetche, T.P.; Raulund-Rasmussen, K.; Borggaard, O.K. (1994b) Stability constants for silicate adsorbed to ferrihydrite. *Clay Miner.* 29(3): 341–350.

Hattori, T.; Saito, T.; Ishida, K.; Scheinost, A.C.; Tsuneda, T.; Nagasaki, S.; Tanaka, S. (2009) The structure of monomeric and dimeric uranyl adsorption complexes on gibbsite: a combined DFT and EXAFS study. *Geochim. Cosmochim. Acta* 73: 5975–5988.

Hauschild, U. (1963) Uber Nordstrandite, γ-$Al(OH)_3$. *Z. Anorg. Allg. Chem.* 324: 15–30.

Hayes, K.F.; Redden, G.; Ela, W.; Leckie, J.O. (1991) Surface complexation models: an evaluation of model parameter estimation using FITEQL and oxide mineral titration data. *J. Colloid Interface Sci.* 142: 448–469.

He, L.M.; Zelazny, L.W.; Baligar, V.C.; Ritchey, K.D.; Martens, D.C. (1997) Ionic strength effects on sulfate and phosphate adsorption on γ-alumina and kaolinite: triple-layer model. *Soil Sci. Soc. Am. J.* 61(3): 784–793.

Helyar, K.R.; Munns, D.N.; Burau, R.G. (1976a) Adsorption of phosphate by gibbsite. I. Effects of neutral chloride salts of calcium, magnesium, sodium and potassium. *J. Soil Sci.* 27: 307–314.

Helyar, K.R.; Munns, D.N.; Burau, R.G. (1976b) Adsorption of phosphate by gibbsite. II. Formation of a surface complex involving divalent cations. *J. Soil Sci.* 27: 315–323.

Hem, J. (1985) *Study and Interpretation of the Chemical Characteristics of Natural Water*, 3rd ed., Water Supply Paper 2254, U.S. Geological Survey, Alexandria, VA.

Hemingway, B.S.; Sposito, G. (1996) Inorganic aluminum-bearing solid phases. In: G. Sposito, editor, *The Environmental Chemistry of Aluminum*, 2nd ed., CRC Press Inc., Boca Raton, FL.

Herbelin, A.L.; Westall, J.C. (1999) *FITEQL: A Computer Program for Determination of Chemical Equilibrium Constants from Experimental Data*, version 4.0, Chemistry Department, Oregon State University, Corvallis, OR.

Herrmann, L.; Anongrak, N.; Zarei, M.; Schuler, U.; Spohrer, K. (2007) Factors and processes of gibbsite formation in northern Thailand. *Catena* 71: 279–291.

Hiemstra, T.; Barnett, M.O.; Van Riemsdijk, W.H. (2007) Interaction of silicic acid with goethite. *J. Colloid Interface Sci.* 310: 8–17.

Hiemstra, T.; De Wit, J.C.M.; Van Riemsdijk, W.H. (1989) Multiple proton adsorption modeling at the solid/solution interface of (hydr)oxides: a new approach. II. Application to various important (hydro)oxides. *J. Colloid Interface Sci.* 133(1): 105–117.

Hiemstra, T.; Van Riemsdijk, W.H. (1991) Physical chemical interpretation of primary charging behavior of metal (hydr)oxides. *Colloids Surf.* 59: 7–25.

Hiemstra, T.; Van Riemsdijk, W.H.; Bruggenwert, M.G.M. (1987) Proton adsorption mechanism at the gibbsite and aluminum oxide solid/solution interface. *Neth. J. Agric. Sci.* 35: 281–293.

Hiemstra, T.; Yong, H.; Van Riemsdijk, W.H. (1999) Interfacial charging phenomena of aluminum (hydr)oxides. *Langmuir* 15: 5942–5955.

Hind, A.R.; Bhargava, S.K.; Grocott, S.C. (1999) The surface chemistry of Bayer process solids: a review. *Colloids Surf. A* 146: 359–374.

Hingston, F.J.; Posner, A.M.; Quirk, J.P. (1971) Competitive adsorption of negatively charged ligands on oxide surfaces. *Discuss. Faraday Soc.* 52: 334–351.

Hingston, F.J.; Posner, A.M.; Quirk, J.P. (1972) Anion adsorption by goethite and gibbsite. I. The role of the proton in determining adsorption envelopes. *J. Soil Sci.* 23 (2): 177–192.

Hongxia, Z.; Zheng, D.; Zuyi, T. (2006) Sorption of thorium(IV) ions on gibbsite: effects of contact time, pH, ionic strength, concentration, phosphate and fulvic acid. *Colloids Surf. A* 278: 46–52.

Houston, J.R.; Herberg, J.L.; Maxwell, R.S.; Carroll, S.A. (2008) Association of dissolved aluminum with silica: connecting molecular structure to surface reactivity using NMR. *Geochim. Cosmochim. Acta* 72: 3326–3337.

Hsi, D.C.; Langmuir, D. (1985) Adsorption of uranyl onto ferric oxyhydroxides: application of the surface complexation site-binding model. *Geochim. Cosmochim. Acta* 49: 1931–1941.

Hsu, P.H. (1966) Formation of gibbsite from aging hydroxy-aluminum solutions. *Soil Sci. Soc. Am. Proc.* 30: 173–76.

Hsu, P.H. (1977) Aluminum hydroxides and oxyhydroxides. In: *Minerals in Soil Environments*, Soil Science Society of America, Madison, WI, Chapter 4.

Hsu, P.H. (1989) Aluminum hydroxides and oxyhydroxides. In: J.B. Dixon, S.B. Weed, J.A. Kittrick, M.H. Milford, and J.L. White, editors, *Minerals in Soil Environments*, 2nd ed., ASA-CSSA-SSSA, Madison, WI, pp. 331–378.

Huang, C.; Shiu, H. (1996) Interactions between alum and organics in coagulation. *Colloids Surf. A* 113: 155–163.

Huang, X.; Foster, G.D.; Honeychuck, R.V.; Schreifels, J.A. (2009) The maximum of phosphate adsorption at pH 4.0: why it appears on aluminum oxides but not on iron oxides. *Langmuir* 25(8): 4450–4461.

Hug, S.H. (1997) *In situ* Fourier transform infrared measurements of sulfate adsorption on hematite in aqueous solutions. *J. Colloid Interface Sci.* 188: 415–422.

Huittinen, N.; Rabung, Th.; Lützenkirchen, J.; Mitchell, S.C.; Bickmore, B.R.; Lehto, J.; Geckeis, H. (2009) Sorption of Cm(III) and Gd(III) onto gibbsite, α-$Al(OH)_3$: a batch and TRLFS study. *J. Colloid Interface Sci.* 332: 158–164.

Huneke, J.T.; Cramer, R.E.; Alvarez, R.; El-Swaify, S.A. (1980) The identification of gibbsite and bayerite by laser Raman spectroscopy. *Soil Sci. Soc. Am. J.* 44: 131.

Ingram-Jones, V.J.; Slade, R.C.T.; Davies, T.W.; Southern, J.C.; Salvador, S. (1996) Dehydroxylation sequences of gibbsite and boehmite: study of differences between soak and flash calcination and of particle size effects. *J. Mater. Chem.* 6: 73–79.

IUPAC (1997) *Compendium of Chemical Terminology*, 2nd ed.

Jampor, R.O.; Park, G.A. (1982) Characterization of aqueous colloids by their electrical double-layer and intrinsic surface chemical properties. In: E. Matijevic, editor, *Surface and Colloid Science*, vol. 12, Plenum Press, New York, pp. 119–216.

Jampor, J.L.; Kovalenker, V.A.; Roberts, A.C. (1998) New mineral names. *Am. Miner.* 83: 1117–1121.

Jardine, P.M.; Zelazny, L.W. (1996) Surface reactions of aqueous aluminum species. In: G. Sposito, editor, *The Environmental Chemistry of Aluminum*, 2nd ed., CRC Press Inc., Boca Raton, FL.

Jepson, W.B.; Jeffs, D.G.; Ferris, A.P. (1976) The adsorption of silica on gibbsite and its relevance to the kaolinite surface. *J. Colloid Interface Sci.* 55(2): 454–461.

Jodin, M.-C.; Gaboriaud, F.; Humbert, B. (2004) Repercussions of size heterogeneity on the measurement of specific surface areas of colloidal minerals: combination of macroscopic and microscopic analyses. *Am. Miner.* 89: 1456–1462.

Jodin, M.C.; Gaboriaud, F.; Humbert, B. (2005) Limitations of potentiometric studies to determine the surface charge of gibbsite γ-$Al(OH)_3$ particles. *J. Colloid Interface Sci.* 287: 581–591.

Johansson, G. (1960) On the crystal structure of some basic aluminium salts. *Acta Chem. Scand.* 14: 771–773.

Johansson, G. (1962) The crystal structures of $[Al_2(OH)_2(H_2O)_8](SO_4)_2 \cdot 2H_2O$ and $[Al_2(OH)_2(H_2O)_8](SeO_4)_2 \cdot 2H_2O$. *Acta Chem. Scand.* 16: 403–420.

Johnson, B.B.; Ivanov, A.V.; Antzutkin, O.N.; Forsling, W. (2002) ^{31}P nuclear magnetic resonance study of the adsorption of phosphate and phenyl phosphates on γ-Al_2O_3. *Langmuir* 18: 1104–1111.

Jolicoeur, S.; Ildefonse, P.; Bouchard, M. (2000) Kaolinite and gibbsite weathering of biotite within saprolites and soils of central Virginia. *Soil Sci. Soc. Am. J.* 64: 1118–1129.

Kaiser, K.; Guggenberger, G. (2003) Mineral surfaces and soil organic matter. *Eur. J. Soil Sci.* 54: 219–236.

Karathanasis, A.D.; Evangelou, V.P.; Thompson, Y.L. (1988) Aluminum and iron equilibria in soil solutions and surface waters of acid mine watersheds. *J. Environ. Qual.* 17(4): 534–543.

Karthikeyan, K.G.; Elliott, H.A.; Cannon, F.S. (1997) Adsorption and coprecipitation of copper with the hydrous oxides of iron and aluminum. *Environ. Sci. Technol.* 31(10): 2721–2725.

Kavanagh, B.V.; Posner, A.M.; Quirk, J.P. (1975) Effect of polymer adsorption on the properties of the electrical double layer. *Faraday Discuss. Chem. Soc.* 59: 242–249.

Kim, C.S.; Rytuba, J.J.; Brown, G.E., Jr. (2004a) EXAFS study of mercury(II) sorption to Fe- and Al-(hydr)oxides. I. Effects of pH. *J. Colloid Interface Sci.* 271: 1–15.

Kim, C.S.; Rytuba, J.J.; Brown, G.E., Jr. (2004b) EXAFS study of mercury(II) sorption to Fe- and Al-(hydr)oxides. II. Effects of chloride and sulfate. *J. Colloid Interface Sci.* 270: 9–20.

Kinniburgh, D.G.; Syers, J.K.; Jackson, M.L. (1975) Specific adsorption of trace amounts of calcium and strontium by hydrous oxides of iron and aluminum. *Soil Sci. Soc. Am. Proc.* 39: 464–470.

Kloprogge, J.T.; Duong, L.V.; Wood, B.J.; Frost, R.L. (2006) XPS study of the major minerals in bauxite: gibbsite, bayerite and (pseudo-)boehmite. *J. Colloid Interface Sci.* 296: 572–576.

Kloprogge, J.T.; Ruan, H.D.; Frost, R.L. (2002) Thermal decomposition of bauxite minerals: infrared emission spectroscopy of gibbsite boehmite and diaspore. *J. Mater. Sci.* 37: 1121–1129.

Kodama, H.; Schnitzer, M. (1980) Effect of fulvic acid on the crystallization of aluminum hydroxides. *Geoderma* 24(3): 195–205.

Koretsky, C.M.; Sverjensky, D.A.; Sahai, N. (1998) A model of surface site types on oxide and silicate minerals based on crystal chemistry: implications for site types and densities, multi-site adsorption, surface infrared spectroscopy, and dissolution kinetics. *Am. J. Sci.* 298(5): 349–438.

Kremleva, A.; Krüger, S.; Rösch, N. (2008) Density functional model studies of uranyl adsorption on (001) surfaces of kaolinite. *Langmuir* 24: 9515–9524.

Kyle, J.H.; Posner, A.M.; Quirk, J.P. (1975) Kinetics of isotopic exchange of phosphate adsorbed on gibbsite. *J. Soil Sci.* 26: 32–43.

Ladeira, A.C.Q.; Ciminelli, V.S.T.; Duarte, H.A.; Alves, M.C.M.; Ramos, A.Y. (2001) Mechanism of anion retention from EXAFS and density functional calculations: arsenic (V) adsorbed on gibbsite. *Geochim. Cosmochim. Acta* 65(8): 1211–1217.

Laiti, E.; Persson, P.; Öhman, L.-O. (1998) Balance between surface complexation and surface phase transformation at the alumina/water interface. *Langmuir* 14: 825–831.

Langmuir, D. (1997) *Aqueous environmental geochemistry*. Prentice Hall, Upper Saddle River, NJ.

Lee, M.Y.; Parkinson, G.M. (1999) Growth rates of gibbsite single crystals determined using *in situ* optical microscopy. *J. Cryst. Growth* 198–199: 270–274.

Lee, Y.-P.; Liu, Y.-H.; Yeh, C.-S. (1999) Formation of bayerite, gibbsite and boehmite particle by laser ablation. *Phys. Chem. Chem. Phys.* 1: 4681–4686.

Lefèvre, G. (2004) *In situ* Fourier-transform infrared spectroscopy studies of inorganic ions adsorption on metal oxides and hydroxides. *Adv. Colloid Interface Sci.* 107:109–123.

Lefèvre, G.; Duc, M.; Fédoroff, M. (2004) Effect of solubility on the determination of the protonable surface site density of oxyhydroxides. *J. Colloid Interface Sci.* 269: 274–282.

Lefèvre, G.; Duc, M.; Lepeut, P.; Caplain, R.; Fédoroff, M. (2002) Hydration of γ-alumina in water and its effects on surface reactivity. *Langmuir* 18: 7530–7537.

Lehl, H. (1934) Boehmite and bayerite. *J. Phys. Chem.* XL(1): 47–54.

Letterman, R.D.; Amirtharajah, A.; O'Melia, C.R. (1999). Coagulation and flocculation. In: *Water Quality and Treatment*, McGraw-Hill, New York.

Levin, I.; Bendersky, L.A.; Brandon, D.G.; Ruhle, M. (1997) Cubic to monoclinic phase transformations in alumina. *Acta Mater.* 45(9): 3659–3669.

Levin, I.; Brandon, D. (1998a) Metastable alumina polymorphs: crystal structures and transition sequences. *J. Am. Ceram. Soc.* 81(8): 1995–2012.

Levin, I.; Brandon, D. (1998b) A new metastable alumina polymorph with monoclinic symmetry. *Philos. Mag. Lett.* 77(2): 117–124.

Li, H.; Addai-Mensah, J.; Thomas, J.C.; Gerson, A.R. (2005) The crystallization mechanism of $Al(OH)_3$ from sodium aluminate solutions. *J. Cryst. Growth* 279: 508–520.

Li, S.-Z.; Xu, R.-K. (2008) Electrical double layers' interaction between oppositely charged particles as related to surface charge density and ionic strength. *Colloids Surf. A* 326: 157–161.

Licsko, I. (1997) Realistic coagulation mechanisms in the use of aluminium and iron(III) salts. *Water Sci. Technol.* 36(4): 103–110.

Lin, T.-F.; Wu, J.-K. (2001) Adsorption of arsenite and arsenate within activated alumina grains: equilibrium and kinetics. *Water Res.* 35(8): 2049–2057.

Lippens, B.C. (1961) Structure and textures of aluminas, Proefschrift, Delft, Netherlands.

Liu, H.; Hu, J.; Xu, J.; Liu, Z.; Shu, J.; Mao, H.K.; Chen, J. (2004) Phase transition and compression behavior of gibbsite under high-pressure. *Phys. Chem. Miner.* 31: 240–246.

Liu, Y.; Ma, D.; Blackley, R.A.; Zhou, W.; Han, X.; Bao, X. (2008) Synthesis and characterization of gibbsite nanostructures. *J. Phys. Chem. C* 112: 4124–4128.

Liu, Y.T.; Wang, M.K.; Chen, T.Y.; Chiang, P.N.; Huang, P.M.; Lee, J.F. (2006) Arsenate sorption on lithium/aluminum layered double hydroxide intercalated by chloride and on gibbsite: sorption isotherms, envelopes, and spectroscopic studies. *Environ. Sci. Technol.* 40: 7784–7789.

Lloyd, S.; Thurgate, S.M.; Cornell, R.M.; Parkinson, G.M. (1998) Atomic force microscopy of gibbsite. *Appl. Surf. Sci.* 135: 178–182.

Lodding, W. (1961) Gibbsite vermiforms in the Pensauken formation of New Jersey. *Am. Miner.* 46: 394–401.

Lützenkirchen, J. (2001) A discussion of the surface complexation modeling in the paper by Sarkar et al. (1999). Comments & Letters to the Editor. *Soil Sci. Soc. Am. J.* 65: 1348–1350.

Macias Vazquez, F. (1981) Formation of gibbsite in soils and saprolites of temperate-humid zones. *Clay Miner.* 16: 43–52.

Manceau, A.; Combes, J. (1988) Structure of Mn and Fe oxides and hydroxides: a topological approach by EXAFS. *Phys. Chem. Miner.* 15: 283–295.

Manning, B.A.; Goldberg, S. (1996) Modeling competitive adsorption of arsenate with phosphate and molybdate on oxide minerals. *Soil Sci. Soc. Am. J.* 60(1): 121–131.

Manning, B.A.; Goldberg, S. (1997) Adsorption and stability of arsenic(III) at the clay mineral–water interface. *Environ. Sci. Technol.* 31(7): 2005–2011.

Marboe, E.C.; Bentur, S. (1961) A new interpretation of aging of aluminum hydroxide gel. *Silicates Ind.* 26: 389–399.

Martell, A.E.; Hancock, R.D.; Smith, R.M.; Motekaitis, R.J. (1996) Coordination of Al(III) in the environment and in biological systems. *Coord. Chem. Rev.* 149: 311–328.

Martinez, C.E.; McBride, M.B. (1999) Dissolved and labile concentrations of Cd, Cu, Pb, and Zn in aged ferrihydrite–organic matter systems. *Environ. Sci. Technol.* 33: 745–750.

Masue, Y.; Loeppert, R.H.; Kramer, T.A. (2009) Arsenate and arsenite adsorption and desorption behavior on coprecipitated aluminum:iron hydroxides. *Environ. Sci. Technol.* 41: 837–842.

Mathur, S.S. (1995) Development of a database for ion sorption on goethite using surface complexation modeling, M.S. thesis, Department of Civil and Environmental Engineering, Carnegie Mellon University, Pittsburgh, PA.

Mathur, S.S.; Dzombak, D.A. (2006) Surface complexation modeling: goethite. In: J. Lutzenkirchen, editor, *Surface Complexation Modeling*, American Press, Chapter 16, pp. 443–468.

Maurice, P.A. (2009) Some key techniques for investigating surfaces and interfaces. In: *Environmental Surfaces and Interfaces from the Nanoscale to the Global Scale*, Wiley, Hoboken, NJ, Chapter 4.

May, H.M.; Helmke, P.A.; Jackson, M.L. (1979) Gibbsite solubility and thermodynamic properties of hydroxyl-aluminum ions in aqueous solution at 25°C. *Geochim. Cosmochim. Acta* 43: 861–868.

McBride, M.B. (1982) Cu^{2+}-adsorption characteristics of aluminum hydroxide and oxyhydroxides. *Clays Clay Miner.* 30(1): 21–28.

McBride, M.B.; Fraser, A.R.; McHardy, W.J. (1984) Cu^{2+} interaction with microcrystalline gibbsite evidence for oriented chemisorbed copper ions. *Clays Clay Miner.* 32(1): 12–18.

McBride, M.B.; Wessellnk, L.G. (1988) Chemisorption of catechol on gibbsite, boehmite, and noncrystalline alumina surfaces. *Environ. Sci. Technol.* 22: 703–708.

McHardy, W.J.; Thomson, A.P. (1971) Conditions for the formation of bayerite and gibbsite. *Miner. Mag.* 38: 358–368.

McKinley, J.P.; Zachara, J.M.; Smith, S.C.; Turner, G.D. (1995) The influence of uranyl hydrolysis and multiple site-binding reactions on adsorption of U(VI) to montmorillonite. *Clays Clay Miner.* 43(5): 586–598.

McNeill, L.S.; Edwards, M. (1999) Predicting As removal during metal hydroxide precipitation. *J. Am. Water Works Assoc.* 89(1): 75–86.

Mehta, S.K.; Kalsotra, A. (1991) Kinetics and hydrothermal transformation of gibbsite. *J. Therm. Anal.* 367: 267–275.

Meng, X.; Letterman, R.D. (1993a) Effect of component oxide interaction on the adsorption properties of mixed oxides. *Environ. Sci. Technol.* 27: 970–975.

Meng, X.; Letterman, R.D. (1993b) Modeling ion adsorption on aluminum hydroxide modified silica. *Environ. Sci. Technol.* 27: 1924–1929.

Mercury, J.M.R.; Pena, P.; de Aza, A.H.; Sheptyakov, D.; Turrillas, X. (2006) On the decomposition of synthetic gibbsite studied by neutron thermodiffractometry. *J. Am. Ceram. Soc.* 89(12): 3728–3733.

Micera, G.; Gessa, C.; Melis, P.; Premoli, A.; Dallocchio, R.; Deiana, S. (1986) Zinc(II) adsorption on aluminum hydroxide. *Colloids Surf. A* 17: 389–394.

Mishra, S.P.; Tiwari, D.; Prasad, S.K.; Dubey, R.S.; Mishra, M. (2007) Inorganic particulates in removal of toxic heavy metal ions. Part X. Removal behavior of aluminum hydroxide for Hg(II): a radiotracer study. *J. Radioanal. Nucl. Chem.* 274(2): 257–263.

Misra, C. (2003) Aluminum oxide (alumina), hydrated. In: *Kirk–Othmer Encyclopedia of Chemical Technology*, vol. 2, Wiley.

Mitchell, S.C. (2005) An improved music model for gibbsite, M.Sc. thesis, Department of Geology, Brigham Young University, Idaho.

Molis, E.; Barrès, O.; Marchand, H.; Sauzéat, E.; Humbert, B.; Thomas, F. (2000) Initial steps of ligand-promoted dissolution of gibbsite. *Colloids Surf. A* 163: 283–292.

Morel, F.M.M.; Hering, J.G. (1993) *Principles and Applications of Aquatic Chemistry*, Wiley–Interscience, New York.

Muljadi, D.; Posner, A.M.; Quirk, J.P. (1966) The mechanism of phosphate adsorption by kaolinite, gibbsite, and pseudoboehmite. Part I. The isotherms and the effect of pH on adsorption. *J. Soil Sci.* 17(2): 212–229.

Murray, J.W. (1974) The surface chemistry of hydrous manganese dioxide. *J. Colloid Interface Sci.* 46: 357–371.

Music, S. (1986) Sorption of chromium(VI) and chromium(III) on aluminum hydroxide. *J. Radioanal. Nucl. Chem.* 100(1): 185–196.

Nordstrom, D.K.; Archer, D.G. (2002) Arsenic thermodynamic chemistry and environmental geochemistry. In: A.H. Welch and K.G. Stollenwerk, editors, *Arsenic in Ground Water. Geochemistry and Occurrence*, Kluwer Academic Publishers, pp. 1–26.

Nordstrom, D.K.; Ball, J.W. (1986) The geochemical behavior of aluminum in acidified surface waters. *Science* 232(4746): 54–56.

Okazaki, M.; Sakaidani, K.; Saigusa, T.; Sakaida, N. (1989) Ligand exchange of oxyanions on synthetic hydrated oxides of iron and aluminum. *Soil Sci. Plant Nutr.* 35(3): 337–346.

Okazaki, M.; Takamidoh, K.; Yamane, I. (1986) Adsorption of heavy metal cations on hydrated oxides and oxides of iron and aluminum with different crystallinities. *Soil Sci. Plant Nutr.* 32 (4):523–533.

Oliveira, A.F.; Ladeira, A.C.Q.; Ciminelli, V.S.T.; Heine, T.; Duarte, H.A. (2006) Structural model of arsenic(III) adsorbed on gibbsite based on DFT calculations. *J. Mol. Struct.* 762: 17–23.

Östhols, E. (1995) Thorium sorption on amorphous silica. *Geochem. Cosmochim. Acta* 59(7): 1235–1249.

Ottaviani, M.F.; Martini, G. (1980) Adsorption of the $Cu(OH)_4^{2-}$ complex on aluminas studied by electron spin resonance. *J. Phys. Chem.* 84: 2310–2315.

Palmer, S.J.; Frost, R.L.; Nguyen, T. (2009) Hydrotalcites and their role in coordination of anions in Bayer liquors: anion binding in layered double hydroxides. *Coord. Chem. Rev.* 253: 250–267.

Papée, D.; Tertian, R.; Biais, R. (1958) Recherches sur la constitution des gels et des hydrates crystallines d'aluminin. *Bull. Soc. Chim. Fr.* 81: 1301–1310.

Papelis, Ch.; Brown, G.E., Jr.; Parks, G.A.; Leckie, J.O. (1995) X-ray absorption spectroscopic studies of cadmium and selenite adsorption on aluminum oxides. *Langmuir* 11: 2041–2048.

Parfitt, R.L.; Fraser, A.R.; Russell, J.D.; Farmer, V.C. (1977) Adsorption on hydrous oxides. II. Oxalate, benzoate and phosphate on gibbsite. *J. Soil Sci.* 28: 40–47.

Parks, G.A.; DeBruyn, P.L. (1962) The zero point of charge of oxides. *J. Phys. Chem.* 66: 967–973.

Pascal, P.; Chretien, A.; Trambouze, Y.; Hutter, J.C.; Freundlich, W. (1963) In: *Nouveau Traité de Chimie Minérale*, vol. 6, Masson et Cie, Paris, p. 574.

Paul, K.W.; Kubicki, J.D.; Sparks, D.L. (2007) Sulphate adsorption at the Fe (hydr)oxide–H_2O interface: comparison of cluster and periodic slab DFT predictions. *Eur. J. Soil Sci.* 58: 978–988.

Paulson, A.J.; Balistrieri, L. (1999) Modeling removal of Cd, Cu, Pb, and Zn in acidic groundwater during neutralization by ambient surface waters and groundwaters. *Environ. Sci. Technol.* 33(21): 3850–3856.

Peacock, C.L.; Sherman, D.M. (2005) Surface complexation model for multisite adsorption of copper(II) onto kaolinite. *Geochim. Cosmochim. Acta* 69: 3733–3745.

Peak, D. (2006) Adsorption mechanisms of selenium oxyanions at the aluminum oxide/water interface. *J. Colloid Interface Sci.* 303: 337–345.

Phambu, N.; Humbert, B.; Burneau, A. (2000) Relation between the infrared spectra and the lateral specific surface areas of gibbsite samples. *Langmuir* 16: 6200–6207.

Pivovarov, S. (1997) Surface structure and site density of the oxide–solution interface (Letter to the Editor). *J. Colloid Interface Sci.* 196: 321–323.

Pokrovski, G.S.; Schott, J.; Farges, F.; Hazemann, J.-L. (2003) Iron (III)–silica interactions in aqueous solution: insights from X-ray absorption fine structure spectroscopy. *Geochim. Cosmochim. Acta* 67: 3559–3573.

Pokrovski, G.S.; Schott, J.; Harrichoury, J.-C.; Sergeyev, A.S. (1996) The stability of aluminum silicate complexes in acidic solutions from 25 to 150°C. *Geochim. Cosmochim. Acta* 60 (14): 2495–2501.

Pokrovsky, O.S.; Pokrovski, G.S.; Schott, J.; Galy, A. (2006) Experimental study of germanium adsorption on goethite and germanium coprecipitation with iron hydroxide: X-ray absorption fine structure and macroscopic characterization. *Geochim. Cosmochim. Acta* 70(13): 3325–3341.

Pokrovsky, O.S.; Viers, J.; Emnova, E.E.; Kompantseva, E.I.; Freydier, R. (2008) Copper isotope fractionation during its interaction with soil and aquatic microorganisms and metal oxy(hydr)oxides: possible structural control. *Geochem. Cosmochim. Acta* 72: 1742–1757.

Pokrovsky, O.S.; Viers, J.; Freydier, R. (2005) Zinc stable isotope fractionation during its adsorption on oxides and hydroxides. *J. Colloid Interface Sci.* 291: 192–200.

Prestidge, C.A.; Ametov, I.; Addai-Mensah, J. (1999) Rheological investigations of gibbsite particles in synthetic Bayer liquors. *Colloids Surf. A* 157: 137–145.

Prodromou, K.P.; Pavlatou-Ve, A.S. (1995) Formation of aluminum hydroxides as influenced by aluminum salts and bases. *Clays Clay Miner.* 43(1): 111–115.

Rajan, S.S.S. (1979) Adsorption of selenite, phosphate and sulphate on hydrous alumina. *J. Soil Sci.* 30: 709–718.

Rakotonarivo, E.; Bottero, J.Y.; Thomas, F.; Poirier, J.E.; Cases, J.M. (1988) Electrochemical modeling of freshly precipitated aluminum hydroxide–electrolyte interface. *Colloids Surf. A* 33: 191–207.

Ramos-Gallardo, A.; Vegas, A. (1996) The cation array in aluminum oxides, hydroxides and oxyhydroxides. *Z. Kristallogr.* 211: 299–303.

Randall, S.R.; Sherman, D.M.; Ragnarsdottir, K.V.; Collins, C.R. (1999) The mechanism of cadmium surface complexation on iron oxyhydroxide minerals. *Geochim. Cosmochim. Acta* 63(19–20): 2971–2987.

Rebhum, M.; Lurie, M. (1993) Control of organic matter by coagulation and floc separation. *Water Sci. Technol.* 27(11): 1–20.

Reynolds, R.C. (1971) Clay mineral formation in an alpine environment. *Clays Clay Miner.* 19: 361–374.

Ridley, M.K.; Wesolowski, D.J.; Palmer, D.A.; Benezeth, P.; Kettler, R.M. (1997) Effect of sulfate on the release rate of Al^{3+} from gibbsite in low-temperature acidic waters. *Environ. Sci. Technol.* 31: 1922–1925.

Roberts, D.R.; Ford, R.G.; Sparks, D.L. (2003) Kinetics and mechanisms of Zn complexation on metal oxides using EXAFS spectroscopy. *J. Colloid Interface Sci.* 263: 364–376.

Rosenqvist, J.; Axe, K.; Sjöberg, S.; Persson, P. (2003) Adsorption of dicarboxylates on nanosized gibbsite particles: effects of ligand structure on bonding mechanisms. *Colloids Surf. A* 220: 91–104.

Rosenqvist, J.; Casey, W.H. (2004) The flux of oxygen from the basal surface of gibbsite (α-$Al(OH)_3$) at equilibrium. *Geochim. Cosmochim. Acta* 68(17): 3547–3555.

Rosenqvist, J.; Persson, P.; Sjoberg, S. (2002) Protonation and charging of nanosized gibbsite (α-$Al(OH)_3$) particles in aqueous suspension. *Langmuir* 18: 4598–4604.

Rothbauer, R.; Zigan, F.; O'Daniel, H. (1967) Verfeinerung der Struktur des Bayerits, $Al(OH)_3$ Einschließlich eines Vorschlags fur die H-Positionen. *Z. Kristallogr.* 125: 317–331.

Ruan, H.D.; Frost, R.L.; Kloprogge, J.T. (2001) Comparison of Raman spectra in characterizing gibbsite, bayerite, diaspore and boehmite. *J. Raman Spectrosc.* 32: 745–750.

Saalfeld, H. (1960) Strukturen von Hydrargillit, Entwasserungsmechanismus und Struktur der Zwischenprodukte (Structures of gibbsite and of intermediate products of its dehydration). *Neues. Jahrb. Miner. Abh.* 95: 1–87.

Saalfeld, H.; Wedde, M. (1974) Refinement of the crystal structure of gibbsite. *Z. Kristallogr.* 139: 129–135.

Saeki, K. (2004) Adsorption of Fe^{2+} and Mn^{2+} on silica, gibbsite, and humic acids. *Soil Sci.* 169 (12):832–840.

Sahai, N.; Sverjensky, D.A. (1997) Evaluation of internally consistent parameters for the triple-layer model by the systematic analysis of oxide surface titration data. *Geochim. Cosmochim. Acta* 61(14): 2801–2826.

Sainz-Diaz, C.I.; Timon, V.; Botella, V.; Hernandez-Laguna, A. (2000) Isomorphous substitution effect on the vibration frequencies of hydroxyl groups in molecular cluster models of the clay octahedral sheet. *Am. Miner.* 85(7–8): 1038–1045.

Sarkar, D.; Essington, M.E.; Misra, K.C. (1999) Adsorption of mercury(II) by variable charge surfaces of quartz and gibbsite. *Soil Sci. Soc. Am. J.* 63: 1626–1636.

Scancar, J.; Milacic, R. (2006) Aluminum speciation in environmental samples: a review. *Anal. Bioanal. Chem.* 386: 999–1012.

Schacklette, H.T.; Boerngen, J.G. (1984) Element concentrations in soils and other surfacial materials of the conterminous United States, U.S. Geological Survey Professional Paper 1270.

Schecher, W.D.; Driscoll, C.T. (1988) An evaluation of the equilibrium calculations within acidification models: the effect of uncertainty in measured chemical components. *Water Resour. Res.* 24(4): 533–540.

Scheckel, K.G.; Sparks, D.L. (2000) Kinetics of the formation and dissolution of Ni precipitates in a gibbsite/amorphous silica mixture. *J. Colloid Interface Sci.* 229: 222–229.

Scheinost, A.C.; Sparks, D.L. (2000) Formation of layered single- and double-metal hydroxide precipitates at the mineral/water interface: a multiple-scattering XAFS analysis. *J. Colloid Interface Sci.* 223: 167–178.

Schindler, P.W.; Furst, B.; Dick, R.; Wolf, P.U. (1976) Ligand properties of surface silanol groups. I. Surface complex formation with Fe^{3+}, Cu^{2+}, Cd^{2+} and Pb^{2+}. *J. Colloid Interface Sci.* 55: 469–475.

Schindler, P.W.; Kamber, H.R. (1968) Die Acidität von Silanolgruppen. *Helv. Chim. Acta* 51: 1781–1786.

Schoen, R.; Robertson, C.E. (1970) Structures of aluminum hydroxide and geochemical implications. *Am. Miner.* 55: 43–77.

Schulthess, C.P.; McCarthy, J.F. (1990) Competitive adsorption of aqueous carbonic and acetic acids by an aluminum oxide. *Soil Sci. Soc. Am. J.* 54: 688–694.

Schulthess, C.P.; Sparks, D.L. (1988) A critical assessment of surface adsorption models. *Soil Sci. Soc. Am. J.* 52: 92–97.

Shchukarev, A.; Sjöberg, S. (2005) XPS with fast-frozen samples: a renewed approach to study the real mineral/solution interface. *Surf. Sci.* 584: 106–112.

Sing, K. (2001) The use of nitrogen adsorption for the characterization of porous materials. *Colloids Surf. A* 187–188: 3–9.

Singh, T.S.; Pant, K.K. (2004) Equilibrium, kinetics and thermodynamic studies for adsorption of As(III) on activated alumina. *Sep. Sci. Technol.* 36: 139–147.

Skoog, D.A.; West, D.M. (1982) *Fundamental of Analytical Chemistry*, 4th ed. W.B. Saunders, Philadelphia, PA, pp. 99–148.

Slade, R.C.T.; Southern, J.C.; Thompson, I.M. (1991) ^{27}Al nuclear magnetic resonance spectroscopy investigation of thermal transformation sequences of alumina. Part 1. Gibbsite, γ-$Al(OH)_3$. *J. Mater. Chem.* 1(4): 563–568.

Smith, R.M.; Martell, A.E. (1976) Critical stability constants, Vol. 4, *Inorganic Complexes*, Plenum Press, New York.

Smit, W.; Holten, C.L.M. (1980) Zeta-potential and radiotracer adsorption measurements on EFG α-Al_2O_3 single crystals in NaBr solutions. *J. Colloid Interface Sci.* 78(1): 1–14.

Smith, R.M.; Martell, A.E.; Motekaitis, R.J. (2001) *NIST Critically Selected Stability Constants of Metal Complexes Database*, NIST Standard Reference Database 46, version 6.0, NIST, Gaithersburg, MD.

Smith, R.M.; Martell, A.E.; Motekaitis, R.J. (2003) *NIST Critically Selected Stability Constants of Metal Complexes Database*, NIST Standard Reference Database 46, version 7.0, NIST, Gaithersburg, MD.

Smith, R.S. (1996) Kinetic aspects of aqueous aluminum chemistry: environmental implications. *Coord. Chem. Rev.* 149: 81–93.

Snoeyink, V.L.; Schock, M.R.; Sarin, P.; Wang, L.; Chen, A.S.-C.; Harmon, S.M. (2003) Aluminium-containing scales in water distribution systems: prevalence and composition. *J. Water Supply Res. Technol. AQUA* 52(7): 455–474.

Souza Santos, H.; Kiyohara, P.K.; Souza Santos, P. (2000) A route for synthesis of microcrystalline gibbsite in Na^+ free medium. *J. Mater. Sci. Lett.* 19: 1525–1527.

Souza Santos, P.; Valleijo-Freire, A.; Souza Santos, H.L. (1953) Electron microscope studies on the aging of amorphous colloidal aluminium hydroxide. *Kolloid Z.* 133(2–3): 101–107.

Sposito, G. (1984) *The Surface Chemistry of Soils*, Oxford University Press, New York.

Sposito, G. (1989) *The Chemistry of Soils*, Oxford University Press, New York, p. 6.

Sposito, G., editor (1996) *The Environmental Chemistry of Aluminum*, 2nd ed., CRC Press Inc., Boca Raton, FL.

Srinivasan, P.T.; Viraraghavan, T.; Subramanian, K.S. (1999) Aluminium in drinking water: an overview. *Water SA* 25(1): 47–55.

Struempler, A. (1973) Adsorption characteristics of silver, lead, cadmium, zinc, and nickel on borosilicate glass, polyethylene, and polypropylene surfaces. *Anal. Chem.* 45: 2251–2254.

Stumm, W.; Morgan, J. (1981) *Aquatic Chemistry*, 2nd ed. Wiley, New York.

Stumpf, H.C.; Russell, A.S.; Newsome, J.W.; Tucker, C.M. (1950) Thermal transformations of aluminas and alumina hydrates. *Ind. Eng. Chem.* 42: 1398–1403.

Su, T.; Guan, X.; Gub, G.; Wang, J. (2008) Adsorption characteristics of As(V), Se(IV), and V(V) onto activated alumina: effects of pH, surface loading, and ionic strength. *J. Colloid Interface Sci.* 326: 347–353.

Su, C.; Suarez, D.L. (1997) *In situ* infrared speciation of adsorbed carbonate on aluminum and iron oxides. *Clays Clay Miner.* 45(6): 814–825.

Sullivan, T.J.; Cosby, B.J. (1998) Modeling the concentration of aluminium in surface waters. *Water Air Soil Pollut.* 105: 643–659.

Sverjensky, D.A. (2006) Prediction of the speciation of alkaline earths adsorbed on mineral surfaces in salt solutions. *Geochim. Cosmochim. Acta* 70: 2427–2453.

Sweegers, C.; Boerrigter, S.X.M.; Grimbergen, R.F.P.; Meekes, H.; Fleming, S.; Hiralal, I.D.K.; Rijkeboer, A. (2002a) Morphology prediction of gibbsite crystals: an explanation for the lozenge-shaped growth morphology. *J. Phys. Chem. B* 106: 1004–1012.

Sweegers, C.; de Coninck, H.C.; Meekes, H.; van Enckevort, W.J.P.; Hiralal, I.D.K.; Rijkeboer, A. (2001) Morphology, evolution and other characteristics of gibbsite crystals grown from pure and impure aqueous sodium aluminate solutions. *J. Cryst. Growth* 233: 567–582.

Sweegers, C.; Meekes, H.; van Enckevort, W.J.P.; Bennema, P.; Hiralal, I.D.K.; Rijkeboer, A. (1999) The impact of twinning on the morphology of γ-$Al(OH)_3$ crystals. *J. Cryst. Growth* 197(1–2): 244–253.

Sweegers, C.; Meekes, H.; van Enckevort, W.J.P.; Hiralal, I.D.K.; Rijkeboer, A. (2004) Growth rate analysis of gibbsite single crystals growing from aqueous sodium aluminate solutions. *Cryst. Growth Des.* 4(1): 185–198.

Sweegers, C.; Plomp, M.; de Coninck, H.C.; Meekes, H.; van Enckevort, W.J.P.; Hiralal, I.D. K.; Rijkeboer, A. (2002b) Surface topography of gibbsite crystals grown from aqueous sodium aluminate solutions. *Appl. Surf. Sci.* 187: 218–234.

Swistock, B.R.; Dewalle, D.R.; Sharpe, W.E. (1989) Sources of acid storm flow in an Appalachian headwater stream. *Water Resour. Res.* 25: 2139–2147.

Tabor, N.J.; Yapp, C.J. (2005) Incremental vacuum dehydration–decarbonation experiments on a natural gibbsite (α-$Al(OH)_3$): CO_2 abundance and ^{13}C values. *Geochim. Cosmochim. Acta* 69(3): 519–527.

Tan, K.H. (1998) *Principals of Soil Chemistry*, 3rd ed. Marcel Dekker Inc., New York.

Tanabe, K. (1981) Solid acid and base catalysis. In: J.R. Anderson and M. Boudart, editors, *Catalysis: Science and Technology*, Springer, New York, p. 231.

Tarassoff, P.; Horváth, L.; Pfenninger-Horváth, E. (2006) Famous mineral localities: the Francon Quarry, Montréal, Québec, *Canada. Miner. Rec.* 37(1): 5–60.

Thomas, S.; Sherwood, P.M.A. (1992) Valence band spectra of aluminum oxides, hydroxides, and oxyhydroxides interpreted by Xα calculations. *Anal. Chem.* 64(21): 2488–2495.

Tochiyama, O.; Yamazaki, H.; Li, N. (1996) Effect of the concentration of metal ions on their adsorption on various hydrous iron and aluminum oxides. *J. Nucl. Sci. Technol.* 33(11): 846–851.

Tonkin, J.W.; Balistrieri, L.S.; Murray, J.W. (2004) Modeling sorption of divalent metal cations on hydrous manganese oxide using the diffuse double layer model. *Appl. Geochem.* 19: 29–53.

Tosterud, M. (1928) Aluminum hydrate production, The Aluminum Company of America, Canadian Patent 287,147.

Tournassat, C.; Neaman, A.; Villéras, F.; Bosbach, D.; Charlet, L. (2003) Nanomorphology of montmorillonite particles: estimation of the clay edge sorption site density by low-pressure gas adsorption and AFM observations. *Am. Miner.* 88 (11–12, part 2): 1989–1995.

Trivedi, P.; Axe, L. (2000) Modeling Cd and Zn sorption to hydrous metal oxides. *Environ. Sci. Technol.* 34: 2215–2223.

Trivedi, P.; Axe, L. (2001) Predicting divalent metal sorption to hydrous Al, Fe, and Mn oxides. *Environ. Sci. Technol.* 35: 1779–1784.

Trolard, F.; Tardy, Y. (1987) The stabilities of gibbsite, boehmite, aluminous goethites and aluminous hematites in bauxites, ferricretes and laterites as a function of water activity, temperature and particle size. *Geochim. Cosmochim. Acta* 51: 945–957.

Tsuchida, T.; Ichikawa, N. (1989) Mechanochemical phenomena of gibbsite, bayerite and boehmite by grinding. *React. Solids* 7: 207–217.

Turner, G.D.; Zachara, J.M.; McKinley, J.P.; Smith, S.C. (1996) Surface-charge properties and UO_2^{2+} adsorption of a subsurface smectite. *Geochim. Cosmochim. Acta* 60(18): 3399–3414.

USDA, (1996) *Keys to Soil Taxonomy*, 7th ed. Soil Survey Staff, U.S. Department of Agriculture, Washington, DC.

U.S. Geological Survey, (2008) *Mineral Commodity Summaries*, January.

Valeton, I. (1972) *Bauxites. In: Developments in Soil Sciences*, Elsevier, Amsterdam.

Van Emmerik, T.J.; Sandström, D.E.; Antzutkin, O.N.; Angove, M.J.; Johnson, B.B. (2007) ^{31}P solid-state nuclear magnetic resonance study of the sorption of phosphate onto gibbsite and kaolinite. *Langmuir* 23: 3205–3213.

Van Geen, A.; Robertson, A.P.; Leckie, J.O. (1994) Complexation of carbonate species at the goethite surface: implications for adsorption of metal ions in natural waters. *Geochim. Cosmochim. Acta* 58: 2073–2086.

Van Nordstrand, R.A.; Hettinger, W.P.; Keith, C.D. (1956) A new alumina trihydrate. *Nature* 177: 713–714.

Van Riemsdijk, W.H.; Bolt, G.H.; Koopal, K.K.; Blaakmeer, J. (1986) Electrolyte adsorption on heterogeneous surfaces: adsorption models. *J. Colloid Interface Sci.* 109(1): 219–228.

Van Riemsdijk, W.H.; Lylkema, J. (1980) Reaction of phosphate with gibbsite ($Al(OH)_3$) beyond the adsorption maxima. *J. Colloid Interface Sci.* 76(1): 55–66.

Van Straten, H.A.; Holtkamp, B.T.W.; de Bruyn, P.L. (1984) Precipitation from supersaturated aluminate solutions. I. Nucleation and growth of solid phases at room temperature. *J. Colloid Interface Sci.* 98(2): 342–362.

Veilly, E.; Roques, J.; Jodin-Caumon, M.-C.; Humbert, B.; Drot, R.; Simoni, E. (2008) Uranyl interaction with the hydrated (001) basal face of gibbsite: a combined theoretical and spectroscopic study. *J. Chem. Phys.* 129: 244704.

Vermöhlen, K.; Lewandowski, H.; Narres, H.-D.; Schwuger, M.J. (2000) Adsorption of polyelectrolytes onto oxides: the influence of ionic strength, molar mass, and Ca^{2+} ions. *Colloids Surf. A* 163: 45–53.

Villalobos, M.; Trotz, M.A.; Leckie, J.O. (2003) Variability in goethite surface site density: evidence from proton and carbonate sorption. *J. Colloid Interface Sci.* 268(2): 273–287.

Villiéras, F.; Cases, J.M.; Francois, M.; Michot, L.J.; Thomas, F. (1992) Texture and surface energetic heterogeneity of solids from modeling of low pressure gas adsorption isotherms. *Langmuir* 8(7): 1789–1795.

Villiéras, F.; Michot, L.J.; Bardot, F.; Cases, J.M.; Francois, M.; Rudzinski, W. (1997) An improved derivative isotherm summation method to study surface heterogeneity of clay minerals. *Langmuir* 13(5): 1104–1117.

Violante, A. (1989) Factors affecting aluminum transformations. *6th International Congress of ICSOBA*, Pocos de Caldas, vol. 19, no. 22, pp. 431–439.

Violante, A.; Huang, P.M. (1985) Influence of inorganic and organic ligands on the formation of aluminum hydroxides and oxyhydroxides. *Clays Clay Miner.* 33: 181–192.

Violante, A.; Huang, P.M. (1993) Formation mechanism of aluminum hydroxide polymorphs. *Clays Clay Miner*. 41(5): 590–597.

Violante, A.; Jackson, M.L. (1981) Clay influence on the crystallization of aluminum hydroxide polymorphs in the presence of citrate, sulfate or chloride. *Geoderma* 25: 199–214.

Violante, A.; Ricciardella, M.; Del Gaudio, S.; Pigna, M. (2006) Coprecipitation of arsenate with metal oxides: nature, mineralogy, and reactivity of aluminum precipitates. *Environ. Sci. Technol.* 40(16): 4961–4967.

Violante, A.; Violante, P. (1980) Influence of pH, concentration and chelating power of organic anions on the synthesis of aluminum hydroxides and oxyhydroxides. *Clays Clay Miner*. 28: 425–434.

Violante, P.; Violante, A.; Tait, J.M. (1982) Morphology of nordstrandite. *Clays Clay Miner*. 30 (6):431–437.

Viraraghavan, T.; Subramanian, K.S.; Aruldoss, J.A. (1999) Arsenic in drinking water: problems and solutions. *Water Sci. Technol.* 40(2): 69–76.

Wada, K.; Aomine, S. (1966) Occurrence of gibbsite in weathering of volcanic material at Kuroshibaru, Kumamoto. *Soil Sci. Plant Nutr.* 12: 25–31.

Waite, T.D.; Davis, J.A.; Payne, T.E.; Waychunas, G.A.; Xu, N. (1994) Uranium(VI) adsorption to ferrihydrite. Application of a surface complexation model. *Geochim. Cosmochim. Acta* 58: 5465–5478.

Wang, S.-L.; Cheng, C.-Y.; Tzou, Y.-M.; Liaw, R.-B.; Chang, T.-W.; Chen, J.-H. (2007) Phosphate removal from water using lithium intercalated gibbsite. *J. Hazard. Mater.* 147: 205–212.

Wang, S.L.; Johnston, C.T. (2000) Assignment of the structural OH stretching bands of gibbsite. *Am. Miner.* 85: 739–744.

Waychunas, G.A. (1991) Crystal chemistry of oxides and hydroxides. In: D.H. Lindsley, editor, *Oxide Minerals: Petrologic and Magnetic Significance*, Mineralogical Society of America, Washington, DC, pp. 11–68.

Weaver, C.E.; Pollard, L.D. (1973) The chemistry of clay minerals. In: *Developments in Sedimentology*, vol. 15, Elsevier, New York.

Weerasooriya, R.; Aluthpatabendi, D.; Tobschall, H.J. (2001a) Charge distribution multi-site complexation (CD-MUSIC) modeling of Pb(II) adsorption on gibbsite. *Colloids Surf. A* 189: 131–144.

Weerasooriya, R.; Dharamasena, B.; Aluthpatabendi, D. (2000) Copper–gibbsite interactions: an application of 1-pK surface complexation model. *Colloids Surf. A* 170(1): 65–77.

Weerasooriya, R.; Seneviratne, W.; Kathriarachchi, H.A.; Tobschall, H.J. (2006) Thermodynamic assessment of Hg(II)–gibbsite interactions. *J. Colloid Interface Sci.* 301: 452–460.

Weerasooriya, R.; Tobschall, H.J.; Bandara, A. (2001b) Modeling interactions of Hg(II) and bauxitic soils. *Chemosphere* 69: 1525–1532.

Weerasooriya, R.; Tobschall, H.J.; Bandara, A. (2007a) Modeling interactions of Hg(II) and bauxitic soils. *Chemosphere* 69: 1525–1532.

Weerasooriya, R.; Tobschall, H.J.; Seneviratne, W.; Bandara, A. (2007b) Transition state kinetics of Hg(II) adsorption at gibbsite–water interface. *J. Hazard. Mater.* 147: 971–978.

Weerasooriya, R.; Tobschall, H.J.; Wijesekara, H.K.D.K.; Arachchige, E.K.I.A.U.K.; Pathirathne, K.A.S. (2003) On the mechanistic modeling of As(III) adsorption on gibbsite. *Chemosphere* 51: 1001–1013.

Weerasooriya, R.; Tobschall, H.J.; Wijesekara, H.K.D.K.; Bandara, A. (2004) Macroscopic and vibration spectroscopic evidence for specific bonding of arsenate on gibbsite. *Chemosphere* 55: 1259–1270.

Weerasooriya, R.; Wijesekara, H.K.D.K.; Bandara, A. (2002) Surface complexation modeling of cadmium adsorption on gibbsite. *Colloids Surf. A* 207: 13–24.

Wefers, K. (1967) Zum system Na_2O–TiO_2–H_2O. *Naturwissenschaften* 54(1): 18–19.

Wefers, K.; Misra, C. (1987) Oxides and hydroxides of aluminum, Technical Paper No. 19, Alcoa Laboratories, Pittsburgh, PA.

Wesolowski, D.J. (1992) Aluminum speciation and equilibria in aqueous solution. I. The solubility of gibbsite in the system Na–K–Cl–OH–$Al(OH)_4$ from 0 to 100°C. *Geochim. Cosmochim. Acta* 56: 1065–1091.

Westall, J.C. (1982) FITEQL: a program for the determination of chemical equilibrium constants from experimental data, Technical Report, Chemistry Department, Oregon State University, Corvallis, OR.

Whittington, B.; Ilievski, D. (2004) Determination of the gibbsite dehydration reaction pathway at conditions relevant to Bayer refineries. *Chem. Eng. J.* 98: 89–97.

Wierenga, A.M.; Lenstra, T.A.J.; Philipse, A.P. (1998) Aqueous dispersions of colloidal gibbsite platelets: synthesis, characterization and intrinsic viscosity measurements. *Colloids Surf. A* 134: 359–371.

Wiesner, A.D.; Katz, L.E.; Chen, C.-C. (2006) The impact of ionic strength and background electrolyte on pH measurements in metal ion adsorption experiments. *J. Colloid Interface Sci.* 310: 329–332.

Wilson, M.J. (1970) A gibbsite soil derived from the weathering of an ultrabasic rock on the island of Rhum. *Scott. J. Geol.* 5: 81–89.

Xu, R.-K.; Kozak, L.M.; Huang, P.M. (2008) Kinetics of phosphate-induced desorption of arsenate adsorbed on crystalline and amorphous aluminum hydroxides. *Soil Sci.* 173(10): 1–11.

Yamada, K.; Harato, T.; Hamano, S.; Horinouchi, K. (1984) Dehydration products of gibbsite in rotary kiln and stational calciner. *Light Met.* 157–171.

Yamaguchi, G.; Sakamoto, K. (1958) Crystal structure of bayerite. *Chem. Soc. Jpn. Bull.* 31: 140–141.

Yamamoto, S.; Bluhm, H.; Andersson, K.; Ketteler, G.; Ogasawara, H.; Salmeron, M.; Nilsson, A. (2008) *In situ* X-ray photoelectron spectroscopy studies of water on metals and oxides at ambient conditions. *J. Phys.: Condens. Matter* 20: 1–14.

Yang, L.; Steefel, C.I. (2008) Kaolinite dissolution and precipitation kinetics at 22 °C and pH 4. *Geochim. Cosmochim. Acta* 72: 99–116.

Yang, X.; Sun, Z.; Wang, D.; Forsling, W. (2007a) Surface acid–base properties and hydration/dehydration mechanisms of aluminum (hydr)oxides. *J. Colloid Interface Sci.* 308: 395–404.

Yang, X.; Wang, D.; Sun, Z.; Tang, H. (2007b) Adsorption of phosphate at the aluminum (hydr)oxides–water interface: role of the surface acid–base properties. *Colloids Surf. A* 297: 84–90.

Yates, D.E. (1975) The structure of the oxide/aqueous electrolyte interface, Ph.D. thesis, University of Melbourne, Melbourne, Australia.

Yopps, J.A.; Fuerstenau, D.W. (1964) The zero point of charge of alpha-alumina. *J. Colloid Sci.* 19: 61–71.

Yu, G.; Saha, U.K.; Kozak, L.M.; Huang, P.M. (2007) Combined effects of tannate and ageing on structural and surface properties of aluminum precipitates. *Clays Clay Miner.* 55(4): 369–379.

Zachara, J.M.; Girvin, D.C.; Schmidt, R.L.; Resch, C.T. (1987) Chromate adsorption on amorphous iron oxyhydroxide in the presence of major groundwater ions. *Environ. Sci. Technol.* 21: 589–594.

Zachara, J.M.; Resch, C.T.; Smith, S.C. (1994) Influence of humic substances on Co^{2+} sorption by a subsurface mineral separate and its mineralogic components. *Geochim. Cosmochim. Acta* 58(2): 553–566.

Zänker, H.; Hüttig, G.; Arnold, T.; Nitsche, H. (2006) Formation of iron-containing colloids by the weathering of phyllite. *Aquat. Geochem.* 12: 299–325.

Zhang, H.; Dong, Z.; Tao, Z. (2006) Sorption of thorium(IV) ions on gibbsite: effects of contact time, pH, ionic strength, concentration, phosphate and fulvic acid. *Colloids Surf. A* 278: 46–52.

Zhang, H.; Xie, Y.; Tao, Z. (2005) Sorption of uranyl ions on gibbsite: effects of contact time, pH, ionic strength, concentration and anion of electrolyte. *Colloids Surf. A* 252: 1–5.

Zigan, F.; Joswig, W.; Burger, N. (1978) Die Wasserstoffpositionen im Bayerit $Al(OH)_3$. *Z. Kristallogr.* 148: 255–273.

APPENDIX A

SUMMARY OF EXPERIMENTAL DETAILS

This appendix provides a summary of key experimental information for all of the data for proton, cation, and anion sorption on gibbsite considered in the study. Experimental information is included for data sets considered but not used for the extraction of surface-complexation constants. Explanations are provided for why data sets were not used. The experimental information provided relates to the criteria for data evaluation which are discussed in Chapter 3. A list of abbreviations used throughout Appendix A is provided in Table A1.

ADU-WUSU AND WILCOX (1991b)

Experiment: Silicate ($H_2SiO_4^{2-}$) Sorption on Gibbsite

Solid prep.: Gibbsite was obtained as commercial C-31 coarse granules from the Aluminum Company of America (Alcoa). X-ray diffraction (XRD) verified its structure as being pure gibbsite with an average particle size of 63 μm and a specific surface area of 0.1 m^2/g.

I = NS

T = 30 to 60°C

Reaction vessel: Glass flasks (sorption), plastic components

CO_2 exclusion: NS

Max. S/S: 2.4×10^{-5}

Equil. time: 2 days

TABLE A.1 List of abbreviations

AAS	Atomic Absorption Spectrophotometry
approx.	Approximately
atm	Atmosphere
conc.	Concentration
coppt.	Coprecipitation
d	Day
equil.	Equilibrium
expt.	Experiment
HAO	Hydrous aluminum oxide
h	Hour
ICP-AES	Inductively coupled plasma atomic emission spectrometry
ICP-MS	Inductively coupled plasma mass spectrometry
max.	Maximum
min.	Minimum
min	Minute
NS	Not specified; no specifications reported
prep.	Preparation
rcf	Relative centrifugal force
rpm	Revolutions per minute
SC	Crystal or liquid scintillation counting
S/L	Solid/liquid
S/S	Sorbate/sorbent ratio (based on initial total concentrations)
temp.	Temperature
TOT	Total analytical concentration

Measurement: Suspensions containing various concentrations of silicate were equilibrated by agitation in a constant temperature bath (±0.2°C) for 2 days. Equilibration time was based on a study of kinetics, and the choice of 2 days was made to provide enough time for sorption but not precipitation. The suspension for each run contained 2.8 g of gibbsite and 14 cm^3 of aluminate solution. Aliquots (10 cm^3 of supernatant) were removed for total silicon and aluminum analyses. The amount of silicate sorbed was obtained from the difference between the initial and final silicate concentrations. Silicon and aluminum concentrations in the supernatants were determined by atomic adsorption spectroscopy.

Note: Data sets from this study were not used, because temperature of each experiment was kept high >30°C.

BAUMANN ET AL. (2005)

Experiment: Uranyl (UO_2^{2+}) Sorption on Gibbsite

Solid prep.: Gibbsite was obtained as commercial hydrargillite and verified as being 99.9% pure gibbsite by XRD spectroscopy.

$I = 0.1$ M $NaClO_4$

$T = 25°C$

Reaction vessel: Polyethylene tubes

CO_2 exclusion: No, batch experiments were conducted in air atmosphere.

Max. S/S: 2.16×10^{-4}

Equil. time: 2 days

Measurement: After equilibration gibbsite was separated by centrifugation (3500 rotations/min, 15 minutes). The concentration of uranium in the centrifugate was measured by ICP-MS. After centrifugation, the vials were cleaned with deionized water. Then 40 mL of 1 N HNO_3 was added to remove any remaining uranium from the container walls (after 48 hours of mixing). The uranium content in this nitric acid solution was determined with ICP-MS. With the obtained uranium contents in solution and on the container walls the amount of uranium sorbed onto gibbsite was calculated.

Note: The data set in Fig. 1 was not considered and fitted, because sorption experiments were conducted in air.

ESSINGTON AND ANDERSON (2008)

Experiment: Phosphate (PO_4^{3-}), Arsenate (AsO_4^{3-}), and Sulfate (SO_4^{2-}) Sorption on Gibbsite

Solid prep.: Gibbsite was obtained from Alcan Chemicals and it was a hydrated alumina (SF-4). XRD showed monoclinic gibbsite without detectable impurities. To remove poorly crystalline $Al(OH)_{3(s)}$, samples of SF-4 were treated with CO_2-free, 0.01 M NaOH for 30 minutes. After this treatment, gibbsite was washed with the background electrolyte until a pH-neutral suspension was obtained.

$I = 0.01$ and 0.001 M NaCl

$T = 20$ to 25°C

Reaction vessel: Polypropylene tubes

CO_2 exclusion: Yes, N_2 atm.

Max. S/S: 0.033

Equil. time: 72 hours

Measurement: Batch equilibrium sorption experiments were performed in triplicate. 50 g/L of gibbsite suspension was vigorously shaken and a 4 mL aliquot (containing 0.2 g of solid) was transferred to a centrifuge tube. Then 15.5 mL of NaCl, the background electrolyte, and HCl or NaOH were added to adjust the pH from 3 to 10. The tubes were vortex mixed and 0.5 mL of the 4 mM of sorptive working solution was added. The solid/solution ratio was 10 g/L. After equilibration for 72 hours at ambient temperature centrifugation at 1000 *g* for 20 minutes was followed separating the liquid phase from solid.

Note: The authors assumed a site density of singly coordinated aluminal groups at the edge of the gibbsite crystal of 8 sites/nm^2, a relative contribution of edge area to the total surface area of 20%, and a concentration of reactive aluminal groups on gibbsite of 9.3 mmol/kg. The data sets were used after the data were normalized to the total surface area of gibbsite used in this study.

FARRAH ET AL. (1987)

Experiment: Fluoride (F^-) Sorption on Gibbsite

Solid prep.: Gibbsite was prepared by adding NaOH solution to vigorously stirred 16.7 mM $Al(NO_3)_3$ until the pH was 7. XRD examination confirmed that the produced solid was predominantly gibbsite.

$I = 0.1$ M NaCl

$T = 25°C$

Reaction vessel: Plastic vial

CO_2 exclusion: NS

Max. S/S: 0.125

Equil. time: 1 to 14 days

Measurement: The evaluation of the effect that pH had on the uptake by gibbsite was done by treating aliquots of well-stirred suspension with small volumes of 10 mM HNO_3 or NaOH. The pH range was 3 to 8. Then addition of an aliquot (5 mL) of standard NaF solution followed. The solutions were diluted to a final volume of 25 mL, the vials capped, tumbled, and maintained at a temperature of 25°C. The mixtures were equilibrated for 1 to 14 days. After this period, the solids were separated mainly by filtration (0.45 μm membrane) prior to analysis. The concentration of the soluble aluminum was measured with atomic absorption spectroscopy. The concentration of the free fluoride ion was measured by immersion of an ion-selective electrode and reference half-cell in the isolated aqueous phases, which gave a response directly related to the concentration of free noncomplexed F^-.

Note: The data sets in Fig. 4b were not considered and fitted, because there was no clear indication of the ionic strength at which the experiments took place. Also in some cases, sorption experiments were done on freshly precipitated gibbsite.

FUKUSHI AND SVERJENSKY (2007)

Experiment: Arsenate (AsO_4^{3-}) Sorption on Gibbsite

Solid prep.: Gibbsite was prepared using the Gastuche and Herbillon (1962) method.

$I = 0.005$ *and* 0.1 M $NaNO_3$

$T = 20°C$

Reaction vessel: see Note below

CO_2 exclusion: Yes, N_2 atm.

Max. S/S: see Note below

Equil. time: see Note below

Measurement: see Note below

Note: The data sets used in this study were extracted from Manning and Goldberg's (1996) paper. Thus, the experimental procedure of Manning and Goldberg (1996) is applicable (see details under that entry in Appendix A).

GIVRIN ET AL. (1996)

Experiment: Cobalt (Co^{2+}) Sorption on Gibbsite

Solid prep.: Two types of gibbsite were used for this study: (1) commercially available as *superfine 4 gibbsite* (low surface area; 3.5 m^2/g) and (2) high surface area (16.8 m^2/g) material from ALCOA. Both of these gibbsites were precipitated under highly alkaline conditions ($pH \geq 14$).

$I = 0.01$ to 1 M $NaClO_4$

$T = 25 \pm 1°C$

Reaction vessel: Polypropylene tubes

CO_2 exclusion: Yes, N_2 atm.

Max. S/S: 2.95×10^{-5}

Equil. time: 24 hours (equilibrium reached in 4 hours)

Measurement: The procedures followed in this study were presented in Girvin et al. (1993). The bulk suspension contained 0.5 g gibbsite in 500 mL of 0.01 to 1 M $NaClO_4$. Bulk suspension pH was set at 9.5 and let to stabilize for 30 minutes before it was dual-label spiked. Then it was mixed for 20 minutes. The pH of the bulk suspension was then reduced in 0.5 pH unit steps, by the addition of 0.1 M $HClO_4$. Samples were equilibrated for 24 hours in an end-over-end rotator (at 2 rpm), centrifuged at 27,500*g* for 1 hour and then the final pH of the supernatant was measured in N_2 atmosphere. Sorption was measured as a function of pH ($5 < pH < 10$). Sorption of Co was quantified by dual-labeled scintillation counting of ^{60}Co-labeled remaining in the solution after filtration (0.1 μm). Cobalt was also measured with γ-counting of the 1.17- and 1.33-MeV gamma rays accompanying the ^{60}Co beta decay. Two independent determinations agreed within 3%. Dissolved Al in suspensions was measured by ICP-MS. Sorption was determined from the concentration of sorbate remaining in solution. Loss of Co and Al to container and tube walls was found to be $< 1\%$ of the total dissolved concentration examined across the pH range of the study.

Notes: (1) *Superfine 4 gibbsite* is considered to be an α-$Al(OH)_3$ form. (2) Data sets of Co-sorption edges with high ionic strength (1 M) were not considered, because the Davies equation cannot be applied.

GOLDBERG ET AL. (1993a)

Experiment: Boron (H_3BO_3) Sorption on Gibbsite

Solid prep.: Gibbsite (α-$Al(OH)_3(s)$) was prepared according to the Kyle method (Kyle et al., 1975). The precipitated crystalline gibbsite was verified by XRD. Few trace impurities (e.g., bayerite) were reported.

$I = 0.001$, 0.01, 0.05, 0.1, and 1 M NaCl

$T = 25°C$

Reaction vessel: Polypropylene centrifuge tubes

CO_2 exclusion: Yes, N_2 atm.

Max. S/S: 1.44×10^{-3}

Equil. time: 20 hours

Measurement: Boron-sorption experiments were carried out in batch systems to determine pH sorption edges. 0.375 g of gibbsite were added to 50-mL polypropylene centrifuge tubes and equilibrated with 15 mL of NaCl solution (in various concentrations) by shaking for 20 hours in a reciprocating shaker. This solution contained 5.0 g B m^{-3} and had been adjusted to the desired pH values using 1 M HC1 or 1 M NaOH additions. The total volume changed by <2%. The samples were centrifuged at 7800*g* for 20 minutes. Then pH measurement of the decantates was performed followed by filtration through a 0.45-μm filter, and analysis for B.

Note: Data sets at high ionic strength (1 M) were not considered because the Davies equation could not be applied.

GOLDBERG ET AL. (1993b)

Experiment: Boron (H_3BO_3) Sorption on Gibbsite

Solid prep.: Gibbsite (α-$Al(OH)_3(s)$) was prepared according to the Kyle method (Kyle et al., 1975). The precipitated crystalline gibbsite was verified by XRD. Few trace impurities (e.g., bayerite) were reported.

$I = 0.1$ M NaCl

$T = 25 \pm 0.1°C$

Reaction vessel: Polypropylene centrifuge tubes

CO_2 exclusion: Yes, N_2 atm.

Max. S/S: 1.44×10^{-3}

Equil. time: 20 hours

Measurement: Boron-sorption experiments were carried out in batch systems to determine pH sorption edges. 0.375 g of gibbsite were added to 50-mL polypropylene centrifuge tubes and equilibrated with 15 mL of NaCl solution (in various concentrations) by shaking for 20 hours in a reciprocating shaker. This solution contained 5.0 g B m^{-3} and had been adjusted to the desired pH values using 1 M HC1 or 1 M NaOH additions The total volume changed by <2%. The samples were centrifuged at 7800*g* for 20 minutes. Then pH measurement of the decantates was performed followed by filtration through a 0.45-μm filter, and analysis for B.

Note: Data sets at low (10°C) and high temperature (40°C) were not considered. Isotherms were not considered either, because they were produced at low and high temperature.

GOLDBERG ET AL. (1996)

Experiment: Molybdenum (MoO_4^{2-}) Sorption on Gibbsite

Solid prep.: Gibbsite (α-$Al(OH)_3(s)$) was synthesized according to Kyle et al.'s (1975) procedure. The oxide mineralogy was verified using XRD analysis and found not having any impurities.

$I = 0.1$ M NaCl

$T = 23 \pm 0.5$°C

Reaction vessel: Polypropylene centrifuge tubes

CO_2 exclusion: Yes, N_2 atm.

Max. S/S: 0.01825

Equil. time: 20 hours

Measurement: Sorption experiments were carried out in batches. The amounts of Mo adsorbed as a function of solution pH for different total Mo concentrations were determined. Samples of gibbsite (1.25 g/L) were added to 50-mL polypropylene centrifuge tubes and equilibrated with aliquots of a 0.1 M NaCl solution by shaking for 20 hours in a reciprocating shaker at 23±0.5°C. This solution contained 292 mol Mo/L from $Na_2MoO_4 \cdot 2H_2O$. The pH was adjusted to the desired values using 1 M HCl or 1 M NaOH additions which changed the total volume by $\leq$2%. The samples were centrifuged at 7800*g* for 20 minutes. The decantates were analyzed for pH, filtered through a 0.45 μm filter, and analyzed for Mo concentration using ICP-MS.

GOLDBERG (2005)

Experiment: Boron (H_3BO_3) Sorption on Gibbsite

Solid prep.: Gibbsite (α-Al(OH)$_3$(s)) was prepared according to the Kyle method (Kyle et al. 1975). The precipitated crystalline gibbsite was verified by XRD. Few trace impurities (e.g., bayerite) were reported.

$I = 0.001$, 0.01, 0.05, 0.1, and 1 M NaCl

$T = 25°C$

Reaction vessel: Polypropylene centrifuge tubes

CO_2 exclusion: Yes, N_2 atm.

Max. S/S: 1.44×10^{-3}

Equil. time: 20 hours

Measurement: Boron-sorption experiments were carried out in batch systems to determine pH sorption edges. 0.375 g of gibbsite were added to 50-mL polypropylene centrifuge tubes and equilibrated with 15 mL of NaCl solution (in various concentrations) by shaking for 20 hours in a reciprocating shaker. This solution contained 5.0 g B m^{-3} and had been adjusted to the desired pH values using 1 M HC1 or 1 M NaOH additions. The total volume changed by <2%. The samples were centrifuged at 7800*g* for 20 minutes. Then pH measurement of the decantates was performed followed by filtration through a 0.45-μm filter, and analysis for B using the colorimetric azomethine-H method.

Notes: (1) The data sets at high ionic strength (1 M) in Fig. 3 were not considered because the Davies equation could not be applied. (2) Generalized two-layer model produced poor fitting at data points at pH < 5.5. This was partially attributed to the formation of inner and outer-sphere complexes by boron (personal communication with Dr. Goldberg).

GUAN ET AL. (2005)

Experiment: Phosphate (PO_4^{3-}) Sorption on Aluminum Hydroxide

Solid prep.: The pH of a 0.167 M $AlCl_3$ solution was adjusted to 6 ± 0.1 and maintained by adding NaOH. After the pH of the suspension became stable, the suspension was stirred at room temperature for another 48 hours. The suspension was then filtered, resuspended, and centrifuged, and the supernatant was decanted. The aluminum hydroxide obtained was freeze-dried and stored at room temperature. XRD analysis showed that the solid was amorphous.

$I = 0.05$ M NaCl

$T = 21 \pm 1°C$

Reaction vessel: NS

CO_2 exclusion: NS

Max. S/S: 0.006

Equil. time: 24 hours

Measurement: Sorption isotherms of phosphate on amorphous aluminum hydroxide were obtained at 21 ± 1°C. The aluminum hydroxide suspension contained 285.7 mg $Al(OH)_3$/L and was shaken for 24 hours to disperse the aggregates and hydrate the sorption sites before the sorption test. The pH of the phosphate solutions and the $Al(OH)_3$ suspension was adjusted with NaOH and/or HCl solutions before the sorption tests. The initial pH was 6. A 70-mL $Al(OH)_3$ suspension was mixed with a 30-mL phosphate solution of various concentrations to make the initial P concentration in the mixture vary from 0 to 1.3 mmol P/L (corresponding to an initial loading ranging from 0 to 6500 mmol/kg $Al(OH)_3$). Phosphorus concentration was measured at the end of the sorption test, by filtering aliquots of the mixture through a 0.45-μm cellulose acetate membrane filter. Phosphorus analysis was performed in accordance with the ascorbic acid method (APHA, 1995).

Note: The data sets for phosphorous sorption on aluminum hydroxide were not considered, because the solid used in the experiments was an amorphous aluminum hydroxide.

GUAN ET AL. (2006)

Experiment: Phosphate (PO_4^{3-}) Sorption on Aluminum Hydroxide

Solid prep.: The pH of a 0.167 M $AlCl_3$ solution was adjusted to 6 ± 0.1 and maintained by adding NaOH. After the pH of the suspension became stable, it was stirred at room temperature for another 48 hours. The suspension was then filtered, resuspended, and centrifuged, and the supernatant was decanted. The aluminum hydroxide obtained was freeze-dried and stored at room temperature. XRD analysis showed that the solid was amorphous.

I = 0.05 M NaCl

T = 21 ± 1°C

Reaction vessel: NS

CO_2 exclusion: NS

Max. S/S: 0.1

Equil. time: 24 hours

Measurement: Sorption isotherms of phosphate on amorphous aluminum hydroxide were obtained at 21 ± 1°C. The aluminum hydroxide suspension contained 200 mg $Al(OH)_3$/L and was shaken for 24 hours to disperse the aggregates and hydrate the sorption sites before the sorption test. The pH of the phosphate solutions and the $Al(OH)_3$ suspension was adjusted in the range of 5–8 with NaOH and/or HCl solutions before the sorption tests. Phosphorus concentration was measured at the end of the sorption test, by filtering aliquots of the mixture through a 0.45-μm filter paper. Phosphorus analysis was performed in accordance with the ascorbic acid method (APHA, 1995).

Note: The data sets for phosphorous sorption on aluminum hydroxide were not considered, because the solid used in the experiments was an amorphous aluminum hydroxide.

HELYAR ET AL. (1976a)

Experiment: Phosphate (PO_4^{3-}) Sorption on Gibbsite

Solid prep.: Gibbsite was a commercial synthetic *Hydral 710* (Alcoa). XRD was applied for crystal confirmation. Thermal dehydration patterns and XRD of the solid reported to be typical of gibbsite.

I = 0.001–0.02 M $CaCl_2$, 0.002–0.02 M NaCl, 0.001–0.01 M MgCl, and 0.002–0.02 M KCl

$T = 26 \pm 0.5$°C

Reaction vessel: Glass

CO_2 exclusion: No; a CO_2/N_2 mixture was used. CO_2 was used to adjust the pH at the desired values.

Max. S/S: 1.5×10^{-3}

Equil. time: 1, 3, 6, 12, 48, 72, 96 hours

Measurement: The suspension used for the sorption of phosphate by gibbsite was 242 μM $Ca(H_2PO_4)_2$ in 1 mM $CaCl_2$ at pH 5.5. A water-vapor-saturated mixture of CO_2 and N_2 was bubbled into a 1.2 L $Ca(H_2PO_4)_2$–$CaCl_2$ solution. The pH of the solution was adjusted, by varying the CO_2 partial pressure, between 4.1 to 4.3 (equilibrating the phase for 12 minutes). Then 12 g of gibbsite was added. The pH of the suspension was increased to 5.5 and by using the CO_2 partial pressure the pH was maintained at the desired value. The suspension was equilibrated at various times ranging from 1 to 96 hours. Afterwards, the suspensions were centrifuged at 1100 rcf. The supernatant was kept for analysis (within the 24-hour period).

Sorption isotherms were obtained with the same conditions described above. Initial concentration of P ranged from 0 to 1600 μM, while concentration of Cl was either 2 or 20 mM. Additional Ca and Na isotherms were run at 0 or 40 mM Cl.

Notes: The data sets regarding phosphorous sorption on gibbsite in Fig. 3 were not considered, because the solution contained divalent cations (Ca and Mg). Also, in conjunction with the presence of HCO_3, due to CO_2 inclusion in the solution, these data sets were also not considered.

HELYAR ET AL. (1976b)

Experiment: Phosphate (PO_4^{3-}) Sorption on Gibbsite

Solid prep.: Gibbsite was a commercial synthetic *Hydral 710* (Alcoa). XRD was applied for crystal confirmation. Thermal dehydration patterns and XRD of the solid reported to be typical of gibbsite.

$I = 0.001$–0.02 M $CaCl_2$, 0.002–0.02 M NaCl, 0.001–0.01 M MgCl, and 0.002–0.02 M KCl

$T = 25 \pm 0.5$°C

Reaction vessel: Glass

CO_2 exclusion: No; a CO_2/N_2 mixture was used. CO_2 was used to adjust the pH at the desired values.

Max. S/S: 1.5×10^{-3}

Equil. time: 1, 3, 6, 12, 48, 72, 96 hours

Measurement: The suspension used for the sorption of phosphate by gibbsite was 242 μM $Ca(H_2PO_4)_2$ in 1 mM $CaCl_2$ at pH 5.5. A water-vapor-saturated mixture of CO_2 and N_2 was bubbled into a 1.2 L $Ca(H_2PO_4)_2$–$CaCl_2$ solution. The pH of the solution was adjusted, by varying the CO_2 partial pressure, between 4.1 and 4.3 (equilibrating the phase for 12 minutes). Then 12 g of gibbsite was added. The pH of the suspension was increased to 5.5 and by using the CO_2 partial pressure the pH was maintained at the desired value. The suspension was equilibrated at various times ranging from 1 to 96 hours. Afterwards, the suspensions were centrifuged at 1100 rcf. The supernatant was kept for analysis (within the 24-hour period).Sorption isotherms were obtained with the same conditions described above. Initial concentration of P ranged from 0 to 1600 μM, while concentration of Cl was either 2 or 20 mM. Additional Ca and Na isotherms were run at 0 or 40 mM Cl.

Notes: The data sets for phosphorous sorption on gibbsite in Figs. 1 and 2 were not considered, because the solution contained divalent cations (Zn, Cd, Ca, Sr). The study stated that monovalent cations (Na, K) and Mg did not affect sorption, as some of the divalent metals did. However, in conjunction with the presence of HCO_3, due to CO_2 inclusion in the solution, these data sets were also not considered.

HIEMSTRA ET AL. (1987)

Experiment: Proton Sorption on Gibbsite

Solid prep.: Gibbsite suspensions were prepared according to Gastuche and Herbillon (1962) method. ppt. by addition of NaOH (0.1 M) to $AlCl_3$ concentrated solution (0.33 M) until constant pH = 4.5; ppt. resuspended in distilled water for 4 weeks at 70°C. The suspension was concentrated by flocculation at pH 8 in 1 M NaCl followed by dialysis. The suspension was stored at 4°C during 5 years of experimentation with it.

$I = 0.005$, 0.05, and 0.5 M NaCl

$T = 20 \pm 1$°C

Reaction vessel: polypropylene

CO_2 exclusion: Yes, N_2 atm.

Equil. time: rapid titrations

Measurement: The titration experiments were carried out at the pH range of 4.5–9.5. Fifty milliliters of concentrated suspension (37 g/kg) was put into a vessel along with 1 mL of electrolyte (NaCl). N_2 atm was provided over the suspension in the closed vessel until the pH was stabilized at 7.7. Forward and backward titration was performed for each electrolyte level.

Note: The acid–base titration data were not used in this study because the suspension concentration was high > 30 g/L. Such concentrated suspensions favor agglomeration of the suspension and consequently partial blockage of the active sorption sites (see Chapter 5).

HIEMSTRA ET AL. (1999)

Experiment: Proton Sorption on Gibbsite

Solid prep.: Gibbsite was prepared using the Gastuche and Herbillon (1962) method. An amorphous aluminum hydroxide suspension was produced, followed by dialysis of the suspension at a certain temperature. The am-$Al(OH)_3$ suspensions were obtained by titrating 2 L of an $Al(NO_3)_3$ solution (1 M) with 4 M NaOH in a plastic vessel to a pH of 4.5. The titrant was added at 10 mL/min. Several gibbsite suspensions were synthesized based on two different temperatures of dialysis. Two am-$Al(OH)_3$ suspensions were dialyzed against double-distilled water at 70°C for 4 weeks, resulting in gibbsites GH1 (A_{BET} 18 m^2/g) and GH2 (A_{BET} 40.3 m^2/g), and one at 4°C for 8 weeks, yielding GL1 (A_{BET} 40.8 m^2/g). In all dialysis, the solution was refreshed daily. After preparation of this gibbsite, the sample was dialyzed for another 2 months at 70°C to increase the crystallinity. A fourth gibbsite sample G (A_{BET} 42.1 m^2/g) was prepared, at low temperature. This nonaged sample was used in titration experiments without further dialysis at high temperature. This suspension was aged briefly at 70°C for approximately 2–3 weeks and again titrated.

$I = 0.005$ *and* 0.1 M $NaNO_3$

$T = 20$°C

Reaction vessel: Plastic vessel

CO_2 exclusion: Yes, N_2 atm.

Equil. time: Rapid titration

Measurement: Titration experiments were performed in the pH range of 5–10. A volume of 50 mL of the sample suspension (about 30 g/L) was pipetted together with 1 mL of 0.25 M $NaNO_3$ electrolyte into a 100 mL plastic vessel. N_2 atm was provided over the suspension in the closed vessel to prevent contamination of the suspension with CO_2. The sample was allowed to stand overnight at a pH of about 6–7 to eliminate dissolved and adsorbed (bi) carbonates. Forward and backward titration was performed for each electrolyte level.

Note: The data of this study were not used because of the high suspension concentration. Such concentrated suspensions favor agglomeration of the suspension and consequently partial blockage of the active sorption sites (see Chapter 5). Also for the data of Fig. 7, low (4°C) and high (70°C) temperatures were used.

HINGSTON ET AL. (1971)

Experiment: Arsenate (AsO_4^{3-}) and Phosphate (PO_4^{3-}) Sorption on Gibbsite

Solid prep.: Gibbsite was prepared by precipitation of $Al(OH)_3$ at pH 4.6 from $AlCl_3$ using NaOH. The $Al(OH)_3$ suspension was heated at 40°C for 2 hours and dialyzed with distilled water at 60°C for a period of 23 days with daily changes of water. XRD applied to the gibbsite solid revealed that it is in the hexagonal form.

$I = 0.005$ and 0.1 M $NaNO_3$

$T = 20$°C

Reaction vessel: Polypropylene tubes

CO_2 exclusion: Yes, N_2 atm.

Max. S/S: 2.39×10^{-2}

Equil. time: 24 hours

Measurement: Aliquots of solutions containing the anions of interest were pipetted into a series of tubes and then a portion of gibbsite was added. The pH was adjusted with 0.1 M NaOH or HCl. pH range was 3–10. Ionic strength was maintained constant by NaCl. This electrolyte was also used to facilitate the separation of solids from the solution at centrifugation. The suspension was equilibrated at 20°C for 24 hours. The pH was determined. The amount of each anion adsorbed was estimated from the difference between initial and final concentration at equilibrium. This experimental procedure was followed for anion mixtures or individual anion sorption on gibbsite. Phosphate was determined colorimetrically, while phosphate and arsenate radiometrically.

Note: The data sets of Figs. 3Aii, 3Aiii and 3Aiv were not considered in this study because of the competitive sorption of arsenate with phosphate.

HINGSTON ET AL. (1972)

Experiment: Selenate (SeO_4^{2-}), Molybdate (MoO_4^{2-}), Sulfate (SO_4^{2-}), and Fluoride (F^-) Sorption on Gibbsite

Solid prep.: Gibbsite was prepared by precipitation of $Al(OH)_3$ at pH 4.6 from $AlCl_3$ using NaOH. The $Al(OH)_3$ suspension was heated at 40°C for 2 hours and dialyzed with distilled water at 60°C for a period of 23 days with daily changes of water. Gibbsite was used as a stock suspension containing 25 g/L. A known quantity of adsorbent was shaken with solutions of the anions at a range of pH values and the decrease in anion concentration after 24 hours was used to calculate the amount of adsorption. The pH values were measured at equilibrium and isotherms for a given pH were obtained by interpolation.

$I = 0.1$ M NaCl

$T = 20$–23°C

Reaction vessel: NS

CO_2 exclusion: NS

Max. S/S: 0.078

Equil. time: 24 hours

Measurement: The amount of sorption on gibbsite was estimated at a given initial adsorbate concentration sufficient to give values to complete an isotherm to a well-defined sorption plateau.

JEPSON ET AL. (1976)

Experiment: Silicate ($H_2SiO_4^{2-}$) Sorption on Gibbsite

Solid prep.: A commercial gibbsite was obtained from Martifin. The gibbsite consisted of hexagonal particles up to 2 μm across and with diameter:thickness ratios between 5 : 1 and 10 : 1. XRD analysis showed only gibbsite; other hydrated aluminum solids were less than 1% w/w. The commercial gibbsite used for the experiments was partially unreactive towards silicic acid due to the silica on its surface. Gibbsite was cleaned by leaching in 0.01 M NaOH (at background electrolyte of 1 M NaCl), then in 1 M HCl (at background electrolyte of 0.1 M NaCl), in four solutions of HCl of increasing pH, and finally in water. About 5% of gibbsite was dissolved.

$I = 0.02$ M NaCl

$T = 25$°C

Reaction vessel: Plastic and well-aged containers

CO_2 exclusion: Yes, N_2 atm.

Max. S/S: 0.0125

Equil. time: 30 minutes

Measurement: The silicic solution and the gibbsite suspension were placed into different containers and were adjusted to the same pH using NaOH or HCl. NaCl was added to give a final ionic strength of 0.02 M NaCl. The suspension and solution were then mixed to give a final gibbsite concentration of 10 g/L. Aqueous phases were separated with a membrane filter (porous size 0.05 μm) and the filtrates were analyzed for total silicon and aluminum with AAS. Aluminum was also analyzed colorimetrically by the catechol violet method, whereas silicon was analyzed by the molybdenum blue method.

Note: This study used only the pH edge sorption data extracted from Fig. 1. The data of the two isotherms (Fig. 2) were not used because there was a potential precipitation of $SiO_{2(s)}$.

JODIN ET AL. (2005)

Experiment: Proton Sorption on Gibbsite

Solid prep.: A supersaturated solution of sodium aluminate (Bayer liquor) was aged for 7 to 10 days. Solid separation after aging was done by combining washing in acidic solutions and (A) freeze-drying or (B) dialysis.

$I = 0.001$–0.1 M $NaNO_3$

$T = 25°C$

Reaction vessel: Teflon under Ar

CO_2 exclusion: (A) no, and (B) yes, N_2 atm.

Equil. time: 30 minutes

Measurement: Two methods were used. (A) The precipitate was centrifuged at 5000*g* for 5 minutes, and then washed several times with HNO_3 and once with distilled water. The ppt was filtered and freeze-dried. (B) A cellulose dialysis membrane was filled with acid-washed gibbsite suspension and immersed in distilled water in a PVC tube. Dialysis was stopped when the conductivity was stabilized at 0.5 μS/cm.

Note: The data sets from this study were not considered, because the suspension concentrations used in their experiments were very high (>30 g/L). Such concentrated suspensions favor agglomeration of the suspension and consequent partial blockage of the active sorption sites (see Chapter 5).

KAVANAGH ET AL. (1975)

Experiment: Polymer Sorption on Gibbsite

Solid prep.: Gibbsite was prepared by precipitation by the addition of NaOH to $AlCl_3$ concentrated solution until a constant pH of 4.6 was obtained. The precipitate was then heated for 2 hours and resuspended in distilled water for several weeks at 40°C.

$I = 0.01–1$ M $NaNO_3$

$T = 20$°C

Reaction vessel: Polypropylene tubes

CO_2 exclusion: No

Equil. time: 48 hours

Measurement: After equilibration, a suspension sample was taken and was centrifuged at 30,000*g* for 1 hour. The PVA in the supernatant was determined colorimetrically (back titration).

Note: The data set from this study was not used because the ionic strength of the background electrolyte solution was high (1 M) and the Davies equation correction for ionic strength could not be applied.

KINNIBURGH ET AL. (1975)

Experiment: Calcium (Ca^{2+}) Sorption on Gibbsite

Solid prep.: Gibbsite was prepared by ppt. by addition of NaOH to $Al(NO_3)_3$ concentrated solution until a constant pH of 7 was obtained. The freshly precipitated aluminum gel was amorphous, but after aging it for several months in 0.95 M $NaNO_3$, gibbsite crystals were detected by XRD. The pH after aging the solid dropped from 7 to 4–5. The heated $Al(OH)_3$ powder was originally a mixture of gibbsite and bayerite determined by SEM and XRD. SEM analysis showed that the materials consisted of microporous particles, and that some sintering had occurred. The material had a composition of $Al_2O_3 \cdot 0.46H_2O$.

$I = 1$ M $NaNO_3$

$T =$ NS

Reaction vessel: Polypropylene tube

CO_2 exclusion: NS

Max. S/S: 6.2×10^{-5} for gels, 1.5×10^{-5} for heated gibbsite

Equil. time: 3 hours for gels, 24 hours for heated gibbsite

Measurement: Batch-sorption experiments were conducted by transferring 7 mL of the freshly prepared gel (2.33 mmole of Al) or 0.25 g of heated gibbsite (4.55 mmole of Al) into a 30-mL propylene tube. Then 1 M of NaOH or HNO_3 was added to adjust the pH of each suspension to the desired value. Aliquots of

2 mL of the sorbing solution were then added. The Ca stock solution was prepared by labeling a 1 M $NaNO_3$ with 0.5 μCi/mL of ^{45}Ca. The final sorbing solution of Ca had a concentration of 2×10^{-6} M. The suspensions (gels or heated gibbsite) were equilibrated at room temperature by shaking, and were then centrifuged. An aliquot of 1 mL supernatant solution was measured for ^{45}Ca with a liquid scintillation counter.

Notes: These data were not used because:the ionic strength of 1 M used in the experiments was high and the Davies equation correction for ionic strength could not be applied,freshly precipitated aluminum (hydr)oxide referred as "Al gel" in the study is in amorphous form,the aluminum (hydr)oxide heated at 400°C, referred to as "heated gibbsite" in the study, is probably the χ-Al_2O_3 solid.

LI AND XU (2008)

Experiment: Potentiometric Studies; Surface Charge of Gibbsite

Solid prep.: Gibbsite was prepared according to the Kyle method (Kyle et al., 1975). NaOH (4 M) was added slowly to 1 M $AlCl_3$ solution with stirring until the pH reached 4.6, and the suspension was allowed to stand for 2 hours at 40°C. The product was electrodialyzed. The treated sample was air-dried and ground to pass 100-mesh sieve.

$I = 0.1$ M $NaNO_3$

$T = 20$°C

Reaction vessel: NS

CO_2 exclusion: Yes, N_2 atm.

Equil. time: 24 hours

Measurement: NaCl of 0.1 mM was used as a supporting electrolyte to maintain the ionic strength. A colloid suspension containing 0.1 mM NaCl was prepared with a concentration of gibbsite 0.25 g/L. The suspensions were dispersed ultrasonically at a frequency of 40 kHz and a power of 300 W for 1 hour at 25°C. The pH of the suspension was adjusted to required values with dilute HCl or NaOH solution. Then, the suspensions were shaken for 2 hours at 25°C. pH was checked and adjusted accordingly. Before electrokinetic potential determination, equilibration of 24 hours was achieved with irregular stirring of the suspension.

LIU ET AL. (2006)

Experiment: Arsenate (AsO_4^{3-}) Sorption on Gibbsite

Solid prep.: Gibbsite was commercially obtained (Merck) and its crystal structure was characterized by XRD as α-$Al(OH)_3$(s).

$I = 0.1$ M $NaNO_3$

$T = 20°C$

Reaction vessel: Polyethylene

CO_2 exclusion: Yes, N_2 atm.

Max. S/S: 0.0031

Equil. time: 4 days

Measurement: Sorption isotherm batch experiments were performed at pH 5.0 with arsenate concentrations of 0.02, 0.03, 0.04, 0.05, 0.06, 0.08, 0.1, 0.2, 0.5, and 1 mM. $Na_2HAsO_4 \cdot 7H_2O$ was used as the arsenate source. Twenty milliliters of 2.5 g/L gibbsite suspensions in 0.1 M $NaNO_3$ were added to the 50 mL polyethylene vessels and agitated continuously. The pH of the suspensions were adjusted to 5.0 using HNO_3 or NaOH and maintained at this constant value for 1 week. Twenty milliliters of arsenate solution of different concentrations with preadjusted pH were added to the gibbsite suspension and reacted at 150 rpm on a rotary shaker. After 4 days, the suspensions were sampled with a syringe and filtered through a 0.2 μm Millipore filter membrane. Total arsenic concentration of the filtrates was determined by ICP-AES. Triplicate runs were carried out for each treatment.

Sorption pH edges were obtained via batch experiments in 0.1 M $NaNO_3$ at arsenate concentration of 0.1 mM. Twenty milliliters of 2.5 g/L gibbsite suspension in 0.1 M $NaNO_3$ and 20 mL of stock arsenate solution (0.2 mM), both with preadjusted pH (4–9, constant for a week), were added into the 50-mL polyethylene vessel simultaneously and agitated at 150 rpm for 4 days. At the end of reaction, the suspensions were sampled with a syringe and filtered through a 0.2 μm filter membrane. Total arsenic concentration of the filtrate was determined by ICP-AES. Triplicate runs were carried out for each treatment.

MANNING AND GOLDBERG (1996)

Experiment: Arsenate (AsO_4^{3-}), Phosphate (PO_4^{3-}), and Molybdate (MoO_4^{2-}) Sorption on Gibbsite

Solid prep.: Gibbsite was prepared according to the Kyle (Kyle et al., 1975) method. Slow addition of 4 M NaOH to 1 M $AlCl_3$ while stirring was made until the pH was stabilized at 4.6. The suspension was then heated in a water bath at 40°C for 2 hours. The produced precipitate was transferred to cellulose tubing and dialyzed with deionized water for 36 days. The solids then were dried overnight in a gravity convection oven at 40°C, and ground to pass a 500-μm sieve. XRD analysis was used to confirm crystal identity.

$I = 0.1$ M NaCl

$T = 22 \pm 1°C$

Reaction vessel: Polycarbonate centrifuge tubes

CO_2 exclusion: NS

Max. S/S: 0.0083

Equil. time: 4 hours

Measurement: 20 mL of 133 or 266 μM of As(V), P, and Mo in 0.1 M NaCl were added to 50 mg of gibbsite in 40 mL centrifuge tubes. The pH was adjusted between 3 and 11 using 0.1 M of NaOH or HCl. The tubes were shaken for 4 hours at room temperature and centrifuged at 12,500*g* for 10 minutes. Supernatant pH was measured followed by filtering with 0.1 μm pore size cellulose nitrate membrane filters. The amount of the adsorbed anion was calculated as the difference between the anion concentration before and after reaction with the solids. Anion concentrations were determined by ion chromatography.

Note: Some of the data sets (Fig. 4a, 4b, 4c, 4d) were excluded because of the competitive sorption of phosphate with arsenate or molybdate (binary anion-sorption pH edges).

MASUE ET AL. (2009)

Experiment: Arsenate (AsO_4^{3-}) and Arsenite (AsO_3^{2-}) Sorption/Desorption on Gibbsite

Solid prep.: $Al(NO_3)_3$ solution (0.1 M Al) was hydrolyzed using 1 M NaOH with vigorous stirring. The resulting suspension was adjusted to pH 7.5 and dialyzed against deionized water.

$I = 0.0001$, 0.01, and 0.1 M NaCl

$T = 25°C$

Reaction vessel: NS

CO_2 exclusion: NS

Max. S/S: 0.05

Equil. time: 24 hours

Measurement: Sorption isotherms of As(V) and As(III) on $Al(OH)_3(s)$ were obtained as batch experiments at pH 5 and 8 with the Al concentration fixed at 267 μM as Al hydroxide and arsenic concentrations ranging from 3 to 133 μM. The pH values of individual samples were adjusted by adding HCl or NaOH. The samples were allowed to react for 24 hours on a reciprocating platform shaker.

Sorption pH edges for As(V) and As(III) on $Al(OH)_3(s)$ in 0.1 M NaCl were obtained via batch experiments at an As:Al molar ratio of 1 : 20 (13.4 μM As and 267 μM Al). The pH values of samples were adjusted between 3 and 11 using HCl or NaOH, and samples were allowed to react for 24 hours on a reciprocating platform shaker. The final pH of the supernatant solutions was measured after centrifugation. Arsenic was analyzed by flow-injection hydride-generation flame-atomic-absorption spectrometry.

Note: The crystalline $Al(OH)_3(s)$ was verified as a mixture of gibbsite and bayerite. Coprecipitated Al:Fe hydroxides were also prepared in this study. The data sets found in Figs. 2 and 3 were not considered in this study for the aforementioned reasons. The desorption data for arsenate and arsenite of Fig. 4 were not considered because of the presence of phosphate.

McBRIDE (1982)

Experiment: Copper (Cu^{2+}) Sorption on Bibbsite

Solid prep.: Gibbsite was obtained from Alcoa. The specific surface area was 5.9 m^2/g and was determined by the BET method.

$I = 0.005$ M NaCl

$T = 25°C$

Reaction vessel: NS

CO_2 exclusion: NS

Max. S/S: 0.0078

Equil. time: 24 hours

Measurement: 0.10 g of gibbsite solid was equilibrated with 20 mL of Cu^{2+} solution in 0.05 M NaCl solution at room temperature. The samples were shaken in capped centrifuge tubes for 1 day. The samples were then centrifuged, and the supernatants were analyzed for Cu^{2+} by atomic absorption spectrophotometry.

Note: Equilibration time in some cases was more than 1 day to assess the importance of slow sorption reactions.

McBRIDE ET AL. (1984)

Experiment: Copper (Cu^{2+}) Sorption on Gibbsite

Solid prep.: Gibbsite was prepared following the Gastuche and Herbillon method (1962).

$I =$ NS

$T =$ NS

Reaction vessel: NS

CO_2 exclusion: NS

Max. S/S: 0.003

Equil. time: 1 day

Measurement: Copper titrations were conducted by combining 10 mL of gibbsite suspension (26 mg/mL) with 10 mL of 10^{-3} or 10^{-4} M $Cu(NO_3)_2$ and adjusting the pH with NaOH. The Cu^{2+} activity was monitored with a specific ion electrode. The reversibility of the reaction was also studied. Reference titrations were performed in the same way using 10 mL of H_2O instead of the gibbsite suspension.

Sorption isotherms were obtained by adjusting the $Cu(NO_3)_2$–gibbsite mixtures over a range of pH using NaOH. Equilibration time was set at 1 day. The reaction equilibrated for 1 day was between 10 mL of gibbsite suspension with 10 mL of 10^{-3} or 10^{-4} M $Cu(NO_3)_2$. Measurement of the final pH and Cu^{2+} activity was done using electrodes.

Note: The data from Fig. 2A and 2B were not used, because the study considered only the (001) face of the solid to be reactive.

McKINLEY ET AL. (1995)

Experiment: Uranyl (UO_2^{2+}) Sorption on Gibbsite

Solid prep.: Gibbsite, ALCAN SF-4, was obtained from Alcoa. This material was pretreated by first shaking a suspension of 5 g of gibbsite in 20 mL of 0.01 N NaOH for 30 minutes, followed immediately by seven successive washes with 0.01 M $NaClO_4$.

$I = 0.001$–0.1 M $NaClO_4$

$T = 25°C$

Reaction vessel: NS

CO_2 exclusion: Yes, N_2 atm.

Max. S/S: 1.29×10^{-3}

Equil. time: 20 hours

Measurement: A $UO_2(NO_3)_2 \cdot 6H_2O$ stock solution containing 195 mg U/kg was added by mass to the equilibrated gibbsite suspensions to yield an initial U concentration of 2 mg/L (8.4×10^{-6} mol/L). The pH of the suspension was increased incrementally by the addition of NaOH, and duplicate 15-mL samples were removed at each 0.5-pH-unit step. The samples were equilibrated for 20 hours in a controlled-environment shaker (150 rpm). The equilibrated samples were centrifuged at 5000 rcf for 1 hours, and separate 4-mL samples of the supernatant were removed for determination of pH, solution-phase uranyl concentration, and Na concentration. The final pH of the supernatant was measured under a N_2 atmosphere using a combination electrode. The quantity of uranyl adsorbed was calculated from the difference between initial and final solution phase uranyl concentrations. Uranium concentration was determined using a kinetic phosphorescence analyzer.

MENG AND LETTERMAN (1993a)

Experiment: Cadmium (Cd^{2+}) and Calcium (Ca^{2+}) Sorption on Gibbsite

Solid prep.: Aluminum suspensions were prepared by the addition of acidified (0.001 M HNO_3) 0.25 M $Al(NO_3)_3$ solution to the SiO_2 suspension. The pH of the suspension was adjusted to 8.0 by slow addition of 1 *N* KOH. The suspension then was aged at 22 ± 3°C, under N_2 atmosphere for 2 hours.

$I = 0.04$ M KNO_3

$T = 22 \pm 3$°C

Reaction vessel: Polypropylene bottles

CO_2 exclusion: Yes, N_2 atm.

Max. S/S: 0.02

Equil. time: 2 hours

Measurement: Sorption experiments were performed with total $Cd(NO_3)_2$ or $Ca(NO_3)_2$ in suspensions of 1.78×10^{-5} M or 1.0×10^{-4} M. Kinetic experiments suggested that equilibrium was reached within 2 hours. The pH of each suspension was measured, and the suspensions were centrifuged at 15,600*g* for 15 minutes. Samples of supernatant were analyzed for soluble Cd^{2+} or Ca^{2+} with atomic absorption spectrophotometry.

Notes: (1) Although discrete $Al(OH)_3$ was precipitated in the suspensions of pure aluminum hydroxide, TEM micrographs verified that the solid was in amorphous form. Therefore, sorption data sets obtained with this solid were excluded from the study. (2) In most of the sorption experiments, a mixture of $Al(OH)_3$ with SiO_2 was used.

MENG AND LETTERMAN (1993b)

Experiment: Calcium (Ca^{2+}) and Sulfate (SO_4^{2-}) Sorption on Gibbsite

Solid prep.: Aluminum suspensions were prepared by the addition of acidified (0.001 M HNO_3) 0.25 M $Al(NO_3)_3$ solution to an SiO_2 suspension (in this study most of the suspensions used were mixtures of SiO_2 with Al or Fe oxides, but there were some experiments using only Al or Fe oxides). The pH of the suspension was adjusted to 8.0 by slow addition of 1 *N* KOH, and the suspension was aged at 22 ± 3°C, under N_2 atmosphere for 2 hours.

$I = 0.04$ M KNO_3

$T = 22 \pm 3$°C

Reaction vessel: Polypropylene bottles

CO_2 exclusion: Yes, N_2 atm.

Max. S/S: 2×10^{-5}

Equil. time: 2 hours

Measurement: Sorption experiments were performed with total SO_4^{2-} in suspensions of 1.0×10^{-4} M. Kinetic experiments suggested that equilibrium was reached within 2 hours. The SiO_2 concentration in the mixed oxide suspension was always 1 g/L. Samples of oxide suspension containing adsorbate and 0.04 M KNO_3 were shaken at 22 ± 3°C for 2 hours to reach surface adsorption equilibrium. The pH of each suspension was measured, and the suspensions were centrifuged at 15,600*g* for 15 minutes. Samples of supernatant were analyzed for soluble Ca^{2+} with atomic absorption spectrophotometry.

Note: The characterization of $Al(OH)_3$ used in this study was reported in Meng and Letterman (1993a), where TEM micrographs verified that the $Al(OH)_3$ precipitate was in amorphous form. For that reason, the data sets reported by Meng and Letterman (1993b) were not considered in this study.

MICERA ET AL. (1986)

Experiment: Zinc (Zn^{2+}) Sorption on Gibbsite

Solid prep.: Aluminum hydroxide was prepared by the addition of NH_4OH to a solution of $AlCl_3$ until pH 9. The precipitate was centrifuge-washed until chloride free. XRD showed the sample to be predominantly amorphous with traces of pseudoboehmite and bayerite.

$I = 1$ M $NaClO_4$

$T = 20 \pm 0.1$°C

Reaction vessel: Polypropylene tubes

CO_2 exclusion: Yes, N_2 atm.

Max. S/S: 3×10^{-6} (?)

Equil. time: <24 hours

Measurement: The procedures and instrumentation used in obtaining absorption isotherms and analytical measurements were as described at Gessa et al. (1984). Aliquots of freshly ppt $Al(OH)_3$ containing 2.6 mmol Al were used with sorbate solutions prepared by dissolving weighed amounts of the metal salt. Sodium chloride or $NaClO_4$ were used as background electrolytes for a total ionic strength of 1.0 M. The pH was adjusted by adding NaOH or HCl ($HClO_4$) and kept constant until equilibrium was achieved. After equilibration time, the samples were centrifuged and the solutions were analyzed for using a flame emission spectrophotometer.

Notes: The $Al(OH)_3$ used for the sorption studies was a precipitate of high pH (9) and characterized by XRD as bayerite. Ionic strength was high and the Davies equation for ionic strength correction could not be applied. Therefore none of the data sets of this study were used.

MISHRA ET AL. (2007)

Experiment: Mercury ($Hg(OH)_2$) Sorption on Gibbsite

Solid prep.: The solid gibbsite used for this study was amorphous.

$I = 0.001–0.1$ M KNO_3

$T = 25°C$

Reaction vessel: NS

CO_2 exclusion: Yes, N_2 atm.

Max. S/S: 0.001

Equil. time: 30 minutes

Measurement: The sorption measurements were carried out by taking 0.1000 g of $Al(OH)_3$ in 10.0 mL of Hg(II) solution at the desired concentration, labeled with ^{203}Hg in a centrifuge tube. The system was kept at the desired temperature and the suspensions were stirred during the sorption process. The solution was centrifuged and the aliquot of the supernatant was dried under IR lamp and measured for β-activity. Radioactivities of some samples were measured for their γ-activities.

Notes: The $Al(OH)_3$ was in amorphous form. There was no direct evidence of crystalline gibbsite. Also, the experiment was conducted at a relatively high temperature (29.85°C).

MUSIC (1986)

Experiment: Chromate (CrO_4^{2-}) Sorption on Gibbsite

Solid prep.: NS. There is evidence for freshly precipitated $Al(OH)_{3(s)}$.

$I = 0.05$ M NaCl

$T = 25°C$

Reaction vessel: NS

CO_2 exclusion: NS

Max. S/S: 0.2

Equil. time: 30 minutes

Measurement: The sorption systems were prepared by the addition of NaOH solution into an $AlCl_3$ solution containing the sorbate labeled with $^{51}CrO_4^{2-}$ or $^{51}Cr^{3+}$. Some of the sorption systems contained NaCl, Na_2MoO_4, $FeSO_4$, Na_2SO_3, or sodium citrate. After equilibration for sorption the liquid phase was separated by centrifugation at 5000 rpm. The separation of the colloidal Al $(OH)_3$ from the liquid phase was performed by an ultracentrifuge at 20,000 rpm. The count rates for radiolabel led Cr in sorption were measured by a well-type NaI(Tl) scintillation crystal. The sorption percentage was calculated by comparing the radiolabelled introduced into the sorption system with the radiolabelled Cr in the separated liquid phase.

Notes: The data sets extracted from Figs. 1, 2, 5, and 6 were not considered, because the $Al(OH)_3$ was freshly precipitated. There was no evidence that the material was gibbsite. Also, in some of those data sets there was competitive sorption with other anions such as molybdate and chloride.

OKAZAKI ET AL. (1986)

Experiment: Copper (Cu^{2+}), Zinc (Zn^{2+}), and Magnesium (Mg^{2+}) Sorption on Gibbsite

Solid prep.: Gibbsite crystals were prepared according to Funaki and Shimizu (1960). The crystal structure of the purchased gibbsite was confirmed by X-ray powder diffraction. It was reported to be predominantly gibbsite.

$I = 0.1$ and 1 M $NaNO_3$

$T = 21°C$

Reaction vessel: Polyethylene tubes

CO_2 exclusion: Yes, N_2 atm.

Max. S/S: 0.1

Equil. time: 24 hours

Measurement: Working solutions were prepared for Cu^{2+}, Zn^{2+}, and Mg^{2+} in dilute HNO_3. Sodium nitrate (0.1 and 1 M) was used as electrolyte in all experiments. The concentration of the working metal mixture solutions was 3.33×10^{-4} M (each 1.11×10^{-4} M) to 3.33×10^{-2} M (each 1.11×10^{-2} M). In a 50 mL polyethylene centrifuge tube 1×10^{-3} mol with respect to Al was placed for the sorption experiments. Then 5 mL of 6 M $NaNO_3$ solution and 5 mL of 3.33×10^{-4} M to 3.33×10^{-2} M mixed solution of Cu^{2+}, Zn^{2+}, and Mg^{2+} were added. The solution pH was adjusted by addition of 0.01 to 0.1 N NaOH or 0.01 N HNO_3. The suspension was shaken for 24 hours at 21°C. The pH of the suspension was recorded and the solution was centrifuged at 10,000 rpm and filtered through a 0.22 μm filter. Cu, Zn, and Mg were measured using AAS.

Note: Data sets of Cu from Figure 3 were not considered in this study because the ionic strength was high ($I = 1$ M), and the Davies correction could not be applied.

OKAZAKI ET AL. (1989)

Experiment: Phosphate (PO_4^{3-}), Arsenate (AsO_4^{3-}), Molybdate (MoO_4^{2-}), Selenite (SeO_3^{2-}), and Chromate (CrO_4^{2-}) Sorption on Gibbsite

Solid prep.: The hydrated oxides were prepared according to Funaki and Shimizu (1960). The crystal structure of gibbsite was confirmed by X-ray powder diffraction. It was reported to be predominantly gibbsite.

$I = 0.1$ M $NaClO_4$

$T = 21$°C

Reaction vessel: Polyethylene centrifuge tube

CO_2 exclusion: Yes, N_2 atm.

Max. S/S: 0.1

Equil. time: 48 hours

Measurement: Solutions of the anions of interest were prepared from their respective salts. A sample of gibbsite was placed in a 50 mL polyethylene centrifuge tube and 5 mL of a 0.06 M $NaClO_4$ solution and 10 mL of a 9.9×10^{-4} M oxyanion solution was added. pH was adjusted by the addition of 0.01 N NaOH or 0.005 or 0.05 N $HClO_4$. The suspension was equilibrated at 21°C for 48 hours. After that, pH was measured and the suspension was centrifuged at 22,000*g* for 20 minutes and filtered (0.22 μm). Phosphates were measured by the vanadomolybdate method; arsenate by the molybdate blue method; selenite, molybdate and chromate with AAS. The sorption percentage of the oxyanion was calculated based on the decrease of analyte concentration in the solution when gibbsite was present.

Notes: (1) Gibbsite 2 (Gi-2) has a high surface area and low crystallinity, both properties indicative of an amorphous-like mineral. Therefore, data sets obtained with this solid were not considered in this study. Additional evidence of the amorphous-like character of the Gi-2 solid was that the sorption of oxyanions was more pronounced with this solid than the more crystalline Gi-3 gibbsite.

PARFITT ET AL. (1977)

Experiment: Phosphate (PO_4^{3-}) Sorption on Gibbsite

Solid prep.: Gibbsite was prepared using the Gastuche and Herbillon (1962) method. Electron microscopy was employed for the identification of the precipitates. The crystals consisted of thin six-sided plates with diameters around 200 nm. Crystal thickness was less than 4 nm (in most cases).

$I = 0.1$ M NaCl

$T = 25$°C

Reaction vessel: NS

CO_2 exclusion: Yes, N_2 atm.

Max. S/S: No evidence of the initial phosphate concentration

Equil. time: 16 hours

Measurement: Sorption isotherms were obtained by mixing 5 mL of solution containing 0.11 g of gibbsite and phosphate solution. The solution was placed in tubes and let to equilibrate for 16 hours. The tubes were then centrifuged for 15 minutes at 2000 rpm. Phosphate sorption from solutions containing 0.1 M NaCl was followed by colorimetric analysis of the supernatant solution. The amount of phosphate adsorbed was calculated by difference between the initial and final solution concentrations.

PEAK (2006)

Experiment: Selenate (SeO_4^{2-}) Sorption on Gibbsite

Solid prep.: Hydrous aluminum oxide precipitate was prepared using the following method. An aliquot of 1.5 M $AlCl_3$ (200 mL) was neutralized slowly to an OH/Al ratio of 2.7 with 2.0 M NaOH. The resulting precipitates were aged for 1 hour, resuspended in 1 L of deionized water, centrifuged at 1000*g*, and the supernatant was discarded. This process was repeated five times to remove residual Al^{3+} and Cl^- from the suspension. The final pH of the suspension was ~5.5.

$I = 0.01$ and 0.1 M NaCl

$T = 25°C$

Reaction vessel: NS

CO_2 exclusion: Yes, N_2 atm. (the water used in the study was not purged with N_2, but the carbonate level was likely minimal)

Max. S/S: 7.8×10^{-4}

Equil. time: 48 hours

Measurement: For sorption experiments with HAO, a 0.5 g/L HAO suspension was precipitated in deionized water. The suspension was adjusted to the desired pH, and then equilibrated for 8 hours. Addition of either Na_2SeO_4 or Na_2SeO_3 was followed to reach a final concentration of 100 μM. Sample pH was readjusted twice per day with 0.1 M HCl or NaOH as the reaction proceeded. After 48 hours of reaction, samples were centrifuged at 1000*g* for 15 minutes, and the suspension was analyzed by X-ray absorption spectroscopy.

Notes: The HAO was not verified as gibbsite, and therefore no data sets from this study were studied. High surface area (BET 267 m^2/g) usually implies a nongibbsite solid.

POKROVSKY ET AL. (2008)

Experiment: Copper (Cu^{2+}) Sorption on Gibbsite

Solid prep.: Synthetic gibbsite $Al(OH)_3$ (SSA = 2.87 m^2/g) was used in this study. The crystallinity of minerals was verified using XRD. Prior to experiments, all solids were reacted 30 minutes in 0.01 M HNO_3, thoroughly rinsed in deionized water and conditioned in N_2-saturated, 0.01 M $NaNO_3$ for at least 1 week. Trace element analysis of the gibbsite using ICP-MS showed no impurities within 60.1 wt.%, and Cu content was below 10 ppm.

$I = 0.01$ and 0.1 M $NaNO_3$

$T = 25 \pm 0.2$°C

Reaction vessel: Polypropylene vials

CO_2 exclusion: Yes, N_2 atm.

Max. S/S: 5.867×10^{-5}

Equil. time: 2–3 hours (tests were run from 0.5 minutes to 43 hours)

Measurement: Two types of experiments were performed: (i) pH-dependent sorption edge and (ii) fixed pH sorption isotherm. Sorption experiments were conducted at 25 ± 0.2°C in a continuously stirred suspension of 0.01 or 0.1 M $NaNO_3$ solution using 8 mL polypropylene vials. The copper concentration in solution ([Cu^{2+}]aq) was varied from 0.2 to 50 μM. The pH was adjusted to the desired value using either NaOH or HNO_3. For the sorption isotherm measurements, HEPES buffer (0.003 M), which is known not to form complexes with divalent metals in aqueous solutions, was used to keep pH constant. The suspension was centrifuged at the end of the experiments and the resulting supernatant was filtered through a 0.22 μM Nylon filter, and acidified with HNO_3. The concentration of copper adsorbed to gibbsite was calculated by the difference of initial and final Cu concentration in the supernatant suspension.

Copper concentration for filtered solutions was achieved using flame atomic absorption spectroscopy with an uncertainty of ± 1%. For Cu concentration lower than 0.5 μM, ICP-MS was employed. Cu concentration in the solid phase was measured by AAS after full acid digestion of this phase (concentrated HNO_3 for gibbsite) was performed in a clean room.

Note: Only the data produced from the pH-dependent sorption edge experiments were used in this study. Data obtained from the sorption isotherm experiments might have been biased due to competitive sorption of divalent metals contained in the buffers used.

ROBERTS ET AL. (2003)

Experiment: Zinc (Zn^{2+}) Sorption on Gibbsite

Solid prep.: Gibbsite was prepared using the Kyle et al. (1975) method. A 1.0 M $AlCl_3$ solution was titrated by 4 M NaOH until a gelatinous precipitate appeared followed by dialysis for 36 days in distilled deionized H_2O. Gibbsite ($Al(OH)_3$) was identified by XRD, with all Al in octahedral coordination as determined by Al nuclear magnetic resonance (Al NMR).

$I = 0.005$ and 0.1 M $NaNO_3$

$T = 25$°C

Reaction vessel: NS

CO_2 exclusion: Yes, N_2 atm.

Max. S/S: 0.0078

Equil. time: 24 hours

Measurement: Suspensions of 10 g/L (gibbsite) were equilibrated at pH 4 for 24 hours in background electrolyte with constant stirring. A pre-equilibration step was conducted at different ionic strengths (0.005 and 0.1 M). Initial Zn concentration of 1 mM was used from an acidified 0.1 M $Zn(NO_3)_2$ stock solution. In order to achieve initial Zn concentration, Zn was added in increments to ensure that there was no local oversaturation. Adjustment of pH in the pH edge experiments was done by addition of 0.1 M NaOH. Suspensions were equilibrated in centrifuge tubes on an end-over-end shaker in a glove box for 24 hours. pH ranged from 4.5 to 8.5. The experimental pH was taken to be the pH measured after a 24 hours reaction time. After equilibration, the solids were separated from the suspensions by centrifugation at 12,000 rpm for 15 minutes. The supernatants were filtrated with 0.2 μm cellulose filters, acidified, and analyzed for Zn using atomic absorption spectrometry (AAS). The amount of Zn sorbed was calculated to be the initial Zn concentration minus the Zn concentration in solution.

ROSENQVIST ET AL. (2002)

Experiment: Proton Sorption on Gibbsite

Solid prep.: The gibbsite was prepared using a method similar to that described by Gastuche and Herbillon (1962). A 1 M $AlCl_3$ solution was titrated with 4 M NaOH until pH reached a value of about 4.6 (precipitation of am-$Al(OH)_3$ occurs). The suspension was then heated for 2 hours at 40°C, and dialyzed against deionized water at 50°C for 4 weeks. The dialysis water was refreshed daily during the first 2 weeks and every second day after that. The resulting suspension was 12 g/L with pH of about 4.6. Two batches were prepared. One batch was aged for 9 months and the other was not aged. The amount of chloride remaining in the suspension after the dialysis was determined by titration with silver nitrate. XRD was used to verify crystalline gibbsite formation.

$I = 20$ and 100 mM NaCl

$T = 25 \pm 0.1$°C

Reaction vessel: NS

CO_2 exclusion: Yes, Ar(g) atm.

Equil. time: Up to 14 hours

Measurement: The titrations were performed with an automated system for precise electromotive force (emf) measurements starting with a suspension of pH 4.6 and 8.9. The concentration of free H^+ was determined by measuring the emf (in mV) of the cell: (−) Ag, AgCl(s)/ionic medium// equilibrium suspension/glass electrode (+) or alternatively: (−) Ag, AgCl (s)/ionic medium//equilibrium suspension/Pt, H_2(g) (1 atm) (+). H_2(g) electrodes were used for titrations at pH > 9, where a glass electrode may not be sufficiently accurate. The suspension was kept homogeneous by stirring, and Ar(g) was provided over the suspension. A maximum drift of 0.2 mV/h, corresponding to 0.003 pH units/h, was used as the equilibrium criterion. This drift was usually reached in 8–14 hours after the addition of acid or base. Occasionally, the equilibrium criterion was not met within 14 hours. In that case, the titration was continued without further delay. Rapid titrations were also performed.

Note: One data set extracted from this study was not used, because the gibbsite solid phase used in this particular titration was not aged sufficiently.

SAEKI (2004)

Experiment: Iron (Fe^{2+}) and Manganese (Mn^{2+}), Copper (Cu^{2+}), Calcium (Ca^{2+}), and Cadmium (Cd^{2+}) Sorption on Gibbsite

Solid prep.: Gibbsite was commercially prepared by Showa Dennko. The average particle size was 1 μm.

$I = 0.1$ M NaCl

$T = 25 \pm 1$°C

Reaction vessel: NS

CO_2 exclusion: Yes, Ar atm.

Max. S/S: 2.34×10^{-5}

Equil. time: 10 hours

Measurement: Sorption experiments of Fe and Mn were conducted in a glove box purged with Ar. Adsorbed quantities of Fe^{2+} and Mn^{2+} were calculated from the differences between the amounts of Fe and Mn added and those measured in the equilibrium solutions. For the Fe^{2+} and Mn^{2+} sorption on the gibbsite, 0.2 g portions of the oxides were weighed into centrifuged tubes. To these tubes, < 18.0 mL of deionized water, 1.0 mL of 2.0 mol/L NaCl background electrolyte solution, and 1.0 mL of 60 mmol/L $FeCl_2$ or $MnCl_2$ solutions were added. The solution pH was changed by the addition of dilute HCl or NaOH solutions. The initial concentration of Fe or Mn was 3.0 mmol/L. The tubes were capped and equilibrated for 10 hours with occasional shaking. After equilibration, the final solution pH was measured and a part of each suspension was filtered through a 0.22 μm membrane filter. The collected filtrate was acidified with HCl and analyzed for Fe and Mn concentrations by atomic absorption spectroscopy.

SARKAR ET AL. (1999)

Experiment: Mercury ($Hg(OH)_2$) Sorption on Gibbsite

Solid prep.: The gibbsite used for this study was commercially obtained as *superfine 4 gibbsite.* The solid was treated for 30 minutes with 0.01 M NaOH to remove poorly crystallized coatings. Then it was centrifuge-washed with 0.1 M $NaNO_3$ several times until the pH of the supernatant became ~7. Concentration of the solid was 3.3 g/L and of Hg(II) 0.6 μM.

$I = 0.01$, 0.1, and 0.5 M $NaNO_3$

$T = 20$–25°C

Reaction vessel: Teflon tubes

CO_2 exclusion: NS

Max. S/S: 1.41×10^{-5}

Equil. time: 48 hours

Measurement: pH was adjusted with 0.02 M HNO_3 or 0.02 M NaOH so the equilibrated solution had pH between 3 and 9. After equilibration, the solutions were centrifuged at 1500g for 40 minutes and the supernatant solution was analyzed. Hg(II) concentrations in solutions, for sorption studies, were measured using an atomic absorption spectrometer equipped with flow injection mercury analysis system. Dilution of samples prior to analysis was performed.

Notes: (1) Superfine 4 gibbsite is considered to be an α-$Al(OH)_3$ form. (2) Experimental data above pH 6 have not been considered. The reason is that the shape of the pH edge plot is not justified by the equilibrium system. TOTHg concentration does not justify surface precipitation; if surface precipitation occurred, the % sorption for mercury would have been higher than described by the plot.

TOCHIYAMA ET AL. (1996)

Experiment: Cobalt (Co^{2+}) Sorption on Gibbsite

Solid prep.: The gibbsite used was a commercially available mineral of Brazilian origin with unknown composition. However, XRD was applied and showed that the solid was gibbsite (γ-$Al(OH)_3$).

$I =$ NS

$T = 23 \pm 2$°C

Reaction vessel: Pyrex-glass tube

CO_2 exclusion: No, but proven not to affect

Max. S/S: Unknown initial amount of sorbate

Equil. time: 4 hours

Measurement: Sorption experiments were conducted by mixing 0.1 g of the gibbsite sample in the Pyrex-glass tube with 5 mL of solution containing ^{60}Co, carrier Co(II), and a buffer mixture, which consisted of 0.1 M 2-morpholinoethane sulfonic acid and 0.1 M of tris(hydroxymethyl)aminomethane. The tubes were sealed and shaken for 4 hours at 23 ± 2°C. After equilibration of 4 hours, centrifugation followed at 5000 rpm. A 1-mL aliquot of the supernatant solution was withdrawn and analyzed for its γ-activity with a well-type NaI(Tl) scintillation counter.

Notes: Sorption data of this study were not used because: (1) the ionic strength at which the sorption experiments were conducted is unknown. (2) The initial concentration of Co is unknown. Sorption was expressed as log (distribution ratio) versus pH.

VAN EMMERIK ET AL. (2007)

Experiment: Phosphate (PO_4^{3-}) Sorption on Gibbsite

Solid prep.: Gibbsite was prepared using the Gastuche and Herbillon (1962) method. $Al(NO_3)_3{\cdot}9H_2O$ was dissolved in 2.5 L of water. To this solution, ~1.5 L of 4 M KOH was gradually added over a period of 2–3 hours with constant stirring until the pH reached 3 to 4. The slurry was then aged at 40–45°C for 24 hours before being dialyzed against distilled water at 50–55°C for 6 weeks with the water being refreshed daily. The resulting slurry was freeze-dried. Powder XRD analysis of the sample using Cu Kα radiation showed only *d*-spacing values that corresponded to those of synthetic gibbsite.

$I = 0.01$ M NaCl

$T = 25 \pm 1$°C

Reaction vessel: Polypropylene tube

CO_2 exclusion: Yes, N_2 atm.

Max. S/S: 0.029

Equil. time: 24 hours

Measurement: A 200 mL suspension of gibbsite in 0.01 M NaCl was stirred for more than 18 hours under N_2 atmosphere at 25 ± 1°C to hydrate the mineral surface. The pH of the suspension was adjusted to 3. Phosphate solution was added to the required concentration, and the pH readjusted. For sorption experiments as a function of pH, 4 mL of suspension was transferred into a polypropylene centrifuge tube, flushed with N_2, and sealed. pH was adjusted in the batch reactor tubes to span the range from 3 to 13. The tubes were then placed on an end-on-end shaker and allowed to equilibrate for 1 day. The pH of each sample was measured after equilibration, and the samples were centrifuged at 18,000*g*. The supernatant solutions were analyzed for phosphorus by the Molybdenum Blue method.

Isotherm experiments were similarly conducted. The pH of the suspension was held constant through the addition of HCl and NaOH as increasing amounts of phosphate were added. After each phosphate addition, the pH was readjusted to the chosen value and a 4 mL portion of suspension transferred to a centrifuge tube that was flushed with N_2 and capped. Samples were analyzed by solid-state NMR spectroscopy.

WEERASOORIYA ET AL. (2000)

Experiment: Proton Sorption on Gibbsite; Copper (Cu^{2+}) Sorption on Gibbsite

Solid prep.: Gibbsite was purchased from Alcoa. XRD data showed the samples used were dominated by >95% of gibbsite crystals.

$I = 0.001–0.1$ M $NaNO_3$

$T = 25°C$

Reaction vessel: In a glove box for potentiometric measurement; polypropylene for sorption experiments

CO_2 exclusion: Yes, N_2 atm.

Max. S/S: 0.003

Equil. time: 24 hours prior titration; 24 hours for sorption experiments.

Measurement: Potentiometric titration was conducted using an autotitrator in a glove box under N_2 atmosphere. Sorption experiments: Total gibbsite suspension volume was 150 mL. Gibbsite concentration was in the range of 1–2 g/L. The equilibrated solid suspension was divided into aliquots of 10 ml each and placed in the N_2-filled polypropylene tubes. Aliquots of 5 ml equilibrated sample were centrifuged at 14,000 rpm for 30 minutes and the supernatant was removed for total Cu analysis. Cu was measured by AAS (flame or graphite). Site density was estimated based on the crystallographic data.

WEERASOORIYA ET AL. (2001)

Experiment: Proton Sorption on Gibbsite; Lead (Pb^{2+}) Sorption on Gibbsite

Solid prep.: Gibbsite was commercially prepared (ALCOA).

$I = 0.001–0.1$ M $NaNO_3$ or NaCl

$T = 25°C$

Reaction vessel: Polypropylene

CO_2 exclusion: Yes, N_2 atm.

Max. S/S: 0.003

Equil. time: 2 hours (potentiometric experiment), 24 hours (sorption edges), 24 hours (isotherms)

Measurement: Surface titrations were carried out at different electrolyte concentrations (0.001, 0.01, and 0.1 M NaCl). An 8 g/L gibbsite suspension was equilibrated for 2 hours (pH $\approx$ 6). The pH was raised to $\approx$9 with 0.1 M NaOH. Afterwards, surface titrations were conducted. Blank titrations were also performed on filtrates of 8 g/L gibbsite suspensions prepared under identical experimental conditions. The net titrant volume at a given pH was determined mathematically by subtracting the sample titration curve from the blank curve.

Sorption pH edges: Lead sorption pH edges were obtained in either $NaNO_3$ or NaCl. A gibbsite suspension of 150 or 500 mL of 2 g/L was prepared. The suspension was equilibrated for 2 hours at pH $\approx$ 4.5, and ionic strength of 0.001 to 0.1 M (using 5 M NaCl or $NaNO_3$ to regulate electrolyte concentration). pH adjustments were made with 0.0178 M HCl/0.8044 M or 0.045 M HNO_3 or 1.042, 0.021, or 0.01 M NaOH. Aliquots of 4.831 mM or 0.4831 mM Pb(II) were used to spike the solution after 6 hours. Equilibration took place in a polypropylene tube reactor under N_2 atmosphere for 24 hours at 25°C. Afterwards, the pH of the suspension was measured and recorded. Aliquots of the sample (2–4 mL) were centrifuged at 15,000 rpm for 30 minutes. Lead concentrations were measured in the supernatant by flame atomic absorption spectroscopy or by graphite furnace AAS.

Sorption Isotherms: For the measurement of sorption isotherms, three pHs (5.5, 6.4, and 7.28) were considered in either 0.01 M $NaNO_3$ or NaCl. Lead concentration was varied from 0.5 to 120 μM. pH was adjusted with acid/base. After 24 hours of equilibration, lead concentration in the acidified supernatant was determined by AAS.

Note: (1) Some dissolution of gibbsite was observed (Al^{3+} concentration in solution) at pH > 9 and pH < 4.5. Only the data points in the pH range 4.5–9.0 were used. (2) Experiments to define possible hysteresis were conducted by equilibrating a batch solution of 0.01 M $NaNO_3$ or NaCl with initial Pb(II) 7.78 μM for 24 hours at pH > 8.5. The suspension pH was varied in the range 9–4.5. With pH unit intervals of 0.5, and withdrawing 10 mL sample portions at each titration point an sorption edge was constructed.

WEERASOORIYA ET AL. (2002)

Experiment: Cadmium (Cd^{2+}) Sorption on Gibbsite

Solid prep.: Gibbsite was commercially prepared (ALCOA).

I = 0.001–0.1 M $NaNO_3$

T = 25°C

Reaction vessel: Polystyrene tubes

CO_2 exclusion: Yes, N_2 atm.

Max. S/S: 5.26×10^{-5}

Equil. time: 24 hours (sorption edges), 24 hours (isotherms)

Measurement: Cadmium sorption on gibbsite was determined as a function of pH, ionic strength, and adsorbate loading. 2 g/L gibbsite suspensions were prepared in 0.001 M $NaNO_3$, and equilibrated for 2 h. The suspension pH ranged between 7 and 8. Before the addition of Cd the pH was adjusted to 3.5 with 0.541 M HNO_3. Aliquots of 0.890 mM Cd^{2+} were added to reach concentration of 0.77–7.74 μM. Suspension samples were in polystyrene centrifuge tubes. pH was adjusted between 3.5 and 10. Samples were equilibrated for 24 hours. After phase separation by centrifugation (14,000 rpm), samples were filtered (0.45 μm), and the filtrates were acidified for metal analysis.

Sorption Isotherms: Cadmium-sorption isotherms were obtained in 0.01 M $NaNO_3$ for different pH values. For the development of isotherms, the procedure described in the previous paragraph for sorption edge experiments were also followed here except some changes: The 2 g/L gibbsite suspension was equilibrated for 2 hours at a desired pH. Afterwards, a portion of samples was transferred into centrifuge tubes. The addition of different quantities of cadmium into these tubes was made after matching the solution pH. The total concentration of cadmium was varied between 0.1 and 1000 μM. At the end of the 24 hour equilibration period, the pH of the stirred suspension was measured, and regulated at a desired value typically for 0.5–1 hours until a stable pH reading was achieved. Subsequently solid–solution phases were separated to determine solute concentration.

Cadmium concentrations were measured in the supernatant by flame atomic absorption spectroscopy or by graphite furnace AAS.

WEERASOORIYA ET AL. (2003)

Experiment: Arsenite (AsO_3^{2-}) Sorption on Gibbsite

Solid prep.: Gibbsite was commercially prepared (ALCOA).

$I = 0.001–0.1$ M $NaNO_3$

$T = 25°C$

Reaction vessel: Polypropylene tubes

CO_2 exclusion: Yes, N_2 atm.

Max. S/S: 5.26×10^{-5}

Equil. time: 30 minutes

Measurement: Sorption pH edges were obtained using solutions of at concentrations 2.67 and 13.5 μM. The pH was set at 10 using 0.887 M NaOH. The concentration of the gibbsite suspension was 20 g/L. The solid suspension was equilibrated for 3 hours. Solutions were then divided into 10 mL aliquots, placed in 50 mL polypropylene centrifuge tubes. The pH was adjusted between

3.5 and 9.5. Samples were then equilibrated for 24 hours at 120 oscillations per hour in a temperature-regulated water bath shaker. After equilibration, the pH of suspension was measured. Samples were centrifuged at 20,000 rpm for 10 minutes in a refrigerated centrifuge, and the supernatant solution was removed for arsenic analysis.

Sorption isotherms were determined over As(III) concentration from 0.269 to 10 μM at pH 4, 5, and 8. The solution ionic strength was 0.01 M $NaNO_3$. The experimental procedure was the same as in the sorption pH edge experiments except for the following changes: after a 24 hour period the suspension pH was measured and adjusted with a 0.218 M NaOH/0.887 M HNO_3 before solid–solution phases were separated.

Arsenic measurements were done by continuous flow hydride generation (HG)-AAS.

WEERASOORIYA ET AL. (2004)

Experiment: Arsenate (AsO_4^{3-}) Sorption on Gibbsite

Solid prep.: Gibbsite was commercially prepared (ALCOA). XRD was applied for verification of the gibbsite crystal.

$I = 0.001$, 0.01, and 0.1 M $NaNO_3$

$T = 25°C$

Reaction vessel: Polysterene (adsorption), Teflon (centrifuge)

CO_2 exclusion: Yes, N_2 atm.

Max. S/S: 5.22×10^{-5}

Equil. time: 24 hours

Measurement: Arsenate sorption pH edges were obtained for different $NaNO_3$ background electrolyte concentrations at initial As(V) concentrations of 1.34, 2.67, and 13.4 μM at 25°C. Five milliliters of As(V) stock solution and 20 mL of 20 g/L gibbsite suspension were added to 40 mL centrifuge tubes and purged for 5–10 minutes with N_2. The ionic strength used for these experiments was 0.1, 0.01, and 0.001 M with 5 M $NaNO_3$. The pH of the suspensions was adjusted to 4–10 using either 0.052 M HNO_3 or 0.045 M NaOH. Equilibration time was 24 hours at 25°C. After equilibration, the pH was measured, the suspension was centrifuged, and the supernatant was filtered through a 0.45 μm membrane filter, acidified, and stored at 4°C until analyzed.

Sorption isotherms were obtained at pH 4.3, 7.5, and 9.2 at initial As(V) concentrations between 0.66 and 16 μM at 25°C. The experimental setup was similar to the one used for the sorption pH edge experiments, except for the following changes: the pH of the gibbsite suspension and the As(V) solution was adjusted to the desired value before mixing. The tubes were purged with N_2 and closed for shaking in a rotary shaker at 240 rpm for 24 hours. After 24 hours, the pH of the suspension was readjusted to the initial value, and the quantity of acid/base consumed was recorded.

Arsenic measurements were done by continuous flow HG—AAS.

Note: The interactive effect between As(III) and As(V) sorption onto gibbsite was examined by the simultaneous addition of both the ions at a 1 : 1 ratio in some tests.

WEERASOORIYA ET AL. (2006)

Experiment: Mercury ($Hg(OH)_2$) Sorption on Gibbsite

Solid prep.: Gibbsite was commercially prepared (ALCOA).
$I = 0.01$ M $NaNO_3$, NaCl, or $NaClO_4$
$T = 25°C$
Reaction vessel: Polystyrene tubes
CO_2 exclusion: Yes, N_2 atm.
Max. S/S: 78
Equil. time: 24 hours (hydroxylation), 2 hours (equilibration)
Measurement: Mercury-sorption isotherms were obtained in either 0.01 M NaCl or $NaNO_3$ at pH ~4 and 8 at 25°C. The initial loadings of Hg(II) were varied between ~0.1 and ~12.5 μM for NaCl–Hg(II)–gibbsite system, and between ~0.05 and ~2.00 μM for the $NaNO_3$ or $NaClO_4$–Hg(II)–gibbsite system. After a hydroxylation step, 15-mL aliquots of solid suspension were transferred into polystyrene tubes. Required Hg(II) amounts were then spiked into the system to yield a final concentration and the tubes were equilibrated for a specific time period (as determined by preliminary kinetic studies) at 100 rpm at temperatures of 25, 40, and 50°C. pH was then measured. After equilibration, the tubes were centrifuged at 15,000 rpm at the desired temperature, and the supernatant was preserved and stored for Hg(II) analysis. Mercury measurement was performed with an atomic absorption spectrometer equipped with a hydride generator.

Notes: (1) The data sets presented in Figs. 3A, 3B, 4A, 4B for the experiments conducted at high temperature (40 and 50°C) were not considered. (2) Control experiments assessing the loss of metal ions to container walls were also conducted. A solution of 0.2 mM Hg(II) was used for the control tests. The recovery of Hg(II) at a given period was larger than 98–99%; hence there was negligible loss of Hg(II) to container vessels.

WEERASOORIYA ET AL. (2007b)

Experiment: Mercury ($Hg(OH)_2$) Sorption on Gibbsite

Solid prep.: Gibbsite was commercially prepared (ALCOA).
$I = 0.001$, 0.01, or 0.1 M $NaNO_3$, or NaCl

$T = 25°C$

Reaction vessel: NS

CO_2 exclusion: NS

Max. S/S: 3.9×10^{-6}

Equil. time: 24 hours (hydroxylation), 2 hours (equilibration)

Measurement: Mercury (Hg(II))-sorption experiments were performed either in $NaNO_3$ or NaCl. From a stock solution of 500 μM Hg(II), aliquots were used to spike gibbsite suspension samples to reach final solute concentrations of 0.1 or 10 μM. The concentration of the background electrolyte was adjusted to 0.001, 0.01, or 0.1 M with 5 M $NaNO_3$ or 5 M NaCl. One gram of gibbsite sample was well hydroxylated at the desired ionic strength for 24 hours at 25°C prior to sorption testing. The pH of the suspension was adjusted to pH 3.5 using an acid of similar anionic constituents. The system was spiked with Hg(II) to yield final concentration of either 1 or 10 μM. The pH of the system was varied gradually between 4 and 9. For each experimental condition a 20 mL well-mixed suspension was transferred into a centrifuge tube. The tube was rotated at 100 rpm for 2 hours and the pH of the suspension was measured prior to solid-solution separation by membrane filtration (0.45 μm pore size). Dissolved Hg (II) measurement was performed by hydride generator AAS. The remaining solids were analyzed for Hg(II) with Fourier transformation infrared (FTIR).

WIESNER ET AL. (2006)

Experiment: Cobalt (Co^{2+}) Sorption on Gibbsite

Solid prep.: Gibbsite was commercially prepared (Almatis Inc., Bauxite, AR).

$I = 0.01$–1 M $NaNO_3$, NaCl, or $NaClO_4$

$T = 20°C$

Reaction vessel: NS

CO_2 exclusion: Yes, N_2 atm.

Max. S/S: 7.8×10^{-5}

Equil. time: 24 hours

Measurement: Batch sorption experiments were conducted using 10 g/L gibbsite; 10^{-5} M $Co(NO_3)_2$, $CoCl_2$, or $Co(ClO_4)_2$; and 0.01–1 M $NaNO_3$, NaCl, or $NaClO_4$. The samples were mixed in an end-over-end tumbler for 24 hours and centrifuged at 11,290 rcf for 20 minutes. The pH of the supernatant was then measured. The cobalt concentration in the supernatant of each batch experiment was measured using either ICP-AES or graphite furnace atomic absorption spectroscopy.

Note: The data sets collected at high ionic strength (>0.5 M) Figs. 2A, 2B, and 2C were not considered in this study, because the Davies equation could not be applied for ionic strength correction.

XU ET AL. (2008)

Experiment: Arsenate Desorption from Gibbsite

Solid prep.: Crystalline $Al(OH)_3$ was prepared according to the method described by Kyle et al. (1975). NaOH, 4 M, was added slowly to a 1 M $AlCl_3$ solution with stirring until the pH reached 4.6. The suspension was maintained at 40°C for 2 hours. The product was electrodialyzed at a potential gradient of 15 V/cm until a constant specific electrical conductance was reached. The suspension was then freeze dried. The dried solid was ground to pass a 100-mesh sieve.

$I = 0.01$ M $NaNO_3$

$T = 25$°C

Reaction vessel: NS.

CO_2 exclusion: Yes, N_2 atm.

Max. S/S: 0.038

Equil. time: 48 hours

Measurement: A suspension of 0.25 g of Al and 487.5 mL of 10.26 mM $NaNO_3$ was prepared. The suspension was dispersed by ultrasonification at 150 W and 20 kHz for 4 minutes and then stirred with a magnetic stirrer for 6 hours. During this period, the pH of the suspension was adjusted to 5.0 several times with HCl or NaOH. An aliquot of 195 mL was placed in 500 mL flask and shaken at 200 rpm overnight at 25°C. Then 5 mL of 0.01 M Na_2HAsO_4 was added to the suspension to make the initial concentration of arsenate 0.25 mM, and 0.01 M $NaNO_3$ was added to achieve a concentration of 0.01 M. The suspension was then shaken for 48 hours at 25°C. The pH of the suspension was adjusted to pH 5.0 again during the shaking period. The suspension was vacuum filtered through a 0.1 μm membrane filter. The As in the filtrate was measured, and the amount of As adsorbed was calculated from the difference between the concentration of As added and that remaining in the filtrate. The As in solution was determined by HG AAS.

Notes: These data sets were not considered for three reasons: (1) Only desorption experiments were reported. (2) The solid used was verified as a mixture of gibbsite and Bayerite. (3) These were competitive sorption of arsenate and phosphate.

YANG ET AL. (2007a)

Experiment: Proton Sorption on Gibbsite

Solid prep.: Gibbsite crystals were commercial products. The crystal structure of the purchased gibbsite was confirmed by X-ray powder diffraction. It was reported to be predominantly gibbsite.

$I = 0.001$–0.1 M KNO_3

$T = 25$°C

Reaction vessel: NS

CO_2 exclusion: Yes, N_2 atm.

Equil. time: 30 minutes

Measurement: Potentiometric titrations were performed under CO_2-free conditions. The experiments were carried out using a Metrohm automatic system (716 DMS), with a combination pH electrode calibrated in standard buffer solutions. Equilibration time before titration was set at 1 hour. For the two-side titration, base back-titration and acid back-titration HNO_3 and KOH were used.

ZACHARA ET AL. (1994)

Experiment: Cobalt (Co^{2+}) Sorption on Gibbsite

Solid prep.: The gibbsite was of industrial origin (Alcan) a *superfine-4* with a surface area of 4 m^2/g, and a basal crystallite length of 1.5 μm. The gibbsite was treated with 0.01 mol/L NaOH for 30 minutes and washed in 0.01 mol/L $NaClO_4$ before use as a sorbent.

$I = 0.01$–0.1 M $NaClO_4$

$T = 25°C$

Reaction vessel: Pyrex flask

CO_2 exclusion: Yes, N_2 atm.

Max. S/S: 1.04×10^{-5}

Equil. time: 30 minutes

Measurement: Cobalt sorption was measured on gibbsite over the pH range of 4.5–8.5. The experiments were initiated by titrating a 220-mL sorbent suspension to pH 4.5 with $HClO_4$ under N_2 gas. The suspension was maintained at pH 4.5 for 12 hours and was spiked, by mass, with 10^{-4} mol/L $CoCl_2$ solution labeled with ^{57}Co to yield a final Co concentration of 10^{-6} mol/L. After spiking, the suspension was stirred for 30 minutes and two 10-mL aliquots of the suspension were removed via glass transfer pipet to 25-mL, acid-washed Corex centrifuge tubes. The pH was then adjusted upward approximately 0.5 units with a titrator, for 30 minutes, and two more 10-mL aliquots were removed as described previously. This procedure was repeated until the suspension pH reached 8.5. The 10-mL subsamples of the suspension were shaken overnight under N_2 atmosphere and were then centrifuged at 4800 rcf for 30 minutes. Aliquots was removed for ICP analysis of major cations and counting of ^{57}Co. Two series of six samples that varied in pH from 4.5 to 7.0 were studied.

Notes: (1) *Superfine 4 gibbsite* is considered to be an α-$Al(OH)_3$ form. (2) Two data sets from Fig. 8D were not considered in this study because the experiments were conducted in the presence of leonardite humic acid.

ZHANG ET AL. (2005)

Experiment: Uranyl (UO_2^{2+}) Sorption on Gibbsite

Solid prep.: Gibbsite was prepared using the Gastuche and Herbillon (1962) method.

$I = 0.005$ and 0.1 M KNO_3

$T = 20°C$

Reaction vessel: Polyethylene tubes

CO_2 exclusion: Yes, N_2 atm.

Max. S/S: 3.21×10^{-4}

Equil. time: 36 hours

Measurement: Batch sorption experiments were performed in polyethylene test tubes sealed with screw caps and it was found that the sorption of UO_2^{2+} on the wall of polyethylene test tubes was negligible. The concentration of stock solution was gravimetrically determined as U_3O_8. The concentrations of UO_2^{2+} of aqueous solutions in sorption experiments were measured with spectrophotometry.

Note: The data sets of Fig. 2 were not considered in this study because no attempts were made to exclude CO_2 from the sorption experiments. Uranyl forms complexes with carbonates.

ZHANG ET AL. (2006)

Experiment: Thorium (Th^{4+}) Sorption on Gibbsite

Solid prep.: Gibbsite was prepared using the Gastuche and Herbillon (1962) method. XRD revealed that the crystalline structure was that of gibbsite. Chemical analysis showed that there were small quantities of Fe, Si, and Na (<0.02%).

$I = 0.005$ and 0.1 M KNO_3

$T = 20°C$

Reaction vessel: Polyethylene tubes

CO_2 exclusion: Yes, N_2 atm.

Max. S/S: 4.55×10^{-4}

Equil. time: 24 hours

Measurement: Batch sorption experiments were performed in polyethylene test tubes sealed with screw caps and it was found that the sorption of Th(IV) and phosphate on the wall of polyethylene test tubes was negligible in the experimental conditions used.

Note: The data sets of Figs. 4 and 8 were not considered in this study because: (a) in Fig. 4, the pH changed almost by one full unit during the isotherm experiment, (b) in Fig 4., the sorption isotherm was conducted in the presence of fulvic acid, and (c) in Fig. 8, there was an initial concentration of phosphate 5.84×10^{-5} M.

AUTHOR INDEX

Surface Complexation Modeling: Gibbsite By Athanasios K. Karamalidis and David A. Dzombak

SUBJECT INDEX

Printed and bound by CPI Group (UK) Ltd, Croydon, CR0 4YY
16/09/2026
14963461-0001